Molecular Theory of Electric Double Layers

Molecular Theory of Electric Double Layers

Dimiter N Petsev, Frank van Swol and Laura J D Frink
Department of Chemical and Biological Engineering, Center for Micro-engineered Materials, University of New Mexico, Albuquerque, NM, USA

IOP Publishing, Bristol, UK

ISBN 978-0-7503-2276-8 (ebook)
ISBN 978-0-7503-2274-4 (print)
ISBN 978-0-7503-2277-5 (myPrint)
ISBN 978-0-7503-2275-1 (mobi)

DOI 10.1088/978-0-7503-2276-8

Version: 20211001

IOP ebooks

British Library Cataloguing-in-Publication Data: A catalogue record for this book is available from the British Library.

Published by IOP Publishing, wholly owned by The Institute of Physics, London

IOP Publishing, Temple Circus, Temple Way, Bristol, BS1 6HG, UK

US Office: IOP Publishing, Inc., 190 North Independence Mall West, Suite 601, Philadelphia, PA 19106, USA

To Tanya, Bertha and Nathan

Contents

Appendices

Preface

Interfaces between electrolyte solutions and various materials, typically referred to as electric double layers, occur in a wide range of physical and chemical systems. Colloid science, electrochemistry, material science and biology are a few examples where such interfaces play a crucial role. Research into the topic of electric double layers has been ongoing for more than a hundred years. Parallel to this process is the rapid development of liquid state theory, which nowadays allows for relating molecular interactions to macroscopic thermodynamic properties. The focus of this book is on the application of modern liquid state theories to the properties of electric double layers. It is an attempt to demonstrate the ability of statistical mechanical approaches, such as classical density functional theory (DFT), to provide insights and details that will enable a better and more quantitative understanding of electric double layers.

Many of the results presented in this book were obtained with the help of our graduate students at the University of New Mexico. We are particularly grateful to Mark Fleharty, Raviteja Vangara, Divya J Prakash, Dennis C R Brown, Kahlil Stoltzfus, Mary York, John M Baca and Luke Denoyer. Dimiter N Petsev thanks Ms Nia Petseva for providing editorial assistance on chapters 2, 3 and 8–13.

FvS would also like to thank the many collaborators over the years that contributed to his understanding of DFT and molecular simulation. In particular Les Woodcock for his guidance and efforts to teach MD of continuous and discontinuous interaction potentials. In particular, he would like to thank Jim Henderson for his deep insights, generosity and friendship and collaboration on wetting problems. He is indebted to Umberto Marini Bettolo Marconi for his friendship and introduction into lattice DFT. Thanks also to Anthony Malanoski and Sivakumar Challa for their collaboration on lattice DFT and molecular simulation. Finally the friendship and helpful discussions with John Rowlinson, John Maguire, Keith Gubbins, John Shelnutt, Jim Miller, Andy Kraynik, Craig Medforth, Tom Truskett and Jim Martin are gratefully acknowledged.

LJDF would like to thank Frank van Swol for the introduction to classical-DFT many years ago. She is also grateful to Andrew Salinger, Mike Heroux, and Mark Sears for their friendship and collaboration on numerical methods for DFT. In addition, her thanks go to Jim Willenbring for his software engineering expertise that made an open source release of Tramonto possible. She acknowledges Amalie Frischknecht and Jeff Weinhold for the introduction to coarse-grained polymer systems, their work on numerical implementations, and the many collaborative projects that have been the result. Finally, her heartfelt thanks goes to Dimiter Petsev and Frank van Swol for the opportunity to participate in this project.

We also wish to express our gratitude to the Center for Advanced Research Computing at the University of New Mexico where a significant part of the computations, used in this book, were performed.

Author biographies

Dimiter N Petsev

Dr Dimiter N Petsev received his PhD in Physical Chemistry from the University of Sofia. He holds research positions in Purdue University and the University of Alabama in Huntsville. Dr Petsev's research interests are in the area of colloids, surfaces and interfaces. A significant part of the efforts are focused on the dynamics of charged colloids, self-assembly of ionic surfactants, and emulsion stability against flocculation and coalescence. Other research areas of interest include electrokinetic phenomena and their relation to micro and nanofluidics, and capillary driven flow in porous materials.

Frank van Swol

Dr Frank van Swol received his PhD in Physical Chemistry from the University of Amsterdam, The Netherlands, where he was supervised by Prof L V Woodcock. His research has focused on the statistical mechanics of interfaces using molecular dynamics, Monte Carlo and classical density functional theory techniques. Dr van Swol was awarded a Ramsay Memorial Fellowship, which he held at Oxford University (UK) where he worked with Prof John S Rowlinson, from 1982 until 1985. He then joined the Chemical Engineering Department of Cornell as a visiting faculty member and the University of Illinois at Urbana-Champaign before joining Sandia National Laboratories in 1994, from which he retired in 2015. He has been a research professor at the University of New Mexico since 1996.

Laura J D Frink

Dr Laura J D Frink received her PhD in Chemical Engineering from the University of Illinois at Urbana-Champaign in 1995 where she was advised by Frank van Swol and Charles Zukoski. Her doctoral work involved simulation of colloidal systems with solvation potentials from classical density functional theory (c-DFT). She joined Sandia National Laboratories (SNL) in 1995 to pursue the development of a general purpose c-DFT code capable of two- and three-dimensional calculations using parallel computing platforms. She was a lead developer of the c-DFT code Tramonto, which is currently available as open source software. Scientific interests include colloidal systems, adsorption and wetting, polymer interfaces, and self-assembled systems with applications in nanotechnology and biology. In 2008, she left SNL and started a consulting company. She became a research professor at the University of New Mexico in 2017.

IOP Publishing

Molecular Theory of Electric Double Layers

Dimiter N Petsev, Frank van Swol and Laura J D Frink

Chapter 1

Introduction: a historical overview

1.1 Charges and fields

Manifestations of the electromagnetic fundamental force, along with gravity, have been known to humanity since ancient times. Perhaps the main effect of gravity on everyday life was its contribution to the difficulty of physical labor. Electricity, on the other hand, is responsible for spectacular natural phenomena, such as thunder and lightning, which in ancient times seemed to require a supernatural explanation.[1]

A more rational and quantitative approach towards the understanding of electricity and magnetism became available in the 18th century, when Henry Cavendish showed in 1771 that the electrostatic force drops in an inverse proportion to the square of the distance, thus bearing a remarkable resemblance to the law of gravity [1]. The experimental setup consists of two concentric spheres connected by a conducting wire. The outer sphere was charged and the conducting wire was carefully cut using silk thread. The outer sphere was then disassembled into two hemispheres and removed, while the inner sphere was tested for the presence of residual charge. Since no charge was detected, Cavendish concluded that the electric force inside the larger sphere is zero. Using Newton's shell theorem for the gravitational force inside a hollow sphere, he inferred that the electric force should have the same dependence on distance as gravity, and should drop as the square of the distance.

The analysis of the electric forces between charges was further refined by Charles-Augustine de Coulomb. Coulomb designed a precise calibrated torsion balance, which he used in 1785 to measure the forces between charged spheres. Ultimately he arrived at the law for the force, F_{el}, acting between two point charges, that nowadays carries his name [2, 3]

[1] The remaining two fundamental forces are the strong and the weak nuclear forces. Their effects, while present, are more subtle.

doi:10.1088/978-0-7503-2276-8ch1

$$F_{el} = k_e \frac{q_1 q_2}{r^2}, \tag{1.1}$$

where k_e is a the Coulomb constant, q_1 and q_2 are the charges of the interacting particles 1 and 2, and r is the distance between them.

Coulomb also established that the force is proportional to the product of the charges, as well as the fact that like charges repel, while opposite charges attract each other.

An intriguing experiment, performed by Hans Christian Ørsted implied that the electricity and magnetism are somehow intertwined. In 1820 he designed an experiment that consisted of a compass and an active electric circuit. Placing the compass near the circuit, Ørsted noticed that the compass needle moved away from its default North–South orientation. This experiment clearly demonstrated that electric currents generate magnetic fields [4], which implied that there is an intimate relationship between these seemingly different forces.

The subsequent advancement of our understanding of electricity and magnetism is due to Michael Faraday. He discovered the law of electromagnetic induction, which complemented Ørsted's experiment by demonstrating that not only electric currents generate a magnetic fields, but the opposite effect is also present: the relative motion of a magnet and a circuit generates electric current. Even more importantly Faraday pioneered the concept of fields, which while crucially significant for understanding electromagnetism [5], also transcended into other scientific disciplines.

1.2 Electrostatics of systems with distributed charges

The illuminating experiments, performed by Michael Faraday and his predecessors, offered significant insights. However a rigorous theoretical foundation was still missing. This knowledge gap was successfully filled by James Clerk Maxwell [6, 7]. After carefully studying the works of Faraday, Maxwell came up with a set of 20 equations that provided the much needed theoretical foundation of electromagnetism. Oliver Heaviside managed to reduce the number of the original equations down to four, which nevertheless are still referred to as the Maxwell equations [8, 9]

$$\begin{aligned} \nabla \cdot \mathbf{D}(\mathbf{r}, t) &= \rho_{el}(\mathbf{r}, t) \\ \nabla \times \mathbf{B}(\mathbf{r}, t) - \frac{1}{c^2}\frac{\partial \mathbf{E}(\mathbf{r}, t)}{\partial t} &= \mu_0 \mathbf{J}(\mathbf{r}, t) \\ \nabla \times \mathbf{E}(\mathbf{r}, t) + \frac{\partial \mathbf{B}(\mathbf{r}, t)}{\partial t} &= 0 \\ \nabla \cdot \mathbf{B}(\mathbf{r}, t) &= 0. \end{aligned} \tag{1.2}$$

The system (1.2) describes the spatial and temporal behavior of the electric displacement $\mathbf{D}(\mathbf{r}, t)$, the electric $\mathbf{E}(\mathbf{r}, t)$ and magnetic $\mathbf{B}(\mathbf{r}, t)$ field strengths, the spatial charge density distribution $\rho_{el}(\mathbf{r}, t)$, and the electric current $\mathbf{J}(\mathbf{r}, t)$. The remaining parameters are the speed of light $c = 3 \times 10^8$ m/s^{-1}, and the magnetic permeability in vacuum $\mu_0 = 1.257 \times 10^{-6}$ H/m^{-1}. The displacement $\mathbf{D}$ is proportional to the electric field $\mathbf{E}(\mathbf{r}, t)$ and the polarizability of the dielectric medium $\mathbf{P}(\mathbf{r}, t)$, or

$$\mathbf{D}(\mathbf{r}, \mathbf{t}) = \varepsilon_0 \mathbf{E}(\mathbf{r}, \mathbf{t}) + \mathbf{P}(\mathbf{r}, \mathbf{t}), \tag{1.3}$$

where $\varepsilon_0 = 8.854 \times 10^{-12}$ F m^{-1} is the dielectric constant of vacuum. The polarizability is $\mathbf{P}(\mathbf{r}, \mathbf{t})$ a linear function of the electric field

$$\mathbf{P}(\mathbf{r}, \mathbf{t}) = \varepsilon_0 \chi \mathbf{E}(\mathbf{r}, \mathbf{t}), \tag{1.4}$$

where χ is the dielectric susceptibility of the particular dielectric medium. Introducing equation (1.4) into (1.3) eliminates the displacement $\mathbf{D}(\mathbf{r}, t)$ and polarization $\mathbf{P}(\mathbf{r}, t)$ by expressing them in terms of the electric field $\mathbf{E}(\mathbf{r}, t)$, or

$$\mathbf{D}(\mathbf{r}, \mathbf{t}) = \varepsilon_0 (1 + \chi) \mathbf{E}(\mathbf{r}, \mathbf{t}) = \varepsilon_0 \varepsilon \mathbf{E}(\mathbf{r}, \mathbf{t}). \tag{1.5}$$

The quantity $\varepsilon = 1 + \chi$ is called the relative (to vacuum) permittivity and accounts for inhomogeneities (e.g., free charges) in the medium [9].

The focus of this book is on charged interfaces between two phases with different dielectric permittivities and charge densities, at least one of which is typically a liquid (e.g., electrolyte solution). It is the first equation in (1.2) that is most relevant to such systems. The fields in this case are time-independent (i.e., $\partial \mathbf{E}/\partial t = 0$ and $\partial \mathbf{B}/\partial t = 0$) hence, $\nabla \times \mathbf{E}(\mathbf{r}) = 0$. The electric field is irrotational and can be expressed by

$$\mathbf{E}(\mathbf{r}) = -\nabla \psi(\mathbf{r}). \tag{1.6}$$

Combining equations (1.2), (1.5) and (1.6) leads to

$$\nabla \cdot [\varepsilon \varepsilon_0 \psi(\mathbf{r})] = -\rho_{el}(\mathbf{r}), \tag{1.7}$$

with ψ being the scalar electrostatic potential and $\mathbf{r}$ is the spatial position vector. Assuming that the relative dielectric permittivity ε is position independent simplifies equation (1.7) to

$$\nabla^2 \psi(\mathbf{r}) = -\frac{\rho_{el}(\mathbf{r})}{\varepsilon \varepsilon_0}. \tag{1.8}$$

Due to its particular mathematical form, equation (1.8) is commonly referred to as the Poisson equation [10]. It can also be written in integral form

$$\psi(\mathbf{r}) = \frac{1}{4\pi\varepsilon\varepsilon_0} \iiint_V dV \frac{\rho_{el}(\mathbf{r}')}{|\mathbf{r} - \mathbf{r}'|}, \tag{1.9}$$

which implies that the local electrostatic potential depends on a superposition of the distributed charges. The integral is taken over the volume V, encompassing $\mathbf{r}'$.

Another useful relationship links the volume charge to that on the surrounding surface[2]

[2] Following the traditional notation in statistical mechanics, we will hereafter use the notation $\int_V d\mathbf{r}$, or $\int d\mathbf{r}$ instead of $\iiint_V dV$, to denote a volume integral.

$$\iiint_V dV\ \rho_{el}(\mathbf{r}) = -\iiint_V dV\ \nabla\varepsilon\varepsilon_0\nabla\psi(\mathbf{r}) = -\oiint_S dS\ \varepsilon\varepsilon_0\nabla\psi(\mathbf{r})\cdot\mathbf{n}. \qquad (1.10)$$

The transition from volume to surface integration above is a result of the Gauss divergence theorem [10]. Equation (1.10) expresses the balance between the bulk and surface charges. Thus $\sigma = -[\varepsilon\varepsilon_0\nabla\psi(\mathbf{r})]_S$ defines the surface charge density.

1.3 The concept of electric double layer

The electric double layer (EDL) is a term that refers to the charge and potential distributions that form at the interface between two different phases, at least one of which contains mobile charges. A surface charge typically builds up interface, due to chemical reactions or adsorption from the bulk [11–14] (see chapter 2). The mobile charge density then redistributes in the resultant electrostatic field, and an EDL is formed. A common example for such a system is a liquid electrolyte solution (containing mobile ions) in contact with a dielectric substrate, free from mobile charge carriers (i.e., $\rho_{el} = 0$) [11, 15]. Alternatively, if two phases with fluid-like charge densities are in contact, electrostatic field and volume charge redistribution may occur in both of them. A common example is the oil–water two-phase system, with various ionic species dissolved on both sides of the interface [12]. Another instance for EDLs that form in two adjacent phases is the semiconductor–electrolyte interface [16–19].

The experimental evidence for EDLs was first obtained in 1809 by Reuss [20], who applied electric field to a mixture of water and quartz particles. His experiments led to the discovery of electroosmosis (liquid flow) and electrophoresis (particle motion), thus laying the foundations of a new research field now known as electrokinetic phenomena. More importantly, these experiments proved the presence of charge at the interface between a solid substrate and liquid (in this case water). The electroosmotic flow was first quantitatively analyzed by Wiedemann [21], who demonstrated that the ratio of the volumetric fluid flow rate over the total electric current (U_{eo}/J) does not depend on the cross sectional area or thickness of the porous plug. Around the same time, Quincke [22] demonstrated that electroosmotic flow also occurs in straight capillaries, and not only in disordered porous materials.

The first attempt to provide a theoretical description of EDLs was performed by Helmholtz [23], who considered the interface between the electrolyte and the substrate as a capacitor (see figure 1.1). The two-layered structure of the Helmholtz model is responsible for the historically established term 'electric double layer' (EDL) to describe a charged interface that involves an electrolyte solution.

The Helmholtz model [23] has significant shortcomings. With the advancement of the molecular kinetics theories [24] and a better understanding of Brownian motion [25, 26], it became clear that the thermal energy of the electrolyte solution would not allow the perfectly layered ordering of counterions at the interface as inferred by the Helmholtz model. This issue was addressed by Gouy [27, 28] and Chapman [29], who used a modified version of the Poisson equation (1.8) to find the spatial distribution of the electrostatic potential and charge in the solution (see chapter 3). In addition, it explicitly accounts for the effects of temperature and concentration of

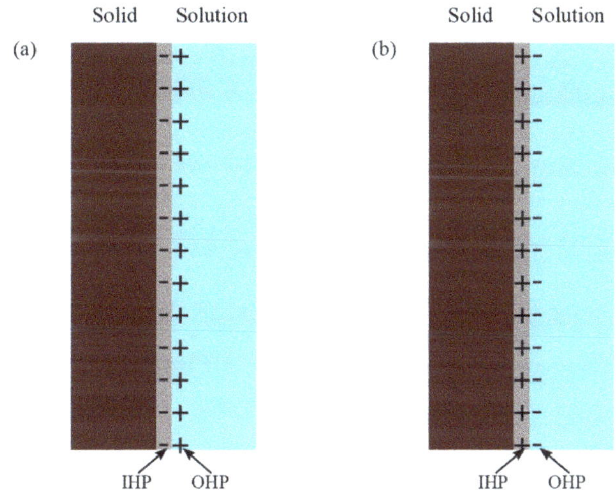

Figure 1.1. A sketch depicting the Helmholtz model of a charged interface between a dielectric substrate (solid) and electrolyte solution. The counterions from the solution are all located on a single plane in the liquid, parallel and adjacent to the interface. (a) A negatively charged interface. (b) A positively charged interface. The inner Helmholtz plane (IHP) and the outer Helmholtz plane (OHP) are indicated by the arrows.

electrolytes. Still, it is an approximation, because in essence, it is a continuum approach, and does not explicitly account for the solution molecular and ionic structure. Despite these shortcomings, the Gouy and Chapman model is very popular. A variation of it was used by Debye and Huckel to develop their celebrated theory of strong electrolytes [30].

EDLs play an important role in the stability and behavior of colloidal systems. A central problem in colloid science is that of the stability (or lack thereof) of dispersions, particularly of those comprised of lyophobic particles. Analysis of this problem was independently offered by Derjaguin and Landau [31], and Verwey and Overbeek [32]. This is the celebrated Derjaguin–Landau–Verwey–Overbeek (or DLVO) theory of colloid stability. The main premise of the DLVO theory is that the stability of lyophobic colloids is determined by the competition between electrostatic repulsion and macroscopic attraction of the van der Waals–Hamaker type [33]. Hence, despite the particle lyophobicity and attraction (due to the van der Waals–Hamaker force), the suspension may be stable because of electrostatic repulsion. The DLVO theory has been remarkably successful, particularly in explaining the salt dependence of the stability of charged colloidal suspensions manifested by empirical observations such as the Schulze–Hardy rule [34, 35]. Charged colloidal suspension can be destabilized and forced to coagulate and precipitate by dissolving more salt (electrolyte) in the solution. The Schulze–Hardy rule empirically established that the critical coagulation concentration (CCC) of counterions 'i' (with charge opposite to that of the colloidal particles) is inversely proportional to the sixth power of their charge number q_i, or CCC $\sim q_i^{-6}$. The DLVO theory not only arrived at the same conclusion for strongly charged particles, but it also correctly predicted a different law (CCC $\sim q_i^{-2}$) in the case of low particle potentials [31, 32, 36, 37].

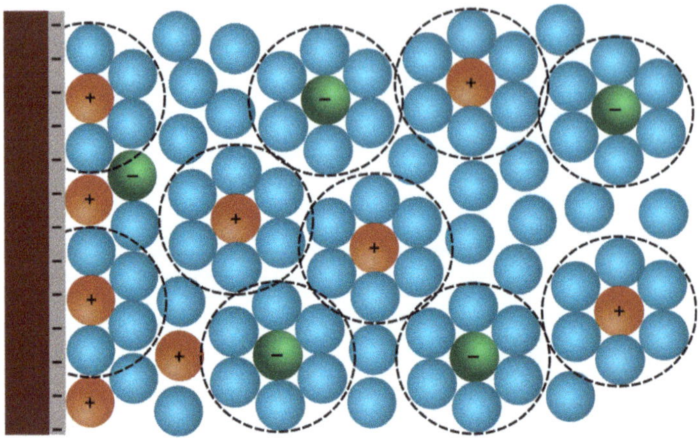

Figure 1.2. A sketch of the molecular structure of an electric double layer. The dashed circles qualitatively illustrate the ionic hydration shell.

The Gouy and Chapman model of the EDL [27–29] and its further developments included in the DLVO theory [31, 32, 36–38] do not account for the molecular structure of the phases in contact. Therefore, the approach is rather approximate. Ignoring the excluded volume of the ions may lead to nonphysically high counterion densities in the vicinity of the charged interface. In addition, it ignores the molecular structure effect due to the solvent molecules (figure 1.2). The structural effects are particularly pronounced in the narrow gap between the interacting macroscopic surfaces, as demonstrated in a series of clever and sophisticated force measurements between surfaces in electrolyte solutions [39, 40]. The results showed that the solution molecular structure has a well-defined and measurable effect on the interaction forces between surfaces in electrolyte solutions that were not present in the original DLVO theory. These additional structural forces are an example of the so-called non-DLVO surface interactions.

An early attempt to correct for some of the inadequacies was made by Stern [41]. His model accounts for the ionic size and non-Coulombic interactions with the substrate in the immediate vicinity (i.e. first ionic layer), but beyond that, the continuum Gouy–Chapman approach is still applied. The solvent molecules contributions were not explicitly included in Stern's analysis. It was Grahame who first analyzed the effects of neutral solvent molecules in the Stern layer [42]. The rest of the EDL in Grahame's model, however, was still represented by a structureless continuum. The solution structure near a charged interface includes effects due to excluded volume and long-range interactions, such as Coulombic, van der Waals, ion–dipole and dipole–dipole. It is often considered that the most important interactions in an electrolyte solution are the Coulombic and all the rest are essentially ignored. Such models are sometimes referred to 'primitive' [43], and are still often used to solve a variety of problems related to EDLs [44–51]. An alternative approach is the 'semi-primitive' model that includes all solution species—ions and neutral (e.g., solvent molecules). The semi-primitive model accounts for the excluded volume and other non-electrostatic interactions (e.g., van der Waals), but not for any dipolar effects [52–54]. The inclusion of the dipolar solvent effects defines a class of models known as 'civilized' [54]. The explicit inclusion of the solvent into the model (semi-primitive or civilized) reveals features and details that otherwise would be entirely overlooked [55–66].

The semi-primitive model is based on modern statistical mechanical theories of fluids. An example is the integral equation approach developed by Ornstein and Zernike, which suggests splitting the molecular interactions into direct and indirect parts. The result is the Ornstein–Zernike (OZ) equation [67] (see chapter 4). The OZ equation is applicable to both pure liquids, and solutions, including electrolytes [68, 69], and has the form

$$h_{ij}(r_{12}) = c_{ij}(r_{12}) + \sum_k \rho_i \int d\mathbf{r}_3 c_{ik}(r_{13}) h_{kj}(r_{23}). \tag{1.11}$$

The functions h_{ij} and c_{ij} correspond to the total and direct correlations between species i and j located at positions $\mathbf{r}_1$ and $\mathbf{r}_2$ ($r_{12} = |\mathbf{r}_1 - \mathbf{r}_2|$). The integral on the right hand side of equation (1.11) takes into account the indirect interactions between species i and j, mediated by a third species k located at $\mathbf{r}_3$. The integration is over the volume $d\mathbf{r}_3$, centered at $\mathbf{r}_3$. The OZ equation contains two unknown functions, which is why an additional (closure) relationship is required [68, 69]. A detailed discussion of different closures is presented in chapter 4.

The OZ method is a significant improvement when compared to the Gouy–Chapman treatment of EDLs, or the Debye–Huckel model of electrolytes. It can be applied to both primitive, semi-primitive and civilized electrolytes [47, 56–59, 70]. The macroscopic charged interface that forms the EDL is usually represented as a solution component with an infinitely large radius [57].

A related approach is offered by the classical density functional theory (c-DFT). This theory has been demonstrated to be a very powerful tool for analyzing inhomogeneous fluids and interfacial systems [69, 71–74]. c-DFT was developed following the blueprints of quantum density functional theory (q-DFT) [75]. Its appeal is due to the fact that it is based on minimizing well-defined and intuitive quantities as the free energy, or the grand thermodynamic potential with respect to the density distributions of all participating components, which provides simple and direct relation to any thermodynamic quantity of interest. In addition, unlike q-DFT, the c-DFT can be easily applied to arbitrarily large systems, which is a great advantage. For systems like the EDL, a grand thermodynamic potential $\Omega[\{\rho_i(\mathbf{r})\}]$ is defined as

$$\Omega\left[\{\rho_i(\mathbf{r})\}\right] = \mathcal{F}\left[\{\rho_i(\mathbf{r})\}\right] - \sum_i \int d\mathbf{r}\, [V_{\text{ext}} - \mu_i \rho_i(\mathbf{r})] \tag{1.12}$$

where $\{\rho_i(\mathbf{r})\}$ is the set of density profiles of all components in the solution, $\mathcal{F}[\{\rho_i(\mathbf{r})\}]$ is the free energy functional, V_{ext} is the external potential that includes all possible interactions with the interface (e.g., excluded volume, Coulombic, van der Waals), and μ_i is the constant chemical potential of species i. All properties of interest can be derived by means of functional differentiation [69, 71–74] (see chapter 5). Hence, c-DFT offer a rigorous, powerful, and computationally efficient method to model and analyze inhomogeneous fluids including EDLs, as long as physically adequate models for $\Omega[\{\rho_i(\mathbf{r})\}]$ (or $\mathcal{F}[\{\rho_i(\mathbf{r})\}]$) are available. Chapter 5 presents a more in-depth discussion of these issues.

It is important to emphasize that c-DFT has been extensively tested against computer simulations [76, 77]. The comparisons showed an excellent agreement between the theory and computational experiments. This is a very encouraging result because c-DFT is very computationally efficient and fast when compared to molecular dynamics (MD) or Monte Carlo (MC) simulations. MD and MC are particularly hard to apply to situations where there is a great disparity in the number densities of different components. An example is a solution, where the physical properties are governed by pH variations. The concentrations of H_3O^+ and OH^- are often much lower than those of other dissolved electrolyte, or even more—the solvent.

An important property of each EDL is the potential and charge at the interface that separates the different phases. These quantities have a major impact on the overall properties of the charged interface. A simple way to introduce either surface potential and/or the surface charge into the model is to treat them as input parameters. In the Gouy and Chapman analysis [27–29], the surface potential is considered constant, which considerably simplifies the mathematical problem and allows for an analytical solution in the case of a flat EDL. Such an approximation, however, is not physical since the actual charge and potential formation at the interface is typically a result of complex surface chemical reactions and physical adsorption/desorption processes [11, 15] (see chapter 2). In 1971, Ninham and Parsegian [13] argued that the surface charge should be determined by the thermodynamic equilibrium between the surface and certain ionic species in the solution, which are referred to as potential determining ions (PDIs). Hence, the surface charge is determined by the specific chemistry at the interface and is known as the charge regulation condition. This condition is able to account for various surface reactions, depending on the system specificity [14]. It may lead to physically different results when compared to the constant surface potential (or charge) based approaches. A very important example is the case of interacting (overlapping) EDLs, which is of crucial importance to colloid stability [78–83].

References

[1] Cavendish H 1771 An attempt to explain some of the principal phaenomena of electricity, by means of an elastic fluid *Phil. Trans. Roy. Soc.* **61** 584–677

[2] de Coulomb C-A 1785 Premier mémoire sur l'electricite et le magnetisme *Histoire de l'Academie Royale des Sciences* pp 569–77

[3] de Coulomb C-A 1785 Second memoire sur l'electricite et le magnetisme *Histoire de l'Academie Royale des Sciences* pp 578–611

[4] Øersted H C 1820 Experiments on the effect of a current of electricity on the magnetic needles *Ann. Philos.* **16** 273–7

[5] Faraday M 1914 *Experimental Researches in Electricity* (New York: J. M. Dent & Sons)

[6] Maxwell J C 1954 *A Treatise on Electricity and Magnetism* vol 1 (New York: Dover)
[7] Maxwell J C 1954 *A Treatise on Electricity and Magnetism* vol 2 (New York: Dover)
[8] Jackson J D 1998 *Classical Electrodynamics* (New York: Wiley)
[9] Landau L D and Lifshitz E M 1984 *Electrodynamics of Continuous Media* (Amsterdam: Pergamon)
[10] Morse P and Feshbach H 1953 *Methods of Theoretical Physics* vol 1 (Minneapolis, MN: Feshbach Publishing)
[11] Overbeek J T G 1952 Electrochemistry of the double layer *Colloid Science* ed H R Kruyt (New York: Elsevier) ch 4 pp 115–93
[12] Davies J T and Rideal E K 1963 *Interfacial Phenomena* (New York: Academic)
[13] Ninham B W and Parsegian V A 1971 Electrostatic potential between surface bearing ionizable groups in ionic equilibrium with physiologic saline solution *J. Theor. Biol.* **31** 405–28
[14] Ettelaie R and Buscall R 1995 Electrical double layer interactions for spherical charge regulating colloidal particles *Adv. Coll. Interf. Sci.* **61** 131–60
[15] Dukhin S S 1974 Development of notions as to mechanism of electrokinetic phenomena and the structure of the colloid micelle ed E Matijevic *Surface and Colloid Science* vol 7 (New York: Wiley Interscience) ch 1 pp 1–49
[16] Zhang X G 2001 *Electrochemistry of Silicon and Its Oxide* (New York: KA/PP)
[17] Konorov P P, Yafyasov A M and Bogevolnov V B 2006 *Field Effect in Semiconductor–Electrolyte Interfaces* (Princeton, NJ: Princeton University Press)
[18] Fleharty M E, van Swol F and Petsev D N 2014 Manipulating semiconductor colloidal stability through doping *Phys. Rev. Lett.* **113** 158302
[19] Fleharty M E, van Swol F and Petsev D N 2015 Electric double layers at the semiconductor–electrolyte interface *J. Colloid Interface Sci.* **449** 409–15
[20] Reuss F 1809 Charge induced flow *Proceedings of the Imperial Society of Naturalists of Moscow* **3** 327–44
[21] Wiedemann G 1852 Ueber die bewegung von flüssigkeiten im kreise der geschlossenen galvanischen säule *Ann. Phys.* **163** 321–52
[22] Quincke G 1861 Ueber die fortführung materieller theilchen durch strömende elektricität *Ann. Phys.* **189** 513–98
[23] Helmholtz H 1853 Ueber einige gesetze der vertheilung elektrischer ströme in kröperlichen leitern mit anwendung auf die thierisch–elektrischen versuche *Ann. Phys. Chem.* **165** 211–33
[24] Maxwell J C 1867 On the dynamical theory of gases *Philos. Trans. Royal Soc. London* **157** 49–88
[25] Einstein A 1905 Über die von der molekularkinetischen theorie der wärme geforderte bewegung von in ruhenden flüssigkeiten suspendierten teilchen *Ann. Phys.* **17** 549–60
[26] von Smoluchowski M 1906 Zur kinetischen theorie der Brownschen molekularbewegung und der suspensionen *Ann. Phys. (Leipzig)* **21** 756–80
[27] Gouy G 1910 Sur la constitution de la charge électrique à la surface d'un électrolyte *J. Physique* **9** 457–68
[28] Gouy G 1917 Sur la fonction electrocapillaire *Ann. Phys.* **7** 129–84
[29] Chapman D L 1913 A contribution to the theory of electrocapillarity *Phil. Mag.* **25** 475–81
[30] Debye P and Huckel E 1923 Zur theorie der elektrolyte. i. gefrierpunktserniedrigung und verwandte erscheinungen *Phys. Ztschr.* **24** 185–206
[31] Derjaguin B and Landau L 1941 Theory of the stability of strongly charged lyophobic sols and of the adhesion of strongly charged particles in solution of electrolytes *Acta. Physicochim. URSS* **14** 633–62

[32] Verwey E J W and Overbeek J T G 1948 *Theory of the Stability of Lyophobic Colloids* (Amsterdam: Elsevier)

[33] Hamaker H C 1937 The London–van der Waals attraction between spherical particles *Physica* **4** 1058–72

[34] Schulze H 1881 Schwefelarsen in wässriger lösung *J. Prakt. Chem.* **25** 431–52

[35] Hardy W B 1900 A preliminary investigation of the conditions which determine the stability of irreversible hydrosols *Proc. Roy. Soc. London* **66** 110–25

[36] Derjaguin B V and Muller V M 1967 Slow coagulation of hydrophobic colloids *Dokl. Akad. Nauk SSSR (in Russian)* **176** 738–41

[37] Overbeek J T G 1952 The interaction between colloidal particles ed H R Kruyt *Colloid Science* (New York: Elsevier) ch 6 pp 245–77

[38] Langmuir I 1938 The role of attractive and repulsive forces in the formation of tactoids, thixotropic gels, protein crystals and coacervates *J. Chem. Phys.* **6** 873–96

[39] Israelachvili J N 2011 *Intermolecular and Surface Forces* 3rd edn (New York: Academic)

[40] Israelachvili J N and Wennerstrom H 1996 Role of hydration and water structure in biological and colloidal interactions *Nature* **379** 219–25

[41] Stern O 1924 Zur theorie der elektrolytischen doppelschicht *Z. Electrochem* **30** 508–16

[42] Grahame D C 1947 The electrical double layer and the theory of electrocapillarity *Chem. Rev.* **41** 441–501

[43] Stokes R H 1972 Debye model and the primitive model for eletrolyte solutions *J. Chem. Phys.* **56** 3382–3

[44] Attard P 1996 Electrolytes and the electric double layer *Adv. Chem. Phys.* **92** 1–159

[45] Voukadinova A and Gillespie D 2019 Energetics of counterion adsorption in the electrical double layer *J. Chem. Phys.* **150** 144703

[46] Valisko M, Kristof T, Gillespie D and Boda D 2018 A systematic monte carlo simulation study of the primitive model planar electrical double layer over an extended range of concentrations, electrode charges, cation diameters and valences *AIP Adv.* **8** 025320

[47] Plischke M 1988 Pair correlation functions and density profiles in the primitive model of the electric double layer *J. Chem. Phys.* **88** 2712–8

[48] Yu Y X, Wu J Z and Gao G H 2004 Density–functional theory of spherical electric double layers and zeta potentials of colloidal particles in restricted–primitive model electrolyte solutions *J. Chem. Phys.* **120** 7223–33

[49] Tang Z, Striven L E and Davis H T 1992 Interactions between primitive electrical double layers *J. Chem. Phys.* **97** 9258–66

[50] Pizio O and S Sokolowski 2006 On the effects of ion–wall chemical association on the electric double layer: A density functional approach for the restricted primitive model at a charged wall *J. Chem. Phys.* **125** 024512

[51] Roth R and Gillespie D 2016 Shells of charge: a density functional theory for charged hard spheres *J. Phys.: Cond. Matter* **28** 244006

[52] Oleksy A and Hansen J-P 2006 Towards a microscopic theory of wetting by ionic solutions. i. surface properties of the semi-primitive model *Mol. Phys.* **104** 2871–83

[53] Oleksy A and Hansen J-P 2009 Microscopic density functional theory of wetting and drying of solid substrate by an explicit solvent model of ionic solutions *Mol. Phys.* **107** 2609–24

[54] Oleksy A and Hansen J-P 2011 Wetting and drying scenarios of ionic solutions *Mol. Phys.* **109** 1275–88

[55] Adelman S A and Deutch J M 1974 Exact solution of the mean spherical model for strong electrolytes in polar solvents *J. Chem. Phys.* **60** 3935–49
[56] Chan D Y C, Mitchell D J, Ninham B W and Pailthorpe B A 1978 On the theory of dipolar fluids and ion–dipole mixtures *J. Chem. Phys.* **69** 691–6
[57] Carnie S and Chan D Y C 1980 The structure of electrolytes at charged surfaces: ion–dipole mixtures *J. Chem. Phys.* **73** 2949–57
[58] Blum L and Henderson D 1981 Mixtures of hard ions and dipoles against a charged wall: The Ornstein–Zernike equation, some exact results, and the mean spherical approximation *J. Chem. Phys.* **74** 1902–10
[59] Dong W, Rosinberg M L, Perera A and Patey G N 1988 A theoretical study of the solid–electrolyte solution interface. i. structure of a hard sphere ion–dipole mixture near an uncharged hard wall *J. Chem. Phys.* **89** 4994–5009
[60] Groot R D 1988 Density-functional theory for inhomogeneous fluids *Phys. Rev.* A **37** 3456–64
[61] Perera L and Berkowitz M L 1992 Dynamics of ion solvation in a stockmayer fluid *J. Chem. Phys.* **96** 3092–101
[62] Fleharty M E, van Swol F and Petsev D N 2016 Solvent role in the formation of electric double layers with surface charge regulation: A bystander or a key participant? *Phys. Rev. Lett.* **116** 048301
[63] Vangara R, Brown D C R, van Swol F and Petsev D N 2017 Electrolyte solution structure and its effect on the properties of electric double layers with surface charge regulation *J. Colloid Interface Sci.* **488** 180–9
[64] Vangara R, van Swol F and Petsev D N 2017 Solvation effects on the potential and charge distributions in electric double layers *J. Chem. Phys.* **147** 214704
[65] Vangara R, van Swol F and Petsev D N 2018 Solvophilic and solvophobic surfaces and non-coulombic surface interactions in charge regulating electric double layers *J. Chem. Phys.* **148** 044702
[66] van Swol F and Petsev D N 2018 Solution structure effects on the properties electric double layers with surface charge regulation assessed by density functional theory *Langmuir* **34** 13808–20
[67] Ornstein L S and Zernike F 1914 Accidental deviations of density and opalescence at the critical point of a sinqle substance *Proc. Acad. Sci. (Amsterdam)* **17** 793–806
[68] McQuarrie D A 2000 *Statistical Mechanics* (Mill Valley, CA: University Science Books)
[69] Hansen J P and McDonald I R 2006 *Theory of Simple Liquids* (New York: Academic)
[70] Blum L and Henderson D 1992 Statistical mechanics of electrolytes at interfaces ed D Henderson *Fundamentals of Inhomogeneous Fluids* (New York: Marcel Dekker, Inc.) ch 6 pp 239–76
[71] Ted Davis H 1996 *Statistical Mechanics of Phases, Interfaces, and Thin Films* (New York: Wiley)
[72] Evans R 1992 Density functionals in the theory of nonuniform fluids ed D Henderson *Fundamentals of Inhomogeneous Fluids* (New York: Marcel Dekker, Inc.) ch 3 pp 85–175
[73] Wu J 2006 Density functional theory for chemical engineering: From capillarity to soft materials *AIChE J.* **52** 1169–91
[74] Roth R 2010 Fundamental measure theory for hard-sphere mixtures: a review *J. Phys.: Condens. Matter* **22** 063102
[75] Honenberg P and Kohn W 1964 Inhomogeneous electron gas *Phys. Rev.* **136** B864–71

[76] Lee J W, Nilson R H, Templeton J A, Griffiths S K, Kung A and Wong B M 2012 Comparison of molecular dynamics with classical density functional and Poisson–Boltzmann theories of the electric double layer in nanochannels *J. Chem. Theory Comput.* **8** 2012–22
[77] Jeanmairet G, Levy N, Levesque M and Borgis D 2014 Introduction to classical density functional theory by a computational experiment *J. Chem. Educ.* **91** 2112–5
[78] Chan D Y C, Perram J W, White L R and Healy T W 1975 Regulation of surface potential at amphoteric surfaces during particle–particle interaction *J. Chem. Soc., Faraday Trans.* I **71** 1046–57
[79] Chan D Y C, Healy T W and White L R 1976 Electrical double layer interactions under regulation by surface ionization equilibria–dissimilar amphoteric surfaces *J. Chem. Soc., Faraday Trans.* I **72** 2844–65
[80] Behrens S H and Borkovec M 1999 Electric double layer interaction of ionizable surfaces: Charge regulation for arbitrary potentials *J. Chem. Phys.* **111** 382–5
[81] Behrens S H and Borkovec M 1999 Electrostatic interaction of colloidal surfaces with variable charge *J. Phys. Chem.* **103** 2918–28
[82] Behrens S H and Borkovec M 1999 Exact Poisson–Boltzmann solution for the interaction of dissimilar charge–regulating surfaces *Phys. Rev.* E **60** 7040–8
[83] Trefalt G, Behrens S H and Borkovec M 2016 Charge regulation in the electrical double layer: Ion adsorption and surface interactions *Langmuir* **32** 380–400

Part I

Theory

Chapter 2

The origin of charge at interfaces involving electrolyte solutions

The interfaces between phases in heterogeneous systems play a crucial role in the thermodynamic analysis of materials [1]. The molecules in the vicinity of an interface are subjected to a very different environment in comparison with those in the bulk fluid. The molecular interactions are between species of a different kind, which leads to a positive interfacial free energy. One way to lower the (unfavorable) positive free energy is by means of chemical (surface chemical reactions) or physical adsorption at the surface. In the case of electrolytes, this often is accomplished by ionic adsorption [2] or surface association/dissociation reactions [3–5]. The result of the ionic adsorption or surface reactions is the accumulation of charges at the interface. The two processes may occur simultaneously, and it is instructive to define them in a more distinctive manner. Examples include amphoteric surfaces, whose surface charge is determined by the pH of the adjacent solution.

The formation of the interfacial charge is an essential prerequisite for existence of electric double layers (EDLs). A proper account for the physical and chemical origin of the surface charges is necessary for a self-consistent thermodynamic analysis of charged interfaces. Relating the charging mechanism to the explicit solution structure is a central theme in this book.

2.1 Effects of the surface chemical reactions and the charge regulation model

The origin of charge at an electrolyte–substrate interface is governed by thermodynamics, which governs the chemical equilibrium between any surface active groups and certain charged species in the solution that are referred to as the potential determining ions (PDIs). This important argument was first formulated and defended by Ninham and Parsegian [3]. The surface chemical charge regulation should be distinguished from the physical adsorption (or desorption) of ions, although the latter also has an effect on

doi:10.1088/978-0-7503-2276-8ch2

the observed surface charge. The surface chemistry is system-specific. Replacing the substrate or certain solutes in the fluid phase can lead to a different set of surface chemical reactions, hence to a different surface charge.

A simple and straightforward charge regulating model, for amphoteric surfaces, was suggested by Chan *et al* [4, 6]. It is relevant to a typical oxide–electrolyte solution interface, and is well-suited to illustrate the charge regulation effect. The model considers two surface chemical reactions

$$\begin{aligned} &\mathrm{AH_2^+ + BH \rightleftharpoons AH + BH_2^+}, \quad \mathrm{pK_+} = -\log_{10}\mathrm{K_+} \\ &\mathrm{AH + BH \rightleftharpoons A^- + BH_2^+}, \quad \mathrm{pK_-} = -\log_{10}\mathrm{K_-}. \end{aligned} \tag{2.1}$$

These reactions are characterized by their respective equilibrium constants $\mathrm{K_+}$ and $\mathrm{K_-}$, which are defined by

$$\mathrm{K_+} = \frac{\rho^{sg}_{\mathrm{AH}}\rho^{s}_{\mathrm{BH_2^+}}}{\rho^{sg}_{\mathrm{AH_2^+}}}, \quad \mathrm{K_-} = \frac{\rho^{sg}_{\mathrm{A^-}}\rho^{s}_{\mathrm{BH_2^+}}}{\rho^{sg}_{\mathrm{AH}}}. \tag{2.2}$$

The superscript *sg* refers to a chemical reactive group that is an integral part of solid oxide surface, while the superscript *s* indicates a solution component in the immediate vicinity of the interface. The surface reactive groups (AH) can release a proton and become negative ($\mathrm{A^-}$), or bind a second one and turn positive ($\mathrm{AH_2^+}$), depending on the concentration of the potential determining ions (PDIs), $\mathrm{BH_2^+}$, in the solution next to the interface. The PDIs are generated by the fluid reaction

$$\mathrm{2BH \rightleftharpoons B^- + BH_2^+}. \tag{2.3}$$

An example for such a system is silica in contact with water, where AH, $\mathrm{A^-}$, and $\mathrm{AH_2^+}$ would correspond to –SiOH, $-\mathrm{SiO^-}$, and $-\mathrm{SiOH_2^+}$ respectively. The fluid species BH, $\mathrm{B^-}$, and $\mathrm{BH_2^+}$ would correspond to H_2O, OH^-, and H_3O^+.

The surface charge σ depends on differences between the relative numbers of reactive surface groups (ρ_i^{sg}), or [4, 6]

$$\sigma = e\Gamma\frac{\rho^{sg}_{\mathrm{AH_2^+}} - \rho^{sg}_{\mathrm{A^-}}}{\rho^{sg}_{\mathrm{AH}} + \rho^{sg}_{\mathrm{AH_2^+}} + \rho^{sg}_{\mathrm{A^-}}}, \tag{2.4}$$

where Γ is the total number of surface ionizable groups per unit area. Combining equations (2.2) and (2.4) leads to the following expression for the surface charge

$$\sigma = e\Gamma\left(\frac{\rho^{s}_{\mathrm{BH_2^+}}}{\mathrm{K_+}} + \frac{\mathrm{K_-}}{\rho^{s}_{\mathrm{BH_2^+}}}\right). \tag{2.5}$$

This expression translates into a relationship between the surface charge σ and surface potential ψ_s [4]

$$\sigma(\psi_s) = e\Gamma\frac{\delta\sinh[e(\psi_N - \psi_s)/k_BT]}{1 + \delta\cosh[e(\psi_N - \psi_s)/k_BT]}. \tag{2.6}$$

The equilibrium constants for the reactions (2.1) are included in the parameter $\delta = 2\sqrt{K_-/K_+}$. Another important quantity in equation (2.5) is the Nernst potential

$$\psi_N = [\ln(10)k_BT/e](\mathrm{pI} - \mathrm{pH}_b) \tag{2.7}$$

where pH_b is defined in the solution bulk, far from the interface, and $\mathrm{pI} = (\mathrm{pK}_- + \mathrm{pK}_+)/2$ is the surface isoelectric point (the number of positively and negatively charged groups is the same, therefore the total charge is zero).

The description of the charged interface provided by equation (2.6) is incomplete. Both σ and ψ_s are unknown, hence an additional relationship between them is necessary. Such a relationship is provided by independently solving for the charge and potential distributions in the fluid, and extrapolating the results to the interface. Still, examining the formal connection between the surface charge and potential in (2.5) and (2.6) is instructive. Specifically, it is worth examining the asymptotic behavior of $\sigma(\psi_s)$ and/or $\psi_s(\sigma)$ in different limits. Thus, for $e(\psi_N - \psi_s)/k_BT \ll 1$, equation (2.5) simplifies to

$$\sigma(\psi_s) = e\Gamma\frac{\delta e(\psi_N - \psi_s)/k_BT}{1 + \delta}, \tag{2.8}$$

and the surface charge is linearly proportional to the surface potential. Naturally, it is also proportional to the total number of surface ionizable groups, while the effect of the surface chemistry is accounted for by the factor $\delta/(1 + \delta)$, as well as by the Nernst potential (see equation (2.7)). The former can be further simplified for $\delta \ll 1$ ($K_- \ll K_+$) or $\delta \gg 1$ ($K_- \gg K_+$) to give

$$\begin{aligned} \sigma(\psi_s) &= e^2\Gamma\delta(\psi_N - \psi_s)/k_BT, \quad \delta \ll 1 \text{ or} \\ \sigma(\psi_s) &= e^2\Gamma(\psi_N - \psi_s)/k_BT, \quad \delta \gg 1 \end{aligned} \tag{2.9}$$

respectively. The second equation (2.9) does not depend on δ, but still may be dependent on the surface chemical equilibrium constants K_- and K_+ (2.1) through ψ_N and more specifically, the point of zero charge pI (see equation (2.7) and the text below).

If $|e(\psi_N - \psi_s)/k_BT| \gg 1$, then equation (2.5) becomes

$$\sigma(\psi_s) = \pm e\Gamma\frac{\delta \exp|e(\psi_N - \psi_s)/k_BT|/2}{1 + \delta \exp|e(\psi_N - \psi_s)/k_BT|/2}. \tag{2.10}$$

The sign will be plus or minus in accordance with the actual sign of $e(\psi_N - \psi_s)/k_BT$. Further simplifications depend on the value of the product $\delta \exp|e(\psi_N - \psi_s)/k_BT|$. The two limiting cases are

$$\begin{aligned} \sigma(\psi_s) &= \pm\, e\Gamma\delta \exp|e(\psi_N - \psi_s)/k_BT|/2, \\ &\qquad \delta \exp|e(\psi_N - \psi_s)/k_BT| \ll 1 \text{ and} \\ \sigma(\psi_s) &= \pm\, e\Gamma, \quad \delta \gtrsim 1. \end{aligned} \tag{2.11}$$

The second equation (2.11) corresponds to fully dissociated surface.

In some cases it may be convenient to invert and solve equation (2.5) for the surface potential. The result is

$$\frac{e[\psi_N - \psi_s(\sigma)]}{k_B T} = \left| \frac{\sigma + \sqrt{\delta^2(e^2\Gamma^2 - \sigma^2) + \sigma^2}}{\delta(e\Gamma - \sigma)} \right|. \tag{2.12}$$

The above example demonstrates that the relationship between the surface charge and potential can be rather complicated and is determined by the specific physical and chemical properties of the phases, forming the EDL. More details are available in the original paper [4].

The model of Chan *et al* [4] discussed above is based on two surface chemical reactions. Depending on the values of the equilibrium constants, it may turn out that one reaction dominates, while the other is much less important and can be ignored. Thus, the first charged regulation model of Ninham and Parsegian [3] is based only on the second reaction in (2.1), assuming that the first one does not contribute much to the surface charge regulation process. The single reaction models are important and were instrumental in obtaining many insightful results regarding the EDL properties, surface interactions, and colloid stability [7–11].

2.2 Effects due to physical adsorption

The physical adsorption of ions at the interface between a substrate and electrolyte solution may also have a profound effect on the overall interfacial charge. The adsorption may be a singular event, or may occur in conjunction with chemical charge regulation reactions. An example of an interface with charge that is governed by ionic adsorption is a silver iodide (AgI) colloidal solution [2]. An AgI colloidal particle suspended in KI solution adsorbs I^- ions. The result is a layer of negative charges, which may be considered as an integral part of the particle. Some of the dissolved K^+ counterions form the dense Stern layer [12] at the surface, as well as the diffuse cloud around the colloid. The I^- ions in this case are the PDIs. This example demonstrates that there is no clear boundary between chemical and physical surface charge regulation. In general, the overall charge of an interface is determined by the specific reactions involving the particular surface groups, as well as by the adsorption of ions in the Stern layer. The true location of the adsorbed ions near the interface is hard to always ascertain, and the solvent effects on the solution structure plays an important role [13]. These difficulties provide clear evidence that a more detailed molecular theory of EDLs would be beneficial.

The charge at the reactive surface (or the IHP) can be experimentally assessed by titration [14–16]. Alternative methods are based on various electrokinetic phenomena, which usually characterize charged surface in terms of the so-called electrokinetic ζ-potential [17, 18]. The electrokinetic phenomena involve the relative motion of fluid and substrate. Thus, electrophoresis refers to colloidal particle migration in electrolyte solutions in the presence of an externally applied electric field. Electroosmosis, on the other hand, describes the motion of fluid in small channels and capillaries that is again driven by an external field. The ζ-potential is defined at the shear plane where the relative motion occurs. This plane is typically not located at the reactive surface (i.e., the IHP), or at the Stern plane (the OHP). In fact, its location is hard to precisely pinpoint. Molecular dynamics analysis [19–23] suggests that the fluid motion at a smooth interface starts immediately after the first layer of adsorbed ions at the interface. This physical picture is consistent with the no-slip boundary condition [24], commonly used in fluid mechanics. The problem becomes even more difficult in the case of more realistic systems, where the surface may exhibit an inherent roughness, the species may not perfectly order in a well-defined plane due to entropy, or the adsorbed species may have different dimensions.

The ion adsorption at the interface between the electrolyte solution and the substrate was first analyzed by Stern [12], who accounted for the effect by using a modified Langmuir adsorption isotherm for each ionic species i in the form

$$\theta_i^{St} = \frac{v_0\rho_i \exp\left[(-\phi_{is} - q_i\psi_{St})/k_B T\right]}{1 + v_0\rho_i \exp\left[(-\phi_{is} - q_i\psi_{St})/k_B T\right]}, \tag{2.13}$$

where θ_i^{St} is the fraction of occupied sites (area) in the Stern plane, v_0 is the solvent molecular volume, ρ_i is the number density of species i in the bulk fluid and q_i is their charge, ϕ_{is} is the non-electrostatic energy of interaction between species i and the surface, and ψ_{St} is the electrostatic potential at the Stern plane. The quantity

$$\rho_s^{St} = \rho_i \exp\left[(-\phi_{is} - q_i\psi_{St})/k_B T\right], \tag{2.14}$$

is the number density of species i in the Stern layer. A major achievement of Stern's model is that takes into account the surface saturation at high ionic concentrations.

The Stern isotherm is commonly used in conjunction with a continuum description of the solution beyond the adsorption layer. The model does not have a rigorous theoretical foundation, and does not offer a self-consistent description of the EDL. This can be accomplished with a molecular level model that accounts for the detailed liquid structure everywhere in the fluid.

2.3 Structural effects on the ionic and solvent concentration at the interface

The surface charging process (chemical or physical) depends on the local concentration of the participating ionic species. For example, the surface chemical equilibria (2.1) is dependent on the concentration of PDIs (e.g., BH_2^+) in the immediate vicinity of the substrate–electrolyte interface. If the species interactions with the interface are of the hard-wall type, then it is the contact density of the PDIs that will enter in the reaction equilibria (2.1). If the interface interactions with the fluid species is through smooth potentials, the distribution can be averaged over a narrow range. For example, the average PDI concentration can be estimated from [25]

$$\langle \rho_{\mathrm{pdi}}^{s} \rangle = \frac{\int_0^{d_{\mathrm{pdi}}/2} dz \rho_{\mathrm{pdi}(z)} w(z)}{\int_0^{d_{\mathrm{pdi}}/2} dz w(z)}, \tag{2.15}$$

where d_{pdi} is the PDI diameter and $w(z)$ is a weighting factor that accounts for the fact that ions have finite sizes, which affects the concentration that enters the chemical equilibria (2.1). The PDIs in the subsurface layer are not perfectly ordered, and some may contribute to the average density $\langle \rho_{\mathrm{pdi}}^{s} \rangle$ despite the fact that they may be slightly shifted away from the reactive surface. Hence, the probability for them to chemically interact with the ionizable groups is less compared to those that are closer (see figure 2.1).

An alternative approach is to define a cavity function [26, 27], which for a simple one-dimensional case reads

$$y_i(z) = \rho_i(z) \exp\left[\frac{\mathscr{V}_{\mathrm{ext}}(z)}{k_B T}\right]. \tag{2.16}$$

The cavity function is not affected by any difficulties that may arise from the excluded volume of the ions and molecules, and discontinuities in the interactions energies upon contact with the charged interface. Instead, it smoothly extends to $z = 0$ and allows to formally define the contact density as

$$\rho_i(0) = y_i(0) \exp\left[\frac{-\mathscr{V}_{\mathrm{ext}}(0)}{k_B T}\right]. \tag{2.17}$$

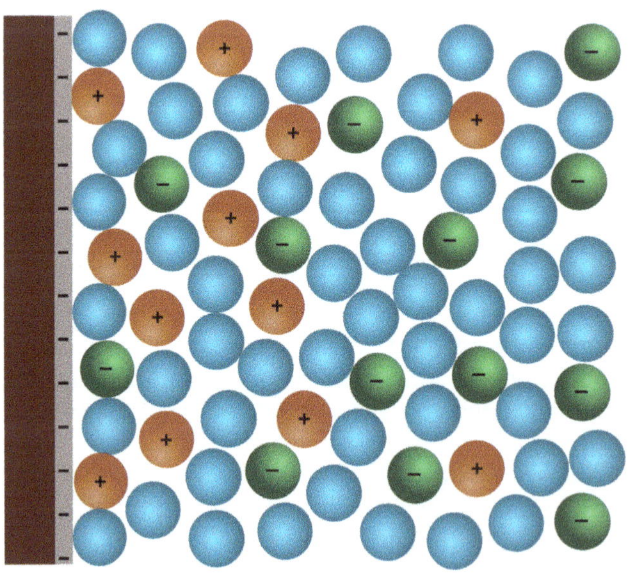

Figure 2.1. Thermal disturbances preventing ions and solvent molecules from ordering on a single plane next to the charged interface.

Equation (2.17) can be used to express the density of the PDIs at the reactive interface and that value can be used to calculate the surface equilibrium dissociation, and therefore the surface charge. Thus, $\rho^{s}_{\mathrm{BH}_2^+}$ in equation (2.2) can be replaced by $\rho_{\mathrm{pdi}}(0)$, which is formally correct for any interactions between the charged interface and the individual fluidic species.

References

[1] Gibbs J W 1993 *The Scientific Papers of J. Willard Gibbs* vol 1 (Woodbridge, CT: Ox Bow)

[2] Dukhin S S 1974 Development of notions as to mechanism of electrokinetic phenomena and the structure of the colloid micelle *Surface and Colloid Science* vol 7 ed E Matijevic (New York: Wiley Interscience) ch 1 pp 1–49

[3] Ninham B W and Parsegian V A 1971 Electrostatic potential between surface bearing ionizable groups in ionic equilibrium with physiologic saline solution *J. Theor. Biol.* **31** 405–28

[4] Chan D Y C, Perram J W, White L R and Healy T W 1975 Regulation of surface potential at amphoteric surfaces during particle–particle interaction *J. Chem. Soc., Faraday Trans.* I **71** 1046–57

[5] Ettelaie R and Buscall R 1995 Electrical double layer interactions for spherical charge regulating colloidal particles *Adv. Coll. Interf. Sci.* **61** 131–60

[6] Chan D Y C, Healy T W and White L R 1976 Electrical double layer interactions under regulation by surface ionization equilibria–dissimilar amphoteric surfaces *J. Chem. Soc., Faraday Trans.* I **72** 2844–65

[7] Behrens S H and Borkovec M 1999 Electric double layer interaction of ionizable surfaces: Charge regulation for arbitrary potentials *J. Chem. Phys.* **111** 382–5

[8] Behrens S H and Borkovec M 1999 Electrostatic interaction of colloidal surfaces with variable charge *J. Phys. Chem.* **103** 2918–28

[9] Behrens S H and Borkovec M 1999 Exact Poisson–Boltzmann solution for the interaction of dissimilar charge–regulating surfaces *Phys. Rev.* E **60** 7040–8

[10] Popa I, Sinha P, Finess M, Maron P, Papastavrou G and Borkovec M 2010 Importance of charge regulation in attractive double–layer forces between dissimilar surfaces *Phys. Rev. Lett.* **104** 228301

[11] Valmacco V, Elzbieciak-Wodka M, Herman D, Trefalt G, Maroni P and Borkovec M 2016 Forces between silica particles in the presence of multivalent cations *J. Colloid Interface Sci.* **472** 108–15

[12] Stern O 1924 Zur theorie der elektrolytischen doppelschicht *Z. Electrochem* **30** 508–16

[13] Grahame D C 1947 The electrical double layer and the theory of electrocapillarity *Chem. Rev.* **41** 441–501

[14] Morgan J W, Forster C F and Evison L 1990 A comparative study of the nature of biopolymers extracted from anaerobic and activated sludges *Wat. Res.* **24** 743–50

[15] Mikkelsen L H 2003 Applications and limitations of the colloid titration method for measuring activated sludge surface charges *Wat. Res.* **37** 2458–66

[16] Garikipati S 2005 Evaluation of colloidal titration for the determination of surface charge of activated sludge flocs *Masters Thesis* Chalmers University of Technology

[17] Hunter R J 1981 *Zeta Potential in Colloid Science* (New York: Academic)

[18] Dukhin S S and Derjaguin B V 1974 Equilibrium double layer and electrokinetic phenomena ed E Matijevic *Surface and Colloid Science* vol 7 (New York: Wiley Interscience) ch 2 pp 50–272

[19] Thompson A 2003 Nonequilibrium molecular dynamics simulation of electro-osmotic flow in a charged nanopore *J. Chem. Phys.* **119** 7503–11
[20] Qiao R and Aluru N R 2003 Ion concentrations and velocity profiles in nanochannel electroosmotic flows *J. Chem. Phys.* **118** 4692–701
[21] Qiao R and Aluru N R 2004 Charge inversion and flow reversal in a nanochannel electro-osmotic flow *Phys. Rev. Lett.* **92** 198301
[22] Qiao R and Aluru N R 2005 Scaling of electrokinetic transport in nanometer channels *Langmuir* **21** 8972–7
[23] Wu P and Qiao R 2011 Physical origins of apparently enhanced viscosity of interfacial fluids in electrokinetic transport *Phys. Fluids* **23** 072005
[24] Landau L D and Lifshitz E M 1987 *Fluid Mechanics* (Amsterdam: Pergamon)
[25] Fleharty M E, van Swol F and Petsev D N 2016 Solvent role in the formation of electric double layers with surface charge regulation: A bystander or a key participant? *Phys. Rev. Lett.* **116** 048301
[26] McQuarrie D A 2000 *Statistical Mechanics* (University Science Books)
[27] Hansen J P and McDonald I R 2006 *Theory of Simple Liquids* (New York: Academic)

IOP Publishing

Molecular Theory of Electric Double Layers

Dimiter N Petsev, Frank van Swol and Laura J D Frink

Chapter 3

Continuum models of the electric double layers

3.1 The Poisson–Boltzmann equation

The relationship between the fluid charge $\rho_{el}(\mathbf{r})$ and potential $\psi(\mathbf{r})$ is given by the Poisson equation (1.8). It can be significantly simplified by assuming that the charge density $\rho_{el}(\mathbf{r})$ follows an equilibrium Boltzmann distribution in the form

$$\rho_{el}(\mathbf{r}) = \sum_i q_i \rho_i^0 \exp\left[\frac{-q_i \psi(\mathbf{r})}{k_B T}\right], \tag{3.1}$$

where ρ_i^0 is the number bulk density (far from any charged surfaces) of ion i, and q_i is the ionic charge number (including its sign) in units of the elementary charge $e = 1.602 \times 10^{-12}$ C. Inserting equation (3.1) into (1.8) results in the Poisson–Boltzmann (PB) equation

$$\nabla^2 \psi(\mathbf{r}) = -\frac{1}{\varepsilon \varepsilon_0} \sum_i q_i \rho_i^0 \exp\left[\frac{-q_i \psi(\mathbf{r})}{k_B T}\right]. \tag{3.2}$$

The PB equation is less general than the Poisson equation (1.8). It only applies to systems in equilibrium. In addition, it does not take into account any interactions between the ions, or between the ions and the solvent molecules. In fact, the solvent molecular structure is not explicitly included. Its only contribution is to the relative dielectric permittivity ε. Hence, the PB model corresponds to an ideal gas of point-like ions in an external (mean) field [1]. Despite the significant approximations in the PB equation, it has been very commonly used in the theoretical analysis of charged interfaces and electrolyte solutions. A good example of both the power and the limitations of the PB approach is the Derjaguin–Landau–Verwey–Overbeek (DLVO) theory of colloid stability [2–6].

Equation (3.2) is a second-order differential equation and therefore requires two boundary conditions. They are usually defined at the charged interfaces, or at a charged interface and the solution bulk, far from any surfaces. As discussed in

doi:10.1088/978-0-7503-2276-8ch3

chapter 2, the thermodynamic consistent surface boundary condition should reflect the chemical reaction (or physical adsorption) equilibrium. However, it is mathematically simpler to use the Dirichlet (fixed surface potential ψ_s) or Neumann (fixed surface charge $\sigma = -\varepsilon\varepsilon_0(\nabla\psi)_s$) boundary conditions at the surface.

The PB equation is nonlinear, and in some cases, it may be challenging to obtain a straightforward analytical solution. The next sections present a brief overview of some of the exact solutions of the PB equation for electric double layers (EDLs), as well as the most commonly used approximations.

3.2 Electric double layer models based on the Poisson–Boltzmann equation: exact and approximate solutions

3.2.1 Single flat electric double layer: the Gouy–Chapman model

The Gouy [7, 8] and Chapman [9] analysis presents a simple example of an exact analytical solution of the nonlinear PB model of an EDL. The system under consideration is a single EDL that is bound on one side by an infinite flat charged interface and propagating to infinity in normal z-direction. Such a geometry simplifies the PB equation (3.2) to a one-dimensional ordinary differential equation. In addition, the electrolyte solution is binary and symmetric (i.e., $q_+ = -q_- = q > 0$ and $\rho_+^0 = \rho_-^0 = \rho^0$). As a result, equation (3.2) simplifies to

$$\frac{d^2\tilde{\psi}(z)}{dz^2} = \kappa^2 \sinh(\tilde{\psi}), \tag{3.3}$$

where $\tilde{\psi} = q\psi/k_BT$ is the dimensionless electrostatic potential and

$$\kappa = \sqrt{\frac{2\rho^0 q^2}{\varepsilon\varepsilon_0 k_B T}} \tag{3.4}$$

is the so-called Debye screening parameter [10]. It offers an estimate of the potential decreasing rate with the distance from the charged interface. The scaling factor k_BT/q has the dimension of volts. For $q = e$, and at room temperature of $T = 298$ K, the scaling factor becomes $k_BT/e = 25.6$ mV.

Equation (3.3) can be easily solved using a simple trick [3, 11]. First, both sides are multiplied by the factor $2d\tilde{\psi}/dz$, which transforms the left hand side of (3.3) into

$$2\frac{d\tilde{\psi}}{dz}\frac{d^2\tilde{\psi}}{dz^2} = \frac{d}{dz}\left(\frac{d\tilde{\psi}}{dz}\right)^2. \tag{3.5}$$

Equation (3.3) then assumes the form

$$d\left(\frac{d\tilde{\psi}}{dz}\right)^2 = 2\kappa^2 \sinh(\tilde{\psi})d\tilde{\psi}. \tag{3.6}$$

Equation (3.6) is integrated by using the fact that at $z \to \infty$ both $\tilde{\psi}$ and $d\tilde{\psi}/dz$ vanish. The result is

$$\frac{d\tilde{\psi}}{dz} = -\kappa\sqrt{2[\cosh(\tilde{\psi}) - 1]} = -2\kappa \sinh\left(\frac{\tilde{\psi}}{2}\right). \tag{3.7}$$

Equation (3.7) allows to calculate the surface charge. From $\sigma = -\varepsilon\varepsilon_0(d\psi/dz)$ follows

$$\tilde{\sigma} = \frac{q\sigma}{\kappa\varepsilon\varepsilon_0 K_B T} = \sqrt{2[\cosh(\tilde{\psi}_s) - 1]} = 2\sinh\left(\frac{\tilde{\psi}_s}{2}\right). \tag{3.8}$$

Rearranging equation (3.7) and integrating from the surface ($z = 0$, and $\tilde{\psi} = \tilde{\psi}_s$) to a location in the EDL (z and $\tilde{\psi}$) leads to

$$\int_{\tilde{\psi}_s}^{\tilde{\psi}} d\tilde{\psi}' \frac{d\tilde{\psi}'}{2\sinh\left(\frac{\tilde{\psi}'}{2}\right)} = -\kappa z. \tag{3.9}$$

Integrating the left hand side results in

$$\tilde{\psi}(z) = 2\ln\left[\frac{1 + \gamma\exp(-\kappa z)}{1 - \gamma\exp(-\kappa z)}\right],\ \gamma = \tanh\frac{\tilde{\psi}_s}{4} \tag{3.10}$$

or alternatively

$$\tilde{\psi}(z) = 4\tanh^{-1}[\gamma\exp(-\kappa x)]. \tag{3.11}$$

The expression for the spatial potential profile $\tilde{\psi}(z)$ is simpler for small surface potentials $\tilde{\psi}_s$, or small γ. Expanding equation (3.10) (or (3.11)) in series around $\tilde{\psi}_s \to 0$ leads to

$$\tilde{\psi}(z) = \tilde{\psi}_s\exp(-\kappa z). \tag{3.12}$$

The same solution could be obtained by the linearization of equation (3.3), expanding the $\sinh(\tilde{\psi})$ function in a series and keeping only the linear term, or

$$\frac{d^2\tilde{\psi}}{dz^2} = \kappa^2\tilde{\psi}. \tag{3.13}$$

Equation (3.13) then can be solved with boundary conditions $\tilde{\psi} = \tilde{\psi}_s$ at the charged interface $z = 0$ and $\tilde{\psi} = 0$, $d\tilde{\psi}/dz = 0$ and $z \to \infty$. The solution is equation (3.12).

The linearization of equation (3.3) works surprisingly well even for moderate potentials, or $\tilde{\psi}_s \sim 1$ instead of $\tilde{\psi}_s \ll 1$. The reason becomes clear after examining the series expansion of $\sinh(\tilde{\psi})$

$$\sinh(\tilde{\psi}) = \tilde{\psi} + \frac{\tilde{\psi}^3}{3!} + \frac{\tilde{\psi}^5}{5!} + \frac{\tilde{\psi}^7}{7!}\cdots, \tag{3.14}$$

which reveals that all even terms cancel. In addition, each term of higher order is much less than the previous one, not only because of the increasing power, but also because of the factorial in the denominator. The latter effect is so powerful that equations (3.10) (or (3.11)) and (3.13) may valid even for $\tilde{\psi}_s \gtrsim 1$.

Figure 3.1 shows the electrostatic potential distribution in the solution near the charged interface. The potential drops with the scaled distance κz because of the charge screening effect of the dissolved ions. The characteristic screening length is given by κ as defined by equation (3.4). The curves in the figure correspond to different values of the surface potentials $\tilde{\psi}_s = q\psi_s/k_B T$. Only the absolute magnitudes of potentials are shown.

A comparison between the exact solutions of the PB equation (3.10) and (3.11) with the linear approximation (3.12), shows a very good agreement even for $\tilde{\psi}_s \gtrsim 1$ in accordance with the above discussion. In fact, a noticeable discrepancy appears around $\tilde{\psi}_s \sim 2$, as evident in figure 3.2. The main difference between the exact and approximate solutions is that the former predicts a steeper decline of the electrostatic potential with distance κz than a simple exponential. There is no discrepancy at $z = 0$ because of the fixed potential $\tilde{\psi}_s$ at the boundary. The difference between the solutions disappears at large distances, because as the potential drops, it better satisfies the linear approximation. Therefore, the greatest difference is at some finite distance from the charged surface. Figure 3.3 presents the relative difference between the exact solution $\tilde{\psi}_{\text{exact}}$ and the linear approximation $\tilde{\psi}_{\text{approx}}$ for $\tilde{\psi}_s = 2$ as in figure 3.2. The error introduced by using the approximate equation (3.12) is

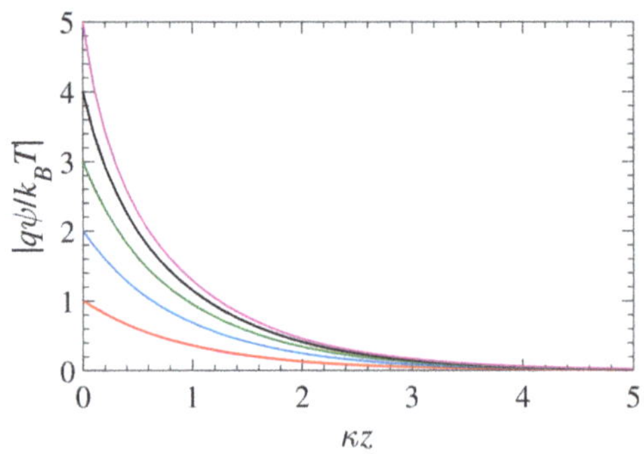

Figure 3.1. Potential magnitude distribution profiles $|\tilde{\psi}| = |q\psi/k_B T|$ vs the scaled distance (see equations (3.10) and (3.11)). The different curves are for different values of the surface potential $\tilde{\psi}_s = q\psi_s/k_B T$. From top to bottom: $\tilde{\psi}_s = 5, 4, 3, 2, 1$.

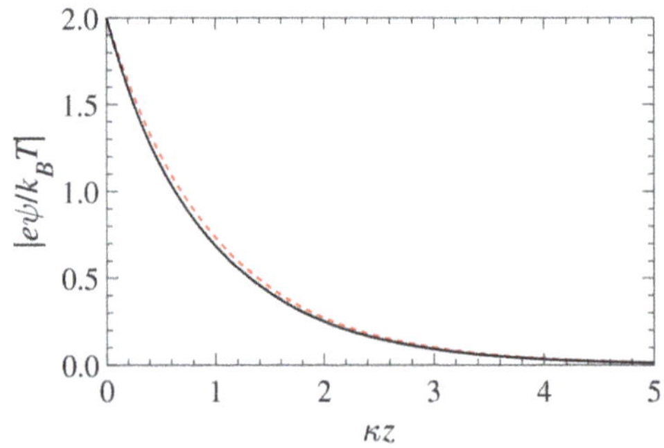

Figure 3.2. Comparison between equations (3.10) (or equivalently (3.11), solid line) and (3.12) (dashed line) for $|\tilde{\psi}| = |q\psi/k_B T| = 2$.

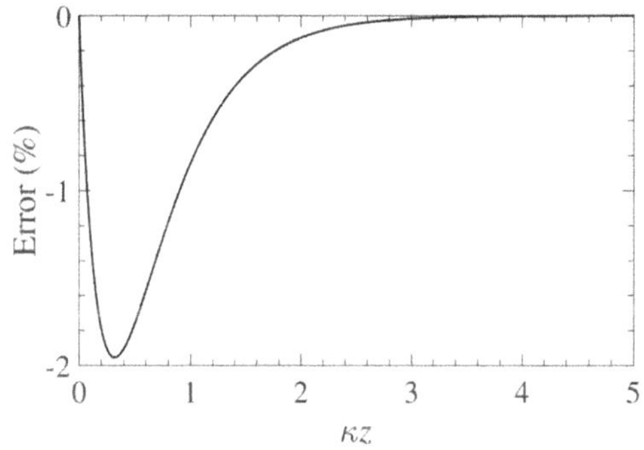

Figure 3.3. Relative error introduced by the linear PB model for surface potential $\tilde{\psi}_s = 2$.

$$\text{Error}(\%) = \frac{\tilde{\psi}_{\text{exact}} - \tilde{\psi}_{\text{approx}}}{\tilde{\psi}_{\text{exact}}} \times 100. \tag{3.15}$$

Hence, for $\tilde{\psi}_s = 2$, the maximum error is about 2 %, and it occurs at $\kappa z \sim 0.3$. Both the magnitude and the location of the maximum error depend on the value of the potential $\tilde{\psi}_s$. For greater surface potentials (e.g., $\tilde{\psi}_s > 2$), the error rapidly increases and moves to greater scaled distances κz. For example, for $\tilde{\psi}_s = 3$ the maximum error magnitudes increases to 9%, while for $\tilde{\psi}_s = 4$ it is over 25%.

An exact analytical solution of the PB equation can also be obtained for asymmetric 2:1 or 1:2 electrolytes (see [4]). This is accomplished by replacing the potential $\tilde{\psi}(z)$ with the following functions:

$$\begin{aligned} v_1(\tilde{\psi}) &= \ln\left[\frac{3}{1 + 2\exp(-\tilde{\psi})}\right], \quad \text{for 2:1 electrolyte} \\ v_2(\tilde{\psi}) &= \ln\left[\frac{1 + 2\exp(\tilde{\psi})}{3}\right], \quad \text{for 1:2 electrolyte.} \end{aligned} \tag{3.16}$$

Using these substitutions allows to write equation (3.3) in the form

$$\frac{d^2\tilde{v}_i(z)}{dz^2} = \kappa^2 \sinh(\tilde{v}_i), \quad i = 1, 2. \tag{3.17}$$

The solution to (3.17) is identical to (3.10) or (3.11), but with v_i instead of ψ. The actual potential will then be recovered by solving (3.16) for $\tilde{\psi}$.

3.2.2 Interaction between two flat electric double layers

The EDL interaction is a key element of the DLVO theory of colloid stability [2, 3]. When two charged surfaces approach each other in an electrolyte solution, the extending EDLs overlap, which leads to an electrostatic interaction. If the charge on both surfaces has the same sign, the interaction is usually repulsive. This repulsion can be expressed as the electrostatic disjoining pressure $\Pi(h)$ between the surfaces, separated by distance h [2, 4].

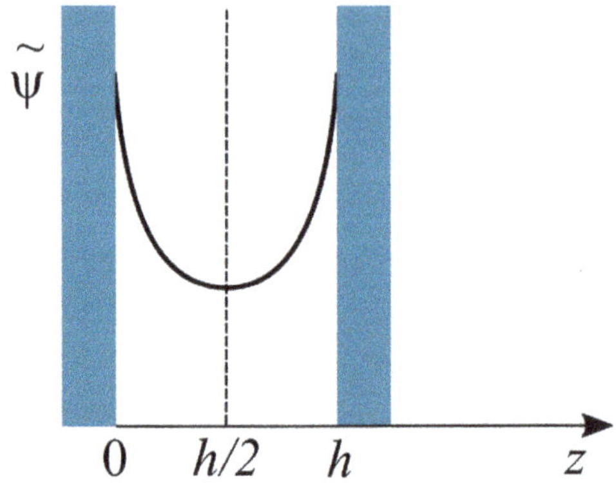

Figure 3.4. A sketch of the electrostatic potential profile between two identically charged surfaces separated by a distance h.

The electrostatic potential in the gap between two charged surfaces in the presence of symmetric binary electrolyte is obtained using equation (3.3). The boundary conditions (in the classical version of the theory) for identically charged surfaces are specified at one of the interfaces and the midplane ($\tilde{\psi} = \tilde{\psi}_s$ at $z = 0$ and $d\tilde{\psi}/dz$ at $z = h/2$, see figure 3.4). The result for the first integral is [3, 12]

$$\frac{d\tilde{\psi}}{dz} = -\kappa\sqrt{\cosh(\tilde{\psi}) - \cosh(\tilde{\psi}_m)}, \ \tilde{\psi}_m = \tilde{\psi}(h/2). \tag{3.18}$$

A second integration leads to the following relationship between electrostatic potential spatial distribution $\tilde{\psi}$ and the running coordinate z:

$$\int_{\tilde{\psi}_s}^{\tilde{\psi}} \frac{d\tilde{\psi}'}{\sqrt{\cosh(\tilde{\psi}') - \cosh(\tilde{\psi}_m)}} = -\int_0^z \kappa dz'. \tag{3.19}$$

The left hand side of (3.19) can be written in the form of an elliptic integral. According to Legendre, a normal incomplete elliptic integral of the first kind is defined as [13]

$$F(k, \varphi) = \int_0^{\varphi} \frac{d\varphi'}{\sqrt{1 - k^2 \sin(\varphi')}}, \tag{3.20}$$

where k is the modulus and ϕ is the amplitude. The complete counterpart of (3.20) is derived by setting $\varphi = \pi/2$, or

$$K(k) = \int_0^{\pi/2} \frac{d\varphi'}{\sqrt{1 - k^2 \sin(\varphi')}} = F(k, \pi/2). \tag{3.21}$$

Substituting $k = \exp(-\tilde{\psi}_m)$ and $\sin(\varphi) = \exp[-(\tilde{\psi}-\tilde{\psi}_m)]$ enables the formal integration of (3.19). The result is [12]

$$2k^{1/2}[K(k) - F(k, \varphi)] = \kappa z. \tag{3.22}$$

Equation (3.22) describes the spatial variation of the electrostatic potential $\tilde{\psi}(z)$. The disjoining pressure $\Pi_{el}(h)$, acting between the charged surfaces separated by a distance h is [4]

$$\Pi_{el}(h) = -\int_0^{\tilde{\psi}} d\tilde{\psi}' \rho_{el}(\tilde{\psi}') - \left(\frac{\varepsilon\varepsilon_0}{2}\nabla\tilde{\psi}\right)^2. \quad (3.23)$$

Applying the Boltzmann approximation to ρ_{el} for binary symmetric electrolyte (cf, equation (3.3)) transforms equation (3.23) into

$$\frac{\Pi_{el}(h)}{\rho^0 k_B T} = 2[\cosh(\tilde{\psi}) - 1] - \kappa^{-2}(\nabla\tilde{\psi})^2. \quad (3.24)$$

Interestingly, the disjoining pressure (3.23) coincides with the electrostatic free energy functional (see section 3.2.6). The disjoining pressure can be calculated by integrating the function $\rho_{el}(\tilde{\psi})$ on any plane located between the charged surfaces that is parallel to them. The potential at that location is determined by equation (3.22). If the two surfaces have identical surface charge and potential, the potential distribution between them is symmetric with respect to the midplane (see figure 3.4). At this location $\nabla\tilde{\psi} = 0$, and the expression for the pressure becomes simpler, or

$$\frac{\Pi_{el}(h)}{\rho^0 k_B T} = 2[\cosh(\tilde{\psi}_m) - 1]. \quad (3.25)$$

Figure 3.5 shows the dependence of electrostatic disjoining pressure on the distance between the charged surfaces. It is calculated using equation (3.22) for two identical charged surfaces with fixed surface potentials. The pressure is repulsive and drops rather steeply with the distance. Therefore, for $z \sim \kappa^{-1}$, the pressure decreases more than three times. This decrease is due to the fact that the hyperbolic cosine is a strong function of $\tilde{\psi}_m$, which subsequently also drops almost exponentially with the distance κz.

A particularly straightforward result is obtained for small surface potentials. Indeed, solving the linear PB equation (3.13) with fixed and low surface potential

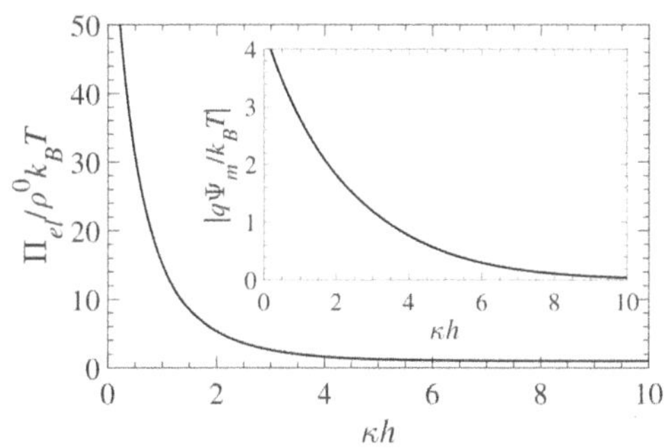

Figure 3.5. Disjoining pressure between two identical surfaces $|\tilde{\psi}_s| = |q\psi_s/k_BT| = 4$ against the distance in units of κ^{-1}. The inset shows the change of the potential in the midplane $|\tilde{\psi}_m| = |q\psi_m/k_BT|$.

$\tilde{\psi}(0) = \tilde{\psi}_s$ and $d\tilde{\psi} = 0$ for $z = h/2$ (see figure 3.4) results in the following potential distribution

$$\tilde{\psi}(z) = \tilde{\psi}_s \frac{\cosh[\kappa(z - h/2)}{\cosh(\kappa h/2)}. \tag{3.26}$$

The midplane potential is the $\tilde{\psi}_m = \tilde{\psi}_s / \cosh(\kappa h/2) \ll 1$ and equation (3.25) can be expanded in series, keeping the quadratic terms, to give

$$\frac{\Pi_{el}^{\psi}(h)}{\rho^0 k_B T} = \tilde{\psi}_m^2 = \frac{\tilde{\psi}_s^2}{\cosh^2(\kappa h/2)}. \tag{3.27}$$

Equation (3.27) is valid for the special case in which the surface potential $\tilde{\psi}_s$ does not change as the distance between the surfaces is varied [4]. This means that the magnitude of the surface charge σ must change, increasing or decreasing with h. For $h = 0$, the surface charge vanishes. An alternative case is maintaining a constant surface charge $\tilde{\sigma}$, and variable surface potential $\tilde{\psi}_s$. The surface charge $\tilde{\sigma} = -(d\tilde{\psi}/\kappa dz)_{z=0}$ is derived using equation (3.27) and the result is

$$\tilde{\sigma} = \tilde{\psi}_s \tanh\left(\frac{\kappa h}{2}\right) = \tilde{\psi}_m \sinh\left(\frac{\kappa h}{2}\right). \tag{3.28}$$

The respective expression for the disjoining pressure at constant surface charge is then

$$\frac{\Pi_{el}^{\sigma}(h)}{\rho^0 k_B T} = \tilde{\psi}_m^2 = \frac{\tilde{\sigma}^2}{\sinh^2(\kappa h/2)}. \tag{3.29}$$

Equation (3.29) is problematic for small distances between the surfaces or $\kappa h < 1$. At such separations, the magnitude of the surface potential increases without a limit, which is clearly nonphysical. It still gives the correct asymptotic dependence at large distances (i.e., $\kappa h \gg 1$), which is $\Pi_{el} \sim h^{-2}$, as shown by Langmuir [12].

Other simplified expressions for the electrostatic potential and disjoining pressure between two flat charged surfaces are based on the nonlinear superposition approximation. It is applicable to arbitrarily high surface potentials but limited to weak EDL overlap, which typically means that $\kappa h > 4$. The idea is that the potential profile can be presented as a superposition of the individual (Gouy–Chapman type) EDLs at the two surfaces. The combined potential in the gap of thickness h is then

$$\tilde{\psi}(z) = 2\ln\left[\frac{1 + \gamma\exp(-\kappa z)}{1 - \gamma\exp(-\kappa z)}\right] + 2\ln\left[\frac{1 + \gamma\exp(-\kappa(h - z))}{1 - \gamma\exp{-\kappa(h - z))}}\right]. \tag{3.30}$$

If the EDL overlap is weak, the potential in the midplane is small and equal to the sum of the contributions of the single layer potentials as seen from (3.30). Replacing this value into equation (3.25) and expanding the latter in series up to the quadratic terms leads to

$$\frac{\Pi_{el}(h)}{\rho^0 k_B T} = 64\gamma^2\exp(-\kappa h). \tag{3.31}$$

The disjoining pressure allows determination of the Helmholtz free energy of interaction f_{el} by integrating the work to bring the two surfaces from infinity to a certain distance h [3], or

$$f_{el}(h) = \int_h^\infty dh' \Pi_{el}(h'). \tag{3.32}$$

Hence, by inserting $\Pi_{el}(h)$ (from equations (3.25), (3.27), or (3.31)) into (3.32), the contribution of electrostatic interactions to the Helmholtz energy can be obtained in exact or approximate form. Thus, the interaction energies per unit area in the Debye and Huckel low potential $\tilde{\psi}_s$ and low charge $\tilde{\sigma}$ limits are (see equations (3.27) and (3.29))

$$\tilde{f}_{el}^{\psi}(h) = \frac{f_{el}^{\psi}(h)\kappa}{\rho^0 k_B T} = 2\tilde{\psi}_s^2\left[1 - \tanh\left(\frac{\kappa h}{2}\right)\right] \tag{3.33}$$

and

$$\tilde{f}_{el}^{\sigma}(h) = \frac{f_{el}^{\sigma}(h)\kappa}{\rho^0 k_B T} = 2\tilde{\sigma}^2\left[\coth\left(\frac{\kappa h}{2}\right) - 1\right]. \tag{3.34}$$

For arbitrarily high surface potential but weak EDL overlap (i.e., $\kappa h > 4$), one can use equation (3.31) to obtain

$$\tilde{f}_{el}(h) = \frac{f_{el}(h)\kappa}{\rho^0 k_B T} = 64\gamma^2 \exp(-\kappa h). \tag{3.35}$$

More details and generalizations in the cases of asymmetric electrolytes are presented in [4].

It is instructive to compare the distance dependence of the free energies of interaction at constant potential (see equation (3.33)) and constant charge (see equation (3.34)), which are shown in figure 3.6. There is a significant discrepancy between the two cases at small distances ($\kappa h \lesssim 3$). In fact, the constant charge curve diverges as κh approaches zero. This behavior is due to the form of the disjoining pressure as defined by equation (3.29). The plot demonstrates that the constant charge expression for the free energy (3.34) is inaccurate even for $\kappa h \gtrsim 1$. For larger separations (e.g., $\kappa h > 3$), the two curves come to a very good agreement. The nonlinear superposition equation (3.35) is also practically indistinguishable from (3.33) and (3.34) for $\kappa h > 4$ and low surface potentials (for the case depicted in figure 3.6 $\tilde{\psi}_s = 1$). In addition, the approximations (for low potentials and/or weak EDL overlap) are also in excellent agreement with the exact results based on equations (3.22), (3.23), and (3.25).

Many systems of practical interest include interactions between surfaces that have different potentials and/or charges. Obviously, the interaction between oppositely charged surfaces is attractive. In those cases, both the potentials and charges satisfy the inequalities $\psi_{s1}\psi_{s2} < 0$ and $\sigma_1\sigma_2 < 0$, where 1 and 2 refer to the two surfaces. However, the situation where the potentials at the two surfaces have the same sign

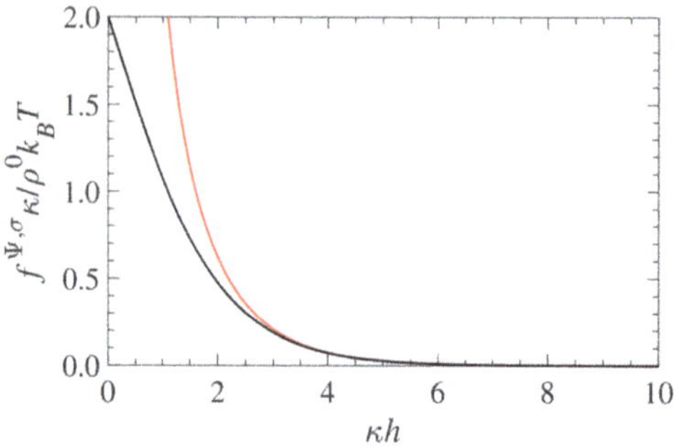

Figure 3.6. A comparison between the free energies of electrostatic interactions vs scaled distance at constant potential $\tilde{\psi}_s = 1$ (equation (3.33), black curve) and constant charge $\tilde{\sigma} = 1$ (equation (3.34), red curve).

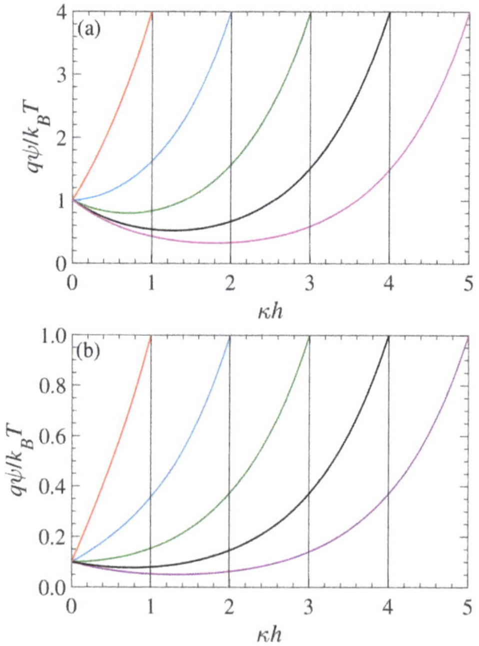

Figure 3.7. Potential distribution between two surfaces having potentials with the same sign, but different magnitudes, $\tilde{\psi}_{s1}\tilde{\psi}_{s2} > 0$. (a) $\tilde{\psi}_{s1} = 1$ and $\tilde{\psi}_{s2} = 4$; (b) $\tilde{\psi}_{s1} = 0.1$ and $\tilde{\psi}_{s2} = 1$. Surface 1 is on the left and surface 2 is on the right.

(both are positive or negative, so $\psi_{s1}\psi_{s2} > 0$) but different magnitude can be more complicated. Simple analysis shows that at certain distances, the surface charge densities may switch from same to opposite charges, thus leading to an effective attraction [4].

Figure 3.7 shows the potential distribution between two differently charged flat surfaces. The top panel (figure 3.7(a)) presents the high surface potential case where the surface potential on the left is $\tilde{\psi}_{s1} = 1$ and the one on the right is $\tilde{\psi}_{s2} = 4$. At larger separations (e.g., $\kappa h = 3$ and $\kappa h = 4$), both the potentials and the charges have the same sign, hence the interactions should be repulsive. Around $\kappa h = 0.2$, the charge at the lower potential surface is $\tilde{\sigma} = -\kappa^{-1}(\nabla\tilde{\psi})_{s1} \approx 0$, as the slope is horizontal. At closer distances, the charge (or the slope) at surface 1 changes sign and becomes opposite to the charge on the right at surface 2, or $\tilde{\sigma}_1\tilde{\sigma}_2 < 0$. Therefore, the two surfaces experience attraction. Note that this interaction depends on the distance, and the attraction appears at smaller separations. Figure 3.7(b), on the other hand, depicts the low potential case, $\tilde{\psi}_{s1} = 0.1$ and $\tilde{\psi}_{s2} = 1.0$. The charge reversal on the left surface occurs at greater distances ($\kappa h \approx 0.3$).

In the general case, the computation of the disjoining pressure requires the full equation (3.23) since the symmetry is broken and the second term is not zero at the midplane. However, the analysis is simple for low potentials where the potential distribution is determined by equation (3.13) with boundary conditions $\tilde{\psi}(0) = \tilde{\psi}_{s1}$ and $\tilde{\psi}(h) = \tilde{\psi}_{s2}$ (see figure 3.7). The result is [4]

$$\frac{\Pi_{el}(h)}{\rho^0 k_B T} = \frac{2\tilde{\psi}_{s1}\tilde{\psi}_{s2}\cosh(\kappa h) - \left(\tilde{\psi}_{s1}^2 + \tilde{\psi}_{s2}^2\right)}{\sinh^2(\kappa h)}. \tag{3.36}$$

Recent measurements of the forces between differently charged surfaces, however, revealed that the assumption of constant potential (or charge) is quantitatively inaccurate and the specific surface chemistry (i.e., charge regulation) needs to be properly taken into account [14].

3.2.3 Curved electric double layers

Electric double layers near curved surfaces present both fundamental and practical interest. Unlike the flat EDL examples discussed above, curved geometries like spherical or cylindrical do not allow for simple analytical solutions. Hence, the choice is between numerically solving equation (3.2) or applying various approximations. Figure 3.8 shows two examples for physical systems where the EDL curvature is important. Others (not shown in the figure) include spherical cavities, rod-shaped colloidal particles, etc.

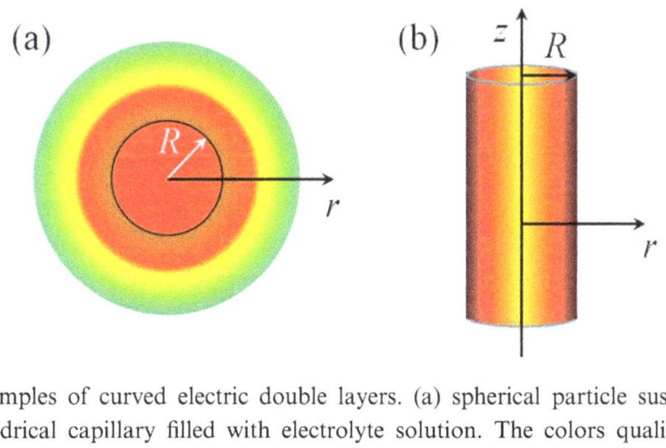

Figure 3.8. Two examples of curved electric double layers. (a) spherical particle suspended in electrolyte solution. (b) A cylindrical capillary filled with electrolyte solution. The colors qualitatively represent the potential magnitude. Red corresponds to high potential, yellow to intermediate and green to low.

Spherical electric double layers: the low surface charge and potential approximation

A celebrated example of the use of the low-potential approximation to describe spherical EDLs is the Debye and Huckel theory of strong electrolytes [10]. A key quantity in this theory is the potential distribution around a spherical ion in the presence of the remaining ions in the electrolyte solution. The potential is assumed to be low everywhere around the central ion and is determined from the linear PB equation (see equation (3.13)) for a spherical geometry with central symmetry

$$\frac{1}{r^2}\frac{d}{dr}r^2\frac{d\tilde{\psi}}{dr} = \kappa^2\tilde{\psi}. \tag{3.37}$$

Inside the ion, the charge density is $\rho_{el} = 0$, and the potential distribution is described by

$$\frac{1}{r^2}\frac{d}{dr}r^2\frac{d\tilde{\psi}}{dr} = 0. \tag{3.38}$$

The boundary conditions necessary to solve equations (3.37) and (3.38) are $(d\tilde{\psi}/dr) = 0$ at $r = 0$, continuity of the potentials $((\tilde{\psi}_s)_{\text{inside}} = (\tilde{\psi}_s)_{\text{outside}} = \tilde{\psi}_s))$ and their derivatives $((d\tilde{\psi}/dr)_{\text{inside}} = (d\tilde{\psi}/dr)_{\text{outside}})$ at ion-solution interface $r = R$, and $\tilde{\psi} = 0$ at $r \to \infty$. A surprisingly well-working additional assumption is that the dielectric constant inside the ion is the same as for the solvent.

The solution for the electrostatic potential is then [10, 15, 16]

$$\tilde{\psi}(r) = \tilde{\psi}_s\frac{R}{r}\exp[-\kappa(r - R)]. \tag{3.39}$$

The surface charge is found from the potential gradient at the particle surface,

$$\tilde{\sigma} = \frac{|q|\sigma}{\kappa\varepsilon\varepsilon_0 k_B T} = -\frac{1}{\kappa}\left(\frac{d\tilde{\psi}}{dr}\right)_{r=R} = \tilde{\psi}_s\frac{1 + \kappa R}{\kappa R}. \tag{3.40}$$

The interactions between charged spheres in electrolyte solutions are not easy to analyze even in the framework of the already simplified Debye and Huckel model. A detailed derivation of the potential around two charged spheres in low-potential limit and the electrostatic energy of interaction Φ_{el} was performed by Verwey and Overbeek [3]. The result is

$$\Phi_{el}(r) = k_B T\left(\frac{e}{q}\right)^2\frac{R^2\tilde{\psi}_s^2}{l_B}\frac{\exp[-\kappa(r - 2R)]}{r}, \tag{3.41}$$

where $l_B = e^2/4\pi\varepsilon\varepsilon_0 k_B T$ is the Bjerrum length. It corresponds to the distance at which the energy between two Coulombic charges is equal to the thermal energy $k_B T$. At room temperature, $l_B \approx 0.7$ nm. Unlike the Debye screening length κ^{-1}, the Bjerrum length is not dependent on the overall electrolyte concentration. Equation (3.41) reasonably approximates the interactions energy between two charged spheres in electrolyte solution as long as the surface potential $\tilde{\psi}_s$ is low.

Equation (3.41) is useful for analyzing the interactions in charged suspension at low volume fractions, where the typical distance between the individual particles is greater than their radii. It has been implemented to model the dynamics of charged suspension, and for the interpretation of dynamic light scattering data from suspensions of charged colloidal particles [17].

It is instructive to compare the potential around a sphere to one in a spherical cavity. The boundary conditions in this case are $\tilde{\psi}(R) = \tilde{\psi}_s$ and $(d\tilde{\psi}/dr)_{r=0} = 0$. The solution of equation (3.37) is

$$\tilde{\psi}(r) = \tilde{\psi}_s \frac{R}{r} \frac{\sinh(\kappa r)}{\sinh(\kappa R)}. \tag{3.42}$$

Hence, the radial dependence inside a spherical cavity is a hyperbolic sine function, as opposed to the outside case where the potential drops as $\exp[-\kappa(r - R)]/r$.

Spherical electric double layers: the thin electric double layer approximation

If both the particle charge and potential are high, the interactions problem becomes considerably more complicated. A tour de force analysis performed by Beresford-Smith *et al* [18] showed that the interaction energy between two charged colloidal spheres in diluted suspensions has the same functional form as equation (3.41), but with a different pre-exponential factor. The interaction energy, obtained by the authors, has the meaning of a potential of mean force and it takes into account not only the effect of the dissolved salt, but also the volume fraction of the suspended colloids and the ions dissociated from each particle. That is why the pre-exponential factor has to be numerically determined in the general case. However, if the electrolyte concentration is high and $\kappa R \gg 1$, an analytical approximation is possible. Such an expression was obtained by Chew and Sen [19], who used a matched asymptotic expansion method [20] to solve the nonlinear PB equation in spherical symmetry,

$$\frac{1}{r^2} \frac{d}{dr} r^2 \frac{d\tilde{\psi}}{dr} = \kappa^2 \sinh(\tilde{\psi}). \tag{3.43}$$

The solution is in the form of series, which are normally truncated at the linear (in $(\kappa R)^{-1}$) term

$$\tilde{\psi}(r) = \tilde{\psi}^0(r) + \frac{1}{\kappa R}\tilde{\psi}^1(r). \tag{3.44}$$

The small parameter is $(\kappa R)^{-1} \ll 1$, which essentially divides the whole domain into inner (close to the particle surface) and outer (away from the surface) subdomains. Hence, the solution (3.44) breaks into inner

$$\tilde{\psi}_{\text{inn}}(r) = \tilde{\psi}^0_{\text{inn}}(r) + \frac{1}{\kappa R}\tilde{\psi}^1_{\text{inn}}(r) \tag{3.45}$$

and outer

$$\tilde{\psi}_{\text{out}}(r) = \tilde{\psi}^0_{\text{out}}(r) + \frac{1}{\kappa R}\tilde{\psi}^1_{\text{out}}(r) \tag{3.46}$$

parts. The potential is strongly screened due the high electrolyte concentration ($\kappa R \gg 1$), so its magnitude is low in the outer region where the two solutions $\tilde{\psi}^0_{\text{out}}$ and $\tilde{\psi}^1_{\text{out}}$ can be obtained from solving the linear equation (3.37). The inner solutions $\tilde{\psi}^0_{\text{inn}}$ and $\tilde{\psi}^1_{\text{inn}}$ are derived by inserting equation (3.45) into equation (3.43). The terms of the same order with respect to $(\kappa R)^{-1}$ are then collected, which leads to two separate differential equations [19]. The zero order equation for $\tilde{\psi}^0_{\text{inn}}$ is identical to (3.3)

$$\frac{d^2\tilde{\psi}^0_{\text{inn}}}{dz^2} = \kappa^2 \sinh\left(\tilde{\psi}^0_{\text{inn}}\right), \quad z = \kappa(r - R) \tag{3.47}$$

while the first order equation is

$$\frac{d^2\tilde{\psi}^1_{\text{inn}}}{dz^2} - \cosh\left(\tilde{\psi}^0_{\text{inn}}\right) = -2\frac{\tilde{\psi}^0_{\text{inn}}}{dz}, \quad z = \kappa(r - R). \tag{3.48}$$

The zero order inner potential $\tilde{\psi}^0_{\text{inn}}$ should satisfy a Dirichlet boundary condition ($\tilde{\psi}^0_{\text{inn}} = \tilde{\psi}_s$) at the surface and vanish at infinity. The first order inner potential $\tilde{\psi}^1_{\text{inn}}$ should be zero on the particle surface and far away from it. The two outer solutions $\tilde{\psi}^0_{\text{out}}$ and $\tilde{\psi}^1_{\text{out}}$ vanish at infinity, while the remaining two constants are determined by the solution matching procedure [19] (for more details see [20]).

The results for $\tilde{\psi}^0$ and $\tilde{\psi}^1$ after matching the inner and outer solutions are

$$\begin{aligned}
\tilde{\psi}^0(r) &= 2\ln\left\{\frac{1 + \gamma\exp[-\kappa(r - R)]}{1 - \gamma\exp[-\kappa(r - R)]}\right\} \\
&\quad + 4R\gamma\frac{\exp[-\kappa(r - R)]}{r} - 4\gamma\exp[-\kappa(r - R)] \\
\tilde{\psi}^1(r) &= \frac{2\gamma\exp[-\kappa(r - R)]}{1 - \gamma^2\exp[-2\kappa(r - R)]} \\
&\quad \{\gamma^2[1 - \exp(-2\kappa(r - R))]2\kappa(r - R)\} \\
&\quad + 2\gamma^3 R\frac{\exp[-\kappa(r - R)]}{r} - 2\gamma[\gamma^2 - 2\kappa(r - R)] \\
&\quad \exp[-\kappa(r - R)],
\end{aligned} \tag{3.49}$$

where $\gamma = \tanh(\tilde{\psi}_s/4)$. The first terms in both equations above are the inner solutions, the second terms correspond to the outer solutions, and the third terms originate from the matching procedure [19]. Introducing (3.49) into (3.44) gives the uniformly valid potential for the whole region around the sphere. The potential distribution derived by Chew and Sen [19] can be combined with the approach outlined by Beresford-Smith *et al* [18] to estimate the interactions between strongly charged particles in diluted suspensions and high electrolyte concentrations [21].

Figure 3.9 shows the electrostatic potential distribution around a charged spherical particle. The potential is high and well beyond the Debye and Huckel

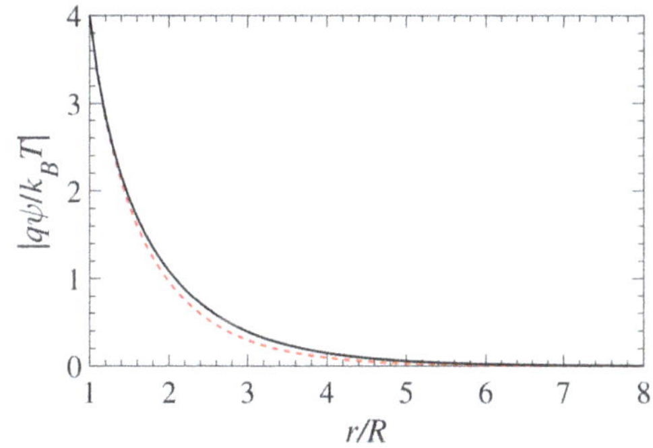

Figure 3.9. Potential distribution around strongly charged spherical particle with surface potential $|\tilde{\psi}_s| = |q\psi_s/K_BT| = 4$ in the thin EDL limit, $\kappa R = 5$. The black solid line represents the uniform solution (3.44). The red dashed line corresponds to the zero order term only.

limit ($\tilde{\psi}_s = 4$). The solid black curve reflects the uniform solution (3.44), which was shown to be identical to the result, obtained by a numerical integration of the PB equation (3.43) [19]. The red dashed curve corresponds to the case where the $(\kappa R)^{-1}$ order term is omitted and only $\tilde{\psi}_0$ is taken into account. The maximum error, introduced by ignoring the $(\kappa R)^{-1}$ term, does not exceed 3.1% for the case presented in figure 3.9. Hence, the matched asymptotic expansion method works well, and even the zero order solution alone is physically reasonable. Chew and Sen tested the method at even more extreme conditions, such as $\tilde{\psi}_s = 10$ and $\kappa R = 1$, and still found very good agreement with the numerical results [19].

The particle charge $\tilde{\sigma} = -\kappa^{-1}(d\tilde{\psi}_{\text{inn}}/dr)_{r=R}$ is (cf equation (3.55)) is

$$\tilde{\sigma} = \frac{|q|\sigma}{\kappa\varepsilon\varepsilon_0 k_B T} = 2\left[\sinh\left(\frac{\tilde{\psi}_s}{2}\right) + \frac{2}{\kappa R}\tanh\left(\frac{\tilde{\psi}_s}{4}\right)\right]. \tag{3.50}$$

Equation (3.50) is identical to the approximation suggested by Loeb *et al* [22]. The first term in the brackets has the same form as the charge for a flat, infinite EDL in the Gouy and Chapman model (see equation (3.8)). The second term is a correction due to the particle curvature. The sign of the curvature term is positive in accordance with that of the particle curvature.

Cylindrical double layers: the low surface charge and potential approximation
The electrostatic potential in cylindrical geometry presents a significant interest. It is used to find the electrostatic energy of rod-shaped molecules [23, 24], or the potential inside a solution-filled charged capillary [25, 26]. The low-potential Debye and Huckel approximation simplifies the problem, and the governing equation is

$$\frac{1}{r}\frac{d}{dr}r\frac{d\tilde{\psi}}{dr} = \kappa^2\tilde{\psi}. \tag{3.51}$$

The potential distribution inside a circular capillary of radius R with low charge and potential at the wall was derived by Rice and Whitehead [25], who showed that the solution is a zero order modified Bessel function of the first kind. The boundary conditions are $\tilde{\psi}(R) = \tilde{\psi}_s$ and $(d\tilde{\psi}/dr)_{r=0} = 0$, which results in

$$\tilde{\psi}(r) = \tilde{\psi}_s \frac{I_0(\kappa r)}{I_0(\kappa R)}. \tag{3.52}$$

The potential around a cylindrical can be found by solving (3.51) with boundary conditions $\tilde{\psi}(R) = \tilde{\psi}_s$ and $\tilde{\psi} = 0$ at $r \to \infty$. The solution is

$$\tilde{\psi}(r) = \tilde{\psi}_s \frac{K_0(\kappa r)}{K_0(\kappa R)}. \tag{3.53}$$

Cylindrical double layers: the thin electric double layer approximation
The thin electrical double layer approximation can be applied to find the potential in charged capillaries, or around cylindrical charged rods. Below, we present the example of a charged cylindrical capillary filled with electrolyte solution. The thin EDL approximation is valid for arbitrarily high surface potentials, but is limited to relatively high electrolyte concentrations, or $\kappa R \gg 1$. The problem can be solved using the matched asymptotic expansions method [27], similarly to the spherical case outlined above. In this case, the PB equation reads:

$$\frac{1}{r}\frac{d}{dr} r \frac{d\tilde{\psi}}{dr} = \kappa^2 \sinh(\tilde{\psi}). \tag{3.54}$$

The solution has the form

$$\tilde{\psi}_{\text{inn}}(r) = \tilde{\psi}^0_{\text{inn}}(r) + \frac{1}{\kappa R}\tilde{\psi}^1_{\text{inn}}(r) \tag{3.55}$$

in the inner subdomain, and

$$\tilde{\psi}_{\text{out}}(r) = \tilde{\psi}^0_{\text{out}}(r) + \frac{1}{\kappa R}\tilde{\psi}^1_{\text{out}}(r) \tag{3.56}$$

in the outer subdomain. The zero order contribution to the potential $\tilde{\psi}^0_{\text{inn}}(r)$ is the solution of the Gouy–Chapman equation (3.47), where $z = \kappa(R - r)$, since we are interested in the capillary interior, and $r = 0$ corresponds to the center. The equation for the first order inner term $\tilde{\psi}^1_{\text{inn}}(r)$ is [27]

$$\frac{d^2\tilde{\psi}^1_{\text{inn}}}{dz^2} - \cosh\left(\tilde{\psi}^0_{\text{inn}}\right) = \frac{\tilde{\psi}^0_{\text{inn}}}{dz}, \quad z = \kappa(r - R), \tag{3.57}$$

which is similar but not identical to equation (3.48). The outer solutions are found from the linear equation (3.51), subject to symmetry boundary conditions in the capillary center $d\tilde{\psi}^0_{\text{out}}/dr)_{r=0} = 0$ and $d\tilde{\psi}^1_{\text{out}}/dr)_{r=0} = 0$. The remaining boundary conditions are determined by the matching of the solutions and the results for $\tilde{\psi}^0$ and $\tilde{\psi}^1$ are [27]

$$\tilde{\psi}^0(r) = 2\ln\left\{\frac{1+\gamma\exp[-\kappa(R-r)]}{1-\gamma\exp[-\kappa(R-r)]}\right\}$$
$$+4\gamma\exp(-\kappa R)\sqrt{2\pi\kappa R}\,I_0(\kappa r) - 4\gamma\exp[-\kappa(R-r)]$$
$$\tilde{\psi}^1(r) = \frac{2\gamma\{\kappa(R-r)\exp[\kappa(R-r)] - \gamma^2\sinh[\kappa(R-r)]\}}{\exp[2\kappa(r-R)]-\gamma^2} \tag{3.58}$$
$$-\gamma^3\exp(-\kappa R)\sqrt{2\pi\kappa R}\,I_0(\kappa r) - \gamma[2\kappa(R-r)-\gamma^2]$$
$$\exp[-\kappa(R-r)].$$

The potential distribution in a cylindrical capillary is shown in figure 3.10. The surface potential is high, $\tilde{\psi}_s = 4$, and the salt concentration is adjusted to $\kappa R = 5$. The black solid curve corresponds to the uniform solution, which for $\kappa R = 5$ and above is practically indistinguishable from the uniform solution or the numerical result from solving (3.54) [27]. Like the case of the charged spherical particle discussed above, even the zero order approximation (red dashed curve) is close to the exact (i.e., numerical) solution.

The charge at the inner capillary surface is

$$\tilde{\sigma} = \frac{|q|\sigma}{\kappa\varepsilon\varepsilon_0 k_B T} = 2\left[\sinh\left(\frac{\tilde{\psi}_s}{2}\right) - \frac{1}{\kappa R}\tanh\left(\frac{\tilde{\psi}_s}{4}\right)\right]. \tag{3.59}$$

It is interesting to compare (3.59) to the spherical particle charge (3.50). The first term in the brackets is the same for both the spherical outer surface and the capillary inner surface. The difference is in the second curvature-related term. The cylindrical capillary curvature is negative and its magnitude is half the one for a sphere of the same radius R.

The Derjaguin Approximation for Interacting Electric Double Layers
An efficient approach to account for the interactions between curved EDLs is offered by the Derjaguin method [4] (see also [5, 6]). The method is particularly

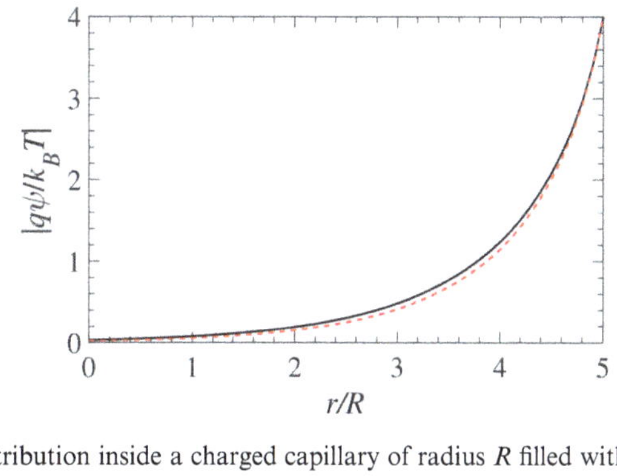

Figure 3.10. Potential distribution inside a charged capillary of radius R filled with electrolyte solution in the thin EDL limit, $\kappa R = 5$. The surface potential is $|\tilde{\psi}_s| = |q\psi_s/K_BT| = 4$. The black solid curve represents the uniform solution. The red dashed curve corresponds to the zero order term only.

valuable when examining particle interactions in suspensions. It is valid at small separations between the surfaces in comparison with the sphere radii (i.e., $h \ll R$). The Derjaguin approximation can be applied to various convex bodies such as spheres, ellipsoids and cylinders [4]. It can also be used to describe the electrostatic interactions between deformable emulsion droplets, where their shapes are approximately represented by truncated spheres [28–31], or obtained by numerically solving the Laplace capillarity equation [32]. These results were then used to examine the flocculation and coalescence dynamics in emulsion systems [33–36]. The overview presented in this section is limited to interacting solid spheres.

The reasoning behind the Derjaguin approximation is that at small separations, the two curved (spherical) surfaces can be represented by parabolas, as illustrated in figure 3.11. Hence, the gap variation $z(r)$ in radial direction is given by

$$z(r) = h + \frac{1}{2}\left(\frac{1}{R_1} + \frac{1}{R_2}\right)r^2. \tag{3.60}$$

For small separation distances and particle curvatures, the surfaces can be treated as locally flat [4–6]. Therefore, the energy of interaction between the two spherical particles can be obtained by integrating the free energy per unit area for flat interacting surfaces [see equation (3.32)], or

$$\Phi_{el}(h) = 2\pi \int_0^\infty drrf_{el}(z) = 2\pi \frac{R_1 R_2}{R_1 + R_2} \int_h^\infty dz\, f_{el}(z). \tag{3.61}$$

The second equality in (3.61) follows from the coordinate change $dz = (R_1^{-1} + R_2^{-1})rdr$ (see equation (3.60)). The upper limit for both integrals is infinity. This is a simplifying but reasonable assumption because the interaction energy drops exponentially with the distance between the surfaces. Hence, the main contribution to the interaction energy comes from surfaces around the fore points of the two spheres, while the remaining portions of the parabolas do not contribute.

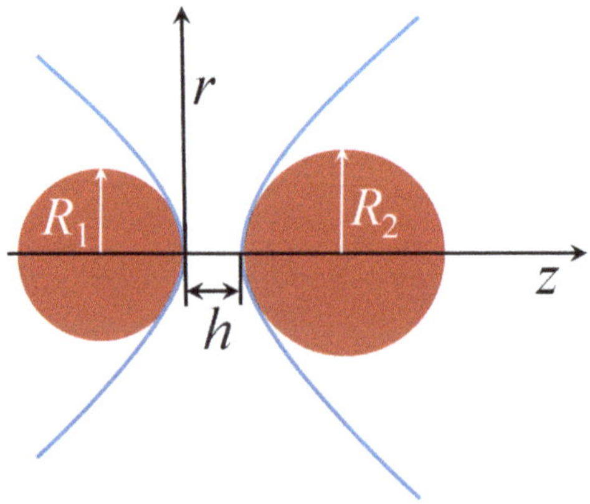

Figure 3.11. A sketch illustrating the Derjaguin approximation for the interaction between two spheres of radii R_1 and R_2, separated by a distance h. See the text for details.

This assumption implies that the range of interactions should be smaller than any of the particle radii.

The corresponding force $F_{el}(h)$ is

$$F_{el}(h) - 2\pi \frac{R_1 R_2}{R_1 + R_2} f_{el}(h). \tag{3.62}$$

For same-size spheres, $R_1 = R_2 = R$, and the geometric factor becomes

$$2\pi \frac{R_1 R_2}{R_1 + R_2} = \pi R, \tag{3.63}$$

which further simplifies equations (3.61) and (3.62).

The Derjaguin approximation allows one to obtain simple and very convenient expressions for the electrostatic interaction energy between charged spheres. However, it is limited to small interparticle distances $h/R \ll 1$, and high electrolyte concentrations $\kappa h \gg 1$.

3.2.4 Effects due to surface charge regulation

The charge regulation chemical equilibrium at the solution interface with the substrate has a significant impact on the interactions between charged surfaces. Varying the distance leads to changes in the surface potential and charge, which affect the disjoining pressure. The charge regulation offers an additional degree of freedom for the system to respond to the changing of the separation distance between the surfaces. As a result, the representative free energy can reach a lower value corresponding to the true minimum, which cannot be accomplished by setting the surface charge or potential to fixed values.

The surface charge regulation condition introduces additional mathematical complexity in comparison with the straightforward Dirichlet (constant potential) or Neumann (constant charge) boundary conditions. Still, for simple chemical equilibria and geometries (e.g., flat EDLs), the nonlinear PB equation can be solved in terms of Jacobian elliptic functions. This reduces the problem to a single transcendental equation [37–46], which can be solved numerically, or further simplified to obtain analytical approximations.

Figure 3.12 shows two examples illustrating the effect of the charge regulation on the disjoining pressure between two identical surfaces. The surface chemical equilibria model corresponds to equation (2.1). The two curves are for pH = 5 (black) and pH = 4 (red), where pH $= -\log_{10}(\rho_{BH_2^+})$. Other parameters include $T = 298$ K, $\varepsilon = 78.5$, $pK_+ = -2$, $pK_- = 6$, $pI = (pK_+ + pK_-)/2 = 2$, surface density of ionizable groups $\Gamma = 75\kappa^2$, and ionic strength $I = 0.01$ M. The surface charge and potential depend on the concentration of the potential determining ions (PDIs), or equivalently, the pH. The greater the difference between the pH and the pI, the greater the surface charge and potential. This is reflected by the disjoining pressure, which is greater for pH = 5 than it is for pH = 4. The lower pH corresponds to a greater number of PDIs, which can neutralize the negatively charged surface groups

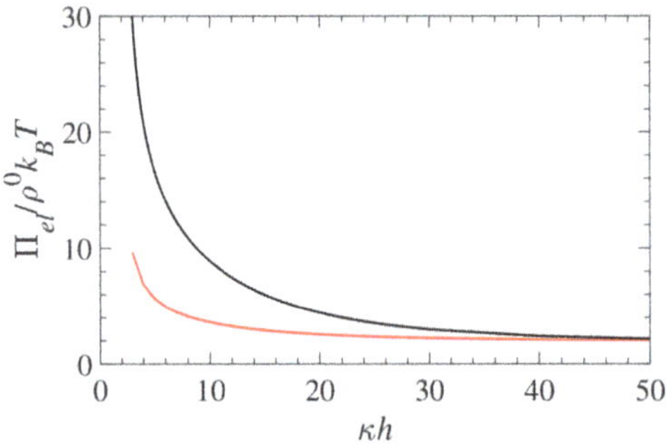

Figure 3.12. Disjoining pressure Π_{el} as a function of distance between two flat, charge regulating surfaces. The top (black) curve corresponds to pH = 5 (higher charge and potential). The lower (red) curve—to pH = 4 (lower charge and potential). The remaining parameters are T = 298 K, ε = 78.5, $pK_+ = -2$, $pK_- = 6$, $pI = (pK_+ + pK_-)/2 = 2$, $\Gamma = 75\kappa^2$, and ionic strength I = 0.01 M.

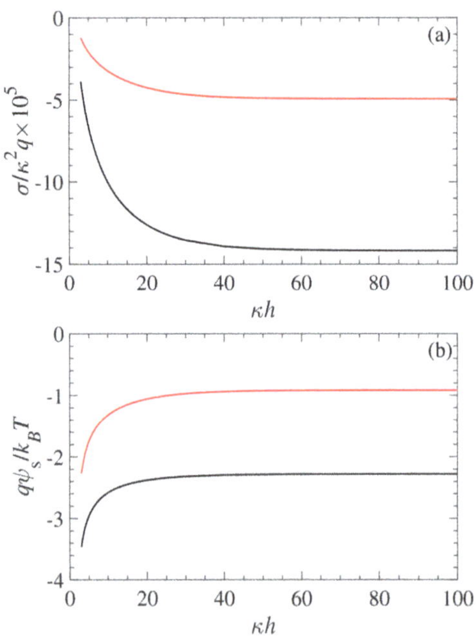

Figure 3.13. (a) Surface charge and (b) potential variation with the distance between the interacting surfaces. The (black) curve corresponds to pH = 5 (higher charge and potential magnitudes). The lower (red) curve—to pH = 4 (lower charge and potential magnitudes). The remaining parameters are T = 298 K, ε = 78.5, $pK_+ = -2$, $pK_- = 6$, $pI = (pK_+ + pK_-)/2 = 2$, $\Gamma = 75\kappa^2$, and ionic strength I = 0.01 M.

and shift the equilibrium to a less charged surface according to (2.1). The pressure indicates that the surfaces experience electrostatic repulsion and it increases with decreasing the distance between them.

The surface charge and potential also vary with the distance between the surfaces as shown in figure 3.13. The overall potential distribution between the surface varies with h, which changes the chemical potential of all ions, including the PDIs. They respond by binding to the reactive surface groups (as h decreases), or dissociating from the surface (as h increases) while remaining in thermodynamic equilibrium with the ions in the reservoir, outside of the gap.

At large separations, the surface charge and potential level off at values that correspond to single, isolated EDLs. The black curves in figure 3.13 correspond to pH = 5, while the red curves correspond to pH = 4. The differences in the surface charges and potentials are significant and noticeable even at relatively large separations up to $\kappa h \gtrsim 20$. In addition, the charge regulation model offers details and insights that are lost by the much simpler constant charge or constant potential approximations discussed above. It also does not suffer from any anomalies and singularities as the constant charge approximation. In fact, the charge regulation model demonstrates that as the surfaces approach, the magnitude of the surface charge decreases. In contact, the surface charge drops to zero, which corresponds to fully neutral surface reactive groups. The magnitude of the surface potential, however, increases with the approach of the surfaces. This high potential drives the repulsion at small distances.

The surface charge regulation is very important in cases where the surface to volume ratio is high. This is the case in narrow micro and nanochannels. The surface acts as a buffer, thus changing the pH of the solution [47], which affects the transport properties of the channels [48].

More detailed examples demonstrating the effect of the chemical charge regulation on the interactions between charged surfaces are available in a number of research articles [37–46].

3.2.5 Electric double layer and colloid stability

The first theoretical analysis of the stability of lyophobic colloidal particles is suspensions was offered by Derjaguin and Landau [2], and independently by Verwey and Overbeek [3]. The result is currently often referred to as the Derjaguin–Landau–Verwey–Overbeek (or DLVO) theory. According to the DLVO model, the stability of the suspended particles is governed by the balance of between the electrostatic repulsion (due to the EDL interactions described above) and the van der Waals attraction, which is a result of the London dispersion interaction energy between the individual atoms and molecules in the macroscopic colloids. The latter was calculated by Hamaker [49] who integrated the pair molecular dispersion energy contributions over the volumes of the interacting particles. The method leads to expressions for the interaction energies between particles and bodies of various shapes. A simple example is the van der Waals interaction energy (per unit area) of two flat, infinitely thick slabs across a distance h, which is

$$f_{vDW}(h) = -\frac{A_H}{12\pi h^2}. \tag{3.64}$$

More importantly, Hamaker derived an expression for the van der Waals energy between two spheres of radii R_1 and R_2

$$\Phi_{vDW}(h) = -\frac{A_H}{6}\left[\frac{2R_1R_2}{(2R_1 + 2R_2 + h)h} + \frac{2R_1R_2}{(2R_1 + h)(2R_2 + h)} + \ln\frac{(2R_1 + 2R_2 + h)}{(2R_1 + h)(2R_2 + h)}\right]. \tag{3.65}$$

The parameter A_H is referred to as the Hamaker constant and it is dependent on the molecular properties of the interacting particles and the solvent. However, due to a phase shift in London fluctuating dipole correlations, the parameter A_H may decrease with the separation, which means it is not truly a constant. This effect is known as electromagnetic retardation, and is related to the finite time necessary for the dipole–dipole communication [50]. It becomes significant for $h > c/\omega$ (c is the speed of light and ω is the frequency of dipole oscillation). The computation of the distance dependent Hamaker 'constant' is formidable task [51], there are simple asymptotic relationships for the non-retarded

$$A_H = \frac{3k_BT}{4}\left(\frac{\varepsilon_1 - \varepsilon_2}{\varepsilon_1 + \varepsilon_2}\right)^2 + \frac{3h_P\omega}{16\sqrt{2}}\frac{\left(n_1^2 - n_2^2\right)^2}{\left(n_1^2 + n_2^2\right)^{3/2}}, \tag{3.66}$$

and the fully retarded

$$A_H = \frac{3k_BT}{4}\left(\frac{\varepsilon_1 - \varepsilon_2}{\varepsilon_1 + \varepsilon_2}\right)^2 + \frac{3h_Pc}{4\pi n_2}\left(\frac{n_1^2 - n_2^2}{n_1^2 + n_2^2}\right)^2\frac{1}{h} \tag{3.67}$$

van der Waals interactions. The parameters ε_1 and ε_2 are the dielectric permittivities of the particles and the solvent, while n_1 and n_2 are the respective refractive indices. The coefficient h_P is the Planck constant.

A simple, uniformly valid expression for A_H can be derived using an interpolation, which reads [6]

$$A_H = \frac{3k_BT}{4}\left(\frac{\varepsilon_1 - \varepsilon_2}{\varepsilon_1 + \varepsilon_2}\right)^2 + \frac{3h_P\omega}{16\sqrt{2}}\frac{\left(n_1^2 - n_2^2\right)^2}{\left(n_1^2 + n_2^2\right)^{3/2}}F(H). \tag{3.68}$$

The interpolation function $F(H)$ is defined as

$$F(H) = \frac{4\sqrt{2}}{\pi}\int_0^\infty dx\frac{(1 + 2Hx)\exp(-2Hx)}{(1 + 2x^2)^2} \approx \left[1 + \left(\frac{\pi H}{4\sqrt{2}}\right)^{3/2}\right]^{-2/3} \tag{3.69}$$

where $H = n_2(n_1^2 - n_2^2)^{1/2}\frac{\omega}{c}h$.

The total DLVO energy of interaction between two colloidal particles in solution is

$$\Phi_{\text{tot}}(h) = \Phi_{vDW}(h) + \Phi_{el}(h). \tag{3.70}$$

A particularly simple case is the interaction between two flat surfaces (i.e., the curvature effects are not important) with low to moderate surface charge and

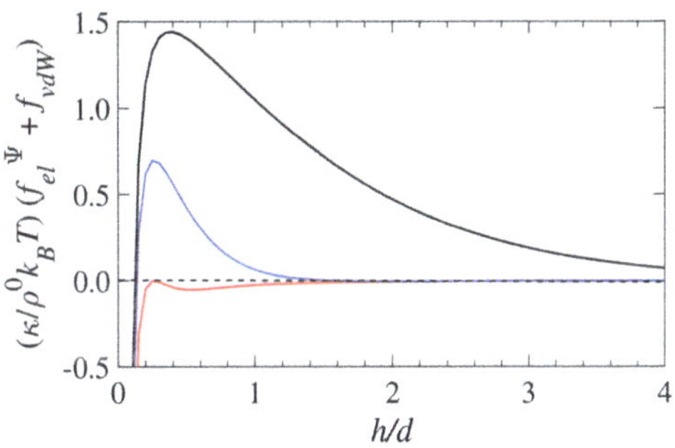

Figure 3.14. DLVO interaction energy between two flat surface as a function of the separation distance h. The surface potential is $\tilde{\psi} = 1$ (see equation (3.33)), and the Hamaker constant is $A_H/k_BT = 1$. The different curves correspond to $\kappa d = 1.00$ (black), $\kappa d = 3.76$ (blue), and $\kappa d = 8.52$ (red).

potential. The van der Waals energy per unit area (see equation (3.64)) may be used in combination with the electrostatic free energies (3.33) or (3.34) as illustrated in figure

The van der Waals and electrostatic contributions have very different dependencies on the separation distance. The van Waals attraction becomes infinite as $h \to \infty$, and decays as h^{-6} at large separations. The electrostatic energy is finite at $h = 0$, and drops exponentially for large h. It may dominate at intermediate distances, which leads to the formation of an energy barrier that may stabilize the suspension. An example is shown in figure 3.14, where the electrostatic barrier depends on the solution ionic strength, or the screening parameter κ. As the ionic strength increases, the barrier height decreases and the suspension becomes less stable. For $\kappa d = 8.52$, the energy maximum becomes zero and the system becomes completely unstable. Further details are available in [2, 3, 5, 6, 52, 53] and others.

3.2.6 Variational approach in the Poisson–Boltzmann limit

The PB equation describes EDL thermodynamic equilibrium. It can be derived by minimizing a free energy functional. Such a functional for a binary symmetric electrolyte was suggested by Sharp and Honig [67] and Reiner and Radke [68]

$$\mathcal{F}_{el}(\mathbf{r}, \tilde{\psi}, \nabla\tilde{\psi}) = \rho^0 k_B T \int d\mathbf{r} \left\{ \frac{1}{\rho^0 |q|} \sum_i \rho_i^{sg} \tilde{\psi}_s - 2[\cosh(\tilde{\psi}(\mathbf{r})) - 1] + \frac{1}{\kappa^2}[\nabla\tilde{\psi}(\mathbf{r})]^2 \right\}. \tag{3.71}$$

The first term under the integral accounts for the boundaries, where ρ_i^{sg} is the number density of fixed, surface charged groups. The functional is then minimized with respect to the potential by taking the functional derivative

$$\frac{\delta \mathcal{F}_{el}[\tilde{\psi}(\mathbf{r})]}{\delta \tilde{\psi}(\mathbf{r})} = 0, \tag{3.72}$$

which defines the equilibrium state. Alternatively, the surface boundary term can be omitted from the functional (3.71) and added as a boundary condition to the differential equation that is a result of the minimization (3.72) [69]. In such cases, equation (3.71) becomes

$$\mathcal{F}_{el}(\mathbf{r}, \tilde{\psi}, \nabla\tilde{\psi}) = -\rho^0 k_B T \int d\mathbf{r} \left\{ 2[\cosh(\tilde{\psi}(\mathbf{r})) - 1] - \frac{1}{\kappa^2}[\nabla\tilde{\psi}(\mathbf{r})]^2 \right\} \qquad (3.73)$$

and after minimization, (see equation (3.73)) leads to the PB equation for symmetric electrolytes,

$$\nabla^2\tilde{\psi}(\mathbf{r}) = \kappa^2 \sinh(\tilde{\psi}). \qquad (3.74)$$

The surface effects are accounted for by the boundary conditions to the problem. They can be of the Dirichlet (constant surface potential), or Neumann (constant surface charge) type. In the more physical case of surface charge regulation, the boundary conditions provide a relationship between the surface charge and potential, $\sigma(\psi_s)$, as demonstrated by equation (2.6).

The variational approach presents a powerful tool to analyze the properties of EDLs in the PB limit [67–70].

3.3 Beyond the Boltzmann distribution: the semiconductor–electrolyte interface

Semiconductors and electrolytes share a common feature: the presence of mobile charges that respond to electric field variations. While the charges in electrolyte solutions are ions, those in the semiconductor materials are electrons or holes. The mobile charge carriers in both phases respond to electric field changes by redistributing in space. Bringing a semiconductor in contact with an electrolyte solution creates an interface with EDLs that forms on both sides. Aqueous electrolyte solutions typically promote the formation of a thin oxide layer that coats the solid semiconductor. This layer chemically interacts with the hydronium ions in the solution, which determines the surface charge and potential (see equation (2.1)). The potential distributions in both phases and the surface charge are interdependent, and any potential perturbation in one phase will propagate into the other. The effect was cleverly utilized by Greenfield and Sivan [54] to design a new type of atomic force microscope that can measure surface electrostatic variations upon the approach of two colloids, or a colloidal sphere to a flat surface. The key idea behind this work is to use a semiconductor substrate that acts as a field effect transistor, modulating the current, as the charged sphere comes closer. The method has great potential for further understanding the interactions between charged colloids and surfaces in electrolyte solutions.

Semiconductor–electrolyte interfaces present a significant practical interest. Examples include catalysis [55, 56], sensing and detection [57, 58], and materials science [59, 60]. There is a fundamental difference between the electrolyte and semiconductor phases. The mobile charges in semiconductors are either all positive or all negative, and their distribution may not follow the exponential Boltzmann law. The rigorous approach to finding the electron (or hole) distribution in the semiconductor phase is based on solving the coupled Poisson and Schrödinger equations [61]. There is a much simpler alternative based on the semi-classical Thomas–Fermi approximation [62]. This approximation was successfully applied to analyzing the electron distribution in confined spaces [61]. The same approach can be applied to the semiconductor–electrolyte interface. Below we present an overview of the properties of the interface between n-doped silicon and aqueous electrolyte solution, separated by thin oxide layer, which typically forms when the two phases are brought in contact [63, 64]. The charge density $\rho_{el}(\mathbf{r})$ in the electrolyte, semiconductor (e.g., silicon) and the thin oxide layer is given by

$$\rho_{el}(\mathbf{r}) = \begin{cases} \sum_i \rho_i^0 q_i \exp\left[\dfrac{-q_i\psi(\mathbf{r})}{k_B T}\right] & \text{electrolyte} \\ eN_d\left\{1 - \mathcal{X}F_{1/2}\left(\dfrac{\mu - e\psi(\mathbf{r})}{k_B T}\right)\right\} & \text{n-doped semiconductor} \\ 0 & \text{oxide,} \end{cases} \tag{3.75}$$

where N_d is the donor number density and

$$F_{1/2}(x) = \frac{2}{\sqrt{\pi}} \int_0^\infty d\xi \frac{\xi^{1/2}}{1 + \exp(\xi - x)} \tag{3.76}$$

is a complete Fermi–Dirac integral of order 1/2 [62, 65]. The parameter $\mathcal{X}$ is defined by

$$\mathcal{X} = \frac{1}{4N_d}\left(\frac{2m^* k_B T}{\pi\hbar^2}\right)^{3/2}, \tag{3.77}$$

and μ is the Fermi level (or the chemical potential of the electrons or holes in the semiconductor). The remaining parameters in (3.78) are the reduced Planck's constant $\hbar$ and the effective mass of the electron m^*. While the oxide layer may also have a non-zero charge density, it is ignored for simplicity (see equation (3.75)). The high temperature limit of the Fermi integral is a Boltzmann exponential, or

$$\mathcal{X}F_{1/2}\left[\frac{\mu - e\psi(\mathbf{r})}{k_B T}\right] \to \exp\left[-\frac{e\psi(\mathbf{r})}{k_B T}\right]. \tag{3.78}$$

The Boltzmann exponential distribution is valid for $T \gg T_F$, where T_F is the Fermi temperature

$$T_F = \frac{\hbar^2(3\pi^2 N_d)^{2/3}}{2k_B m^*}, \tag{3.79}$$

while for $T \lesssim T_F$, the Fermi distribution (see equations (3.75) and (3.76)) or a more rigorous approach, like the Poisson–Schrödinger equation [61], should be used.

The potential distribution in all phases can be found using the Poisson equation (see chapter 1)

$$\nabla^2\psi(\mathbf{r}) = -\frac{\rho_{el}(\mathbf{r})}{\varepsilon\varepsilon_0}, \tag{3.80}$$

where the bulk charge density $\rho_{el}(\mathbf{r})$ is determined by equation (3.75). The boundary conditions at the semiconductor–oxide interface are $(\psi_s)_{sc} = (\psi_s)_{ox}$ and $[\varepsilon_{sc}(\nabla\psi)_{sc} - \varepsilon_{ox}(\nabla\psi)_{ox}] \cdot \mathbf{n} = 0$. The boundary conditions at the oxide–electrolyte interface read $(\psi_s)_{ox} = (\psi_s)_{el}$ and $[\varepsilon_{ox}(\nabla\psi)_{ox} - \varepsilon_{el}(\nabla\psi)_{el}] \cdot \mathbf{n} = \sigma$. The interfacial charge σ is governed by a chemical regulation condition like (2.1) and (2.6).

The solution of equation (3.80), subject to the boundary conditions defined above, implies that the potential and charge distributions in both phases are interrelated, and perturbing any of them in one phase will lead to a response in the other. An example of that effect is the approach of two semiconductor bodies in electrolyte solution [63, 64]. Figure 3.15 shows the potential distribution in two flat

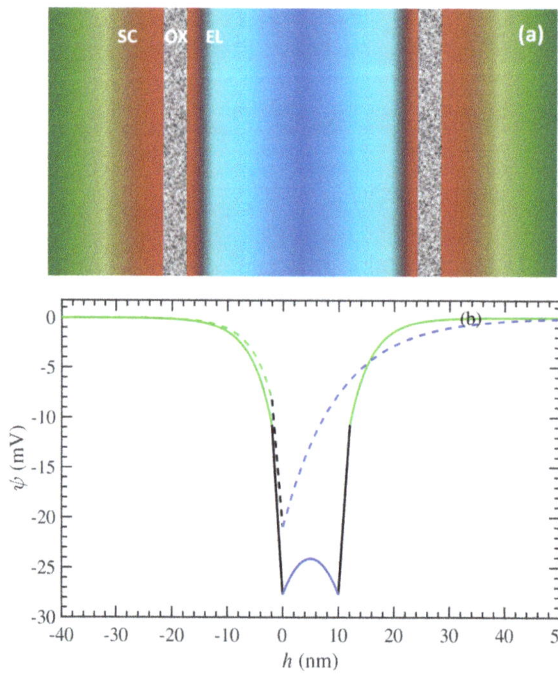

Figure 3.15. Potential distribution in and between two flat macroscopic semiconductor slabs in electrolyte solution. (a) A sketch depicting the potential distribution in the semiconductor (sc), oxide layer (ox) and electrolyte (el). The red color represents higher potential magnitude. (b) Computational results for the potential distribution in all phases. The blue solid curve depicts the potential in the electrolyte filled gap. The black solid lines represent the potential in the thin oxide layers that cover the semiconductor material. The solid green lines describe the potential distribution in the semiconductor phases. The dashed lines refer to infinite separation, or equivalently—a single EDL. The particles are covered with reactive groups with surface density $\Gamma = 8 \times 10^{18}\ \mathrm{m}^{-2}$. These groups may release or attach a proton according to equation (2.1). The parameters for this calculation are: pH = 3.5, the overall electrolyte concentration is 0.925 mM (adjusted by adding symmetric monovalent electrolyte), $\mathrm{pK}_+ = -2$, $\mathrm{pK}_- = 6$. The particle doping is $10^{24}\ \mathrm{m}^{-3}$ and $m^* = 0.28\ m_e$. The dielectric permittivities were $\varepsilon_{el} = 78.5$ for the electrolyte, $\varepsilon_{sc} = 11.7$ for Si, and $\varepsilon_{ox} = 3.9$ for the 2 nm thick layer of SiO_2 [5, 66]. (Reprinted with permission from [63]. Copyright 2014 the American Physical Society.)

semiconductor slabs, covered with thin oxide layers and separated by a gap of thickness h, filled with electrolyte solution. The solid lines correspond to the case where the distance between the surface is $h = 10$ nm. The dashed lines represent a single EDL (infinite separation). The figure shows that as the surfaces approach, the potential distributions change in all phases. The surface potential also undergoes a change with the distance h. It increases as the separation distance decreases. The surface charge, however, is lower at smaller distances. The effect on the potential distribution seems relatively low. However, this small change and the surface charge variation are sufficient to affect the stability of suspension of semiconductor particles in electrolyte solution. A comparison between doped and undoped (while otherwise identical) particles shows that the electrostatic repulsion is strongly affected by the doping. As the mobile charges in the doped semiconductors rearrange upon approach, the repulsion drops significantly, which destabilizes the suspension against coagulation. Balancing the repulsion against the van der Waals attractive energy $\Phi_{vDW} = -A_H d/24h$ (where A_H is the Hamaker constant [49] and d is the particle diameter) and analyzing the coagulation kinetics reveals that the doping can increase the coagulation rate by orders of magnitude [63, 64].

3.4 Electrokinetic phenomena

The understanding of electrokinetic phenomena is intertwined with the development of the notion of the EDL (see chapter 1). Early experimental studies on electro-osmosis and electrophoresis provided crucial insights into the physics and chemistry of charged interfaces involving electrolyte solutions. Informative historical reviews on the development of the theory of electrokinetic phenomena are available in [6, 71–75].

3.4.1 Electrokinetic phenomena in the absence of electric double layer polarization

The EDL is an essential contributor to electrokinetic phenomena. The non-zero charge density $\rho_{el}(\mathrm{r})$ near the charge interface is susceptible to the force due to the externally applied electric field $\mathbf{E}(\mathbf{r})$, and the result is a motion of the fluid relative to the substrate (or vice versa). This motion is viscous-dominated, hence the momentum balance reads [6]

$$\eta \nabla^2 \mathbf{v}(\mathbf{r}) = \nabla p - \rho_{el}(\mathbf{r})\mathbf{E}(\mathbf{r}), \quad \nabla \cdot \mathbf{v}(\mathbf{r}) = 0, \tag{3.81}$$

where $\mathbf{v}(\mathbf{r})$ and $p(\mathbf{r})$ are the pressure and velocity fields, and η is the dynamic viscosity. The first term on the right hand side corresponds to the pressure contribution to the fluid flow, while the second term accounts for the electric force. The charge density $\rho_{el}(\mathrm{r})$ can be expressed in terms of the electrostatic potential using equation (1.8), which rearranges equation (3.81) into

$$\eta \nabla^2 \mathbf{v}(\mathbf{r}) = \nabla p(\mathbf{r}) + \varepsilon\varepsilon_0 \nabla^2 \psi(\mathbf{r})\mathbf{E}(\mathbf{r}). \tag{3.82}$$

In the absence of pressure gradient, the only driving force is due to the electric field and $\mathbf{v}(\mathbf{r}) = \mathbf{v}_{eo}(\mathbf{r})$ is the electroosmotic fluid velocity. Equation (3.82) becomes

$$\nabla \cdot [\eta \nabla \mathbf{v}_{eo}(\mathbf{r}) - \varepsilon\varepsilon_0(\nabla\psi(\mathbf{r}))\mathbf{E}(\mathbf{r})] = \mathbf{0}. \tag{3.83}$$

The expression in the brackets is the tensor of the linear momentum flux. Far from the interface, all gradients vanish, or $\nabla\mathbf{v}(\mathbf{r}) = 0$ and $\nabla\psi(\mathbf{r}) = 0$. Hence, $\eta\mathbf{v}(\mathbf{r}) - \varepsilon\varepsilon_0\psi(\mathbf{r})\mathbf{E}(\mathbf{r}) = \mathbf{const}$. There are many cases of interest that are characterized by very simple geometries. Examples include flat interfaces, or straight channels and capillaries, with electric field that runs parallel to the wall. In those situations, $\mathbf{E}(\mathbf{r}) = \mathbf{E}$ is uniform throughout the EDL, which simplifies the analysis. The fluid motion starts at the shear surface [6, 71–75], where $\mathbf{v} = \mathbf{0}$ and $\psi = \zeta$. Therefore, the unknown constant above equals $-\varepsilon\varepsilon\zeta\mathbf{E}$. The fluid velocity profile in the EDL is then equal to

$$\mathbf{v}_{eo}(\mathbf{r}) = -\frac{\varepsilon\varepsilon_0\mathbf{E}}{\eta}[\zeta - \psi(\mathbf{r})]. \tag{3.84}$$

Equation (3.84) describes the electroosmotic fluid flow velocity profile $\mathbf{v}(\mathbf{r})$ in a straight channel. It implies that the velocity profile $\mathbf{v}(\mathbf{r})$ matches the shape of the electrostatic potential $\psi(\mathbf{r})$. Therefore, one can insert equation (3.52), or equations (3.55)–(3.58) in (3.84) while setting $\psi_s = \zeta$ in order to describe the electroosmotic flow in a cylindrical capillary. Note that for curved geometries (e.g., spherical particles, curves channels), the electric field $\mathbf{E}(\mathbf{r})$ may locally vary, and thus also affect the shape of the velocity profile $\mathbf{v}(\mathbf{r})$ [73, 74].

If the channel is much bigger than the EDL thickness (or κ^{-1}), the electrostatic potential drops to zero at a relatively short distance, and its detailed distribution $\psi(\mathbf{r})$ can be ignored, leading to

$$\mathbf{v}_{eo} = -\frac{\varepsilon\varepsilon_0\zeta\mathbf{E}}{\eta}. \tag{3.85}$$

In the event of electrophoresis (charged particles move relative to the fluid), the fluid velocity profile may not be a match to the electrostatic potential distribution around the charged particle. In the case of a thin EDL (i.e., $\kappa R \gg 1$), it can be considered locally flat, and the particle electrophoretic velocity with respect to the quiescent fluid is

$$\mathbf{v}_{ep} = \frac{\varepsilon\varepsilon_0\zeta\mathbf{E}}{\eta}. \tag{3.86}$$

This is known as the Smoluchowski approximation [76], which is valid for large particles and high electrolyte concentrations.

The opposite limiting case is when the suspended particles are small and the background salt concentration is low (i.e., $\kappa R \ll 1$), often referred to as the Huckel limit [77]. The electrophoretic velocity is derived from the balance between the electric force, acting on the particle $q\mathbf{E}$, and the friction resistance, according to Stokes law $6\pi\eta R\mathbf{v}_{ep}$ [78]. The electric force can be expressed in terms of the particle

ζ-potential using the low-charge approximation $q = 4\pi\varepsilon\varepsilon_0 R\zeta(1 + \kappa R) \approx 4\pi\varepsilon\varepsilon_0 R\zeta$. The electrophoretic velocity is then

$$\mathbf{v}_{ep} = \frac{2\varepsilon\varepsilon_0\zeta\mathbf{E}}{3\eta}. \tag{3.87}$$

The difference between the Smoluchowski and Huckel limiting cases is due to the very different mechanism of energy dissipation associated with the particle motion. The energy dissipation in the Smoluchowski case (equation (3.86)) is a result of the counter electroosmotic flow in the EDL, known as electroosmotic retardation [73]. Therefore, it is spatially limited to a small region around the moving particle with thickness of about κ^{-1}. The energy dissipation in the Huckel case (equation (3.87)) is caused by the Stokes friction force and propagates over large distances, effectively slowing the particle movement much more efficiently. This is why the Huckel electrophoretic velocity is only two thirds of the Smoluchowski velocity.

The intermediate case ($\kappa R \sim 1$) was analyzed by Henry [79], who derived a uniformly valid expression for the electrophoresis of charged spherical particles

$$\begin{aligned} \mathbf{v}_{ep} &= \frac{2\varepsilon\varepsilon_0\zeta\mathbf{E}}{3\eta} f_H(\kappa R), \quad \text{where} \\ f_H(\kappa R) &= 1 + \frac{(\kappa R)^2}{16} - \frac{5(\kappa R)^3}{48} - \frac{(\kappa R)^4}{96} + \frac{(\kappa R)^5}{96} \\ &\quad - \left[\frac{(\kappa R)^4}{8} - \frac{(\kappa R)^6}{96}\right] \exp(\kappa R) \int_{\infty}^{\kappa R} dt \frac{\exp(-t)}{t}. \end{aligned} \tag{3.88}$$

Equation (3.88) is valid for any value of the product κR. In the low salt, small particle (Huckel) limit $\kappa R \to 0$, and $f_H(\kappa) \to 1$. For large particles and high electrolyte concentration, $\kappa R \to \infty$, and $f_H(\kappa) \to 3/2$, approaching the Smoluchowski limit. Henry's correction is important since many systems of practical interest, such as proteins or small colloidal particles, do not conform to the Huckel or Smoluchowski limitations. The integral exponential presents a minor inconvenience, which prompted Ohshima to suggest a much simpler semi-empirical alternative formula to account for the retardation effects [80]

$$f_H(\kappa R) = 1 + \frac{1}{2\left\{1 + \left[\frac{5}{2\kappa R}(1 + 2\exp(-\kappa R))\right]\right\}}. \tag{3.89}$$

The difference between Ohshima (3.89) and Henry's (3.88) corrections is about 1%.

3.4.2 Electrokinetic phenomena in the presence of electric double layer polarization

Externally applied electric fields or fluid motion can perturb the equilibrium distribution of ions (and thus the overall charge and potential) near the charged interface [74]. As a result, the EDL is no longer in equilibrium and the PB equation

(3.2) is no longer valid. Still, the relationship between potential and charge is described by the more general Poisson equation in the form

$$\nabla^2\psi(\mathbf{r}) = -\frac{1}{\varepsilon\varepsilon_0}\sum_i q_i\rho_i(\mathbf{r}), \tag{3.90}$$

where $\rho_i(\mathbf{r})$ and q_i are the density and charge of ionic species 'i'. The summation is performed over all positive and negative ions. The ionic density distributions for all dissolved species, in the non-equilibrium case, are determined by the mass balance equations

$$\begin{aligned}&\nabla\cdot\mathbf{j}_i(\mathbf{r}) = 0,\\ &\mathbf{j}_i = -D_i\nabla\rho_i(\mathbf{r}) - \frac{q_i}{k_BT}D_i\rho_i(\mathbf{r})\mathbf{E}(\mathbf{r}) - \rho_i(\mathbf{r})\mathbf{v}(\mathbf{r}),\end{aligned} \tag{3.91}$$

where D_i is the diffusion coefficient of species 'i'. The field $\mathbf{E}(\mathbf{r}) = \mathbf{E}_{\text{ext}} - \nabla\psi(\mathbf{r})$, where $\mathbf{E}_{\text{ext}}$ is the externally applied contribution, and $\psi(\mathbf{r})$ is the electrostatic potential across the EDL. The fluid velocity $\mathbf{v}(\mathbf{r})$ is obtained from equation (3.81) and generally can be driven by a pressure gradient and/or an external electric field.

In order to account for the EDL polarization effects, equations (3.81), (3.90) and (3.91) have to be solved simultaneously [6, 74]. The boundary conditions at the particle surface are $\mathbf{v} = \mathbf{v}_{ep}$, $-(\varepsilon\varepsilon_0\mathbf{E})_s = \sigma$, and $(\mathbf{j}_i)_s\cdot\mathbf{n} = 0$. Away from the particle, $\mathbf{v} = \mathbf{0}$, $\rho_i = \rho_i^0$ and $\mathbf{E} = \mathbf{E}_{\text{ext}}$. Analytical solutions were obtained by perturbation techniques [81, 82] but subsequent numerical analysis provided more quantitative results [83]. A detailed discussion on the effect of EDL polarization on the electrophoretic mobilities of colloids and particularly their relation to the electrokinetic ζ-potential is available in [6, 74, 75, 80].

3.5 Deficiencies of the continuum approach

The continuum theoretical description of EDLs does not take into account the structure of the solution and the non-Coulombic interactions between all components. A major contributor to the solution structure are the solvent molecules, which are excluded from analyses that are based on the PB equation (3.2). The PB equation represents an ideal gas of charged ions in an external electrostatic field [1]. Hence, its limitations are comparable to those of the ideal gas model in the description of fluids.

The shortcoming of the PB continuum model were recognized by Stern [84] and Grahame [85], but their attempted improvements were limited to the layer next to the charged interface, while the bulk fluid was still considered to be structureless. Subsequent improvements generalized the PB equation to include the ionic excluded volume effects [86–88], but the solvent structure was still ignored. Its importance was recognized much earlier by Stokes, who made a forceful argument in favor of explicitly including the solvent into a comprehensive model of electrolyte solutions [89].

The importance of the solution fluid structure is even greater when the surface charge and potential are determined by a charge regulation mechanism. The surface chemical equilibrium is very sensitive to the fluid structure in the vicinity of the reactive interface, which in turn depends on the surface charge [90, 91]. In addition, the specific interactions between the ions and the solvent molecules (ionic solvation) and/or the surface solvophilicity/solvophobicity properties have a profound effect on the surface chemistry [92, 93]. Hence, the surface charge regulation and the solution structure, including various fluid interactions, are intertwined and have to be all properly accounted for a better quantitative description of EDLs.

References

[1] Attard P 1996 Electrolytes and the electric double layer *Adv. Chem. Phys.* **92** 1–159

[2] Derjaguin B and Landau L 1941 Theory of the stability of strongly charged lyophobic sols and of the adhesion of strongly charged particles in solution of electrolytes *Acta. Physicochim. URSS* **14** 633–62

[3] Verwey E J W and Overbeek J T G 1948 *Theory of the Stability of Lyophobic Colloids* (Amsterdam: Elsevier)

[4] Derjaguin B V, Churaev N V and Muller V M 1987 *Surface Forces* (New York: Plenum)

[5] Israelachvili J N 2011 *Intermolecular and Surface Forces* 3rd edn (New York: Academic)

[6] Russel W B, Saville D A and Schowalter W R 1989 *Colloidal Dispersions* (New York: Cambridge University Press)

[7] Gouy G 1910 Sur la constitution de la charge électrique à la surface d'un électrolyte *J. Physique* **9** 457–68

[8] Gouy G 1917 Sur la fonction electrocapillaire *Ann. Phys.* **7** 129–84

[9] Chapman D L 1913 A contribution to the theory of electrocapillarity *Phil. Mag.* **25** 475–81

[10] Debye P and Huckel E 1923 Zur theorie der elektrolyte. i. gefrierpunktserniedrigung und verwandte erscheinungen *Phys. Ztschr.* **24** 185–206

[11] Overbeek J T G 1952 Electrochemistry of the double layer *Colloid Science* ed H R Kruyt (New York: Elsevier) ch 4 pp 115–93

[12] Langmuir I 1938 The role of attractive and repulsive forces in the formation of tactoids, thixotropic gels, protein crystals and coacervates *J. Chem. Phys.* **6** 873–96

[13] Korn G A and Korn T M 2000 *Mathematical Handbook fdor Scientists and Engineers* (Mineola, NY: Dover)

[14] Ruiz-Cabello F J M, Maroni P and Borkovec M L 2013 Direct measurements of forces between different charged colloidal particles and their prediction by the theory of derjaguin, landau, verwey, and overbeek (dlvo) *J. Chem. Phys.* **138** 234705

[15] Hill T L 1986 *An Introduction to Statistical Thermodynamics* (New York: Dover)

[16] McQuarrie D A 2000 *Statistical Mechanics* (Mill Valley, CA: University Science Books)

[17] Petsev D N and Denkov N D 1992 Diffusion of charged colloidal particles at low volume fraction: Theoretical model and light scattering experiments *J. Colloid Interface Sci.* **149** 329–44

[18] Beresford-Smith B, Chan D Y C and Mitchell D J 1985 The electrostatic inetraction in colloidal systems with low added electrolyte *J. Colloid Interface Sci.* **105** 216–34

[19] Chew W C and Sen P N 1982 Potential of a sphere in an ionic solution in thin double layer approximations *J. Chem. Phys.* **77** 2042–4

[20] Nayfeh A H 2000 *Perturbation Methods* (New York: Wiley-VCH)

[21] Petsev D N, Denkov N D and Nagayama K 1993 Diffusion and light scatteringin dispersions of charged particles with thin electrical double layers *Chem. Phys.* **175** 265–70

[22] Loeb A L, Overbeek J T G and Wiersema P H 1961 *The Electrical Double-Layer Around a Spherical Colloidal Particle* (Boston, MA: MIT Press)
[23] Hill T L 1955 Approximate calculation of the electrostatic free energy of nucleic acids and other cylindrical macromolecules *Arch. Biochem. Biophys.* **57** 229–39
[24] Tracy C A and H Widom 1997 On exact solutions to the cylindrical Poisson–Boltzmann equation with applications to polyelectrolytes *Physica* A **244** 402–13
[25] Rice C L and Whitehead R 1965 Electrokinetic flow in a narrow cylindrical capillary *J. Phys. Chem.* **69** 4017–24
[26] Hildreth D 1970 Electrokinetic flow in fine capillary channels *J. Phys. Chem.* **74** 2006–15
[27] Petsev D N and Lopez G P 2006 Electrostatic potential and electroosmotic flow in a cylindrical capillary filled with symmetric electrolyte: Analytic solutions in thin double layer approximation *J. Colloid Interface Sci.* **294** 492–8
[28] Danov K D, Petsev D N, Denkov N D and Borwankar R 1993 Pair interactions between deformable drops and bubbles *J. Chem. Phys.* **99** 7179–89
[29] Danov K D, Petsev D N, Denkov N D and Borwankar R 1993 Erratum: Pair interactions between deformable drops and bubbles *J. Chem. Phys.* **100** 6104
[30] Denkov N D, Petsev D N and Danov K D 1993 Interaction between deformable brownian droplets *Phys. Rev. Lett.* **71** 3226–9
[31] Petsev D N, Denkov N D and Kralchevsky P A 1995 Flocculation of deformable emulsion droplets: Ii. interaction energy *J. Colloid. Interface Sci.* **176** 201–13
[32] Denkov N D, Petsev D N and Danov K D 1995 Flocculation of deformable emulsion droplets: I. droplet shape and line tension effects *J. Colloid. Interface Sci.* **176** 189–200
[33] Petsev D N 2000 Theoretical analysis of film transitions and coalescence dynamics between charged miniemulsion droplets *Langmuir* **16** 2093–100
[34] Toro-Mendoza J O and Petsev D N 2010 Brownian dynamics of emulsion film formation and droplet coalescence *Phys. Rev.* E **81** 051404
[35] Toro-Mendoza J O and Petsev D N 2010 Erratum: Brownian dynamics of emulsion film formation and droplet coalescence *Phys. Rev.* E **82** 019903
[36] Toro-Mendoza J O, Paredes-Altuve O, Velasquez M A and Petsev D N 2019 Reframing droplet coalescence: Identifying the distinctive dynamics of nanofilm evolution *Phys. Rev. Fluids* **4** 093604
[37] Ninham B W and Parsegian V A 1971 Electrostatic potential between surface bearing ionizable groups in ionic equilibrium with physiologic saline solution *J. Theor. Biol.* **31** 405–28
[38] Carnie S L and Chan D Y C 1993 Interaction free energy between plates with charge regulation: A linearized model *J. Colloid Interface Sci.* **161** 260–4
[39] Chan D Y C, Perram J W, White L R and Healy T W 1975 Regulation of surface potential at amphoteric surfaces during particle–particle interaction *J. Chem. Soc., Faraday Trans.* I **71** 1046–57
[40] Chan D Y C, Healy T W and White L R 1976 Electrical double layer interactions under regulation by surface ionization equilibria–dissimilar amphoteric surfaces *J. Chem. Soc., Faraday Trans.* I **72** 2844–65
[41] Ettelaie R and Buscall R 1995 Electrical double layer interactions for spherical charge regulating colloidal particles *Adv. Coll. Interf. Sci.* **61** 131–60
[42] Behrens S H and Borkovec M 1999 Electric double layer interaction of ionizable surfaces: Charge regulation for arbitrary potentials *J. Chem. Phys.* **111** 382–5

[43] Behrens S H and Borkovec M 1999 Electrostatic interaction of colloidal surfaces with variable charge *J. Phys. Chem.* **103** 2918–28
[44] Behrens S H and Borkovec M 1999 Exact Poisson–Boltzmann solution for the interaction of dissimilar charge–regulating surfaces *Phys. Rev.* E **60** 7040–8
[45] Popa I, Sinha P, Finess M, Maron P, Papastavrou G and Borkovec M 2010 Importance of charge regulation in attractive double–layer forces between dissimilar surfaces *Phys. Rev. Lett.* **104** 228301
[46] Valmacco V, Elzbieciak-Wodka M, Herman D, Trefalt G, Maroni P and Borkovec M 2016 Forces between silica particles in the presence of multivalent cations *J. Colloid Interface Sci.* **472** 108–15
[47] Tseng S, Taiand Y-H and Hsu J-P 2013 Electrokinetic flow in a ph-regulated, cylindrical nanochannel containing multiple ionic species *Microfluid. Nanofluid.* **15** 847–57
[48] Fleharty M E, van Swol F and Petsev D N 2014 The effect of surface charge regulation on conductivity in fluidic nanochannels *J. Colloid Interface Sci.* **416** 105–11
[49] Hamaker H C 1937 The London–van der Waals attraction between spherical particles *Physica* **4** 1058–72
[50] Casimir H B G and Polder D 1948 Influence of retardation on the London–van der Waals forces *Phys. Rev.* **73** 360–72
[51] Dzyaloshinskii I E, Lifshitz E M and Pitaevskii L P 1961 The general theory of van der Waals forces *Adv. Phys.* **10** 165–209
[52] Derjaguin B V and Muller V M 1967 Slow coagulation of hydrophobic colloids *Dokl. Akad. Nauk SSSR (in Russian)* **176** 738–41
[53] Evans D F and Wennerström H 1999 *The Colloidal Domain* (New York: Wiley–VCH)
[54] Greenfield E and Sivan U 2009 Measuring changes in surface potential as two charged bodies approach in electrolyte solution *Phys. Rev. Lett.* **102** 106101
[55] Trotochaud L, Mills T J and S W Boettcher 2013 An optocatalytic model for semi-conductorcatalyst water-splitting photoelectrodes based on *in situ* optical measurements on operational catalysts *J. Phys. Chem. Lett.* **4** 931–5
[56] Nurlaela E, Ould-Chikh S, Harb M, del Gobbo S, Aouine M, Puzenat E, Sautet P, Domen K, Basset J-M and Takanabe K 2013 Critical role of the semiconductorelectrolyte interface in photocatalytic performance for water-splitting reactions using Ta3N5 particles *J. Phys. Chem. Lett.* **4** 931–5
[57] Cui Y, Wei Q, Park H and Lieber C M 2001 Nanowire nanosensors for highly sensitive and selective detection of biological and chemical species *Science* **293** 1289–92
[58] Patolsky F, Zheng G, Hayden O, Lakadamyali M, Zhuang X and Lieber C M 2004 Electrical detection of single viruses *Proc. Natl Acad. Sci. (USA)* **101** 14017–22
[59] Talapin D V, Lee J-S, Kovalenko M V and Shevchenko E V 2010 Prospects of colloidal nanocrystals for electronic and optoelectronic applications *Chem. Rev.* **110** 389–458
[60] Meseguer F, Fenollosa R, Rodriguez I, Xifré-Perez E, Ramiro-Manzano F, Garín M and Tymczenko M 2011 Silicon colloids: A new enabling nanomaterial *J. Appl. Phys.* **109** 102424
[61] Luscombe J H, Bouchard A M and Luban M 1992 Electron confinement in quantum nanostructures: Self-consistent Poisson–Schrodinger theory *Phys. Rev.* B **46** 10262–8
[62] Feynman R P, Metropolis N and Teller E 1949 Equations of state of elements based on the generalized Fermi–Thomas theory *Phys. Rev.* **75** 1561
[63] Fleharty M E, van Swol F and Petsev D N 2014 Manipulating semiconductor colloidal stability through doping *Phys. Rev. Lett.* **113** 158302

[64] Fleharty M E, van Swol F and Petsev D N 2015 Electric double layers at the semiconductor–electrolyte interface *J. Colloid Interface Sci.* **449** 409–15
[65] Kubo R 2004 *Statistical Mechanics, An Advanced Course With Problems and Solutions* (Amsterdam: Elsevier)
[66] Fritz J, Cooper E B, Gaudet S, Sorge P K and Manalis S R 2010 Electronic detection of dna by its intrinsic molecular charge *Proc. Natl Acad. Sci. USA* **110** 389–458
[67] Sharp K A and Honig B 1990 Calculating total electrostatic energies with the nonlinear Poisson–Boltzmann equation *J. Phys. Chem.* **94** 3901–12
[68] Reiner E S and Radke C J 1990 Variational approach to the electrostatic free energy in charged colloidal suspensions : General theory for open systems *J. Phys. Soc. Faraday Trans.* **86** 7684–92
[69] Bozic A L and Podgornik R 2018 Anomalous multipole expansion: Charge regulation of patchy inhomogeneously charged spherical particles *J. Chem. Phys.* **149** 163307
[70] Majee A, Bier M and Podgornik R 2018 Spontaneous symmetry breaking of charge-regulated surfaces *Soft Matter* **14** 985–91
[71] Overbeek J T G 1952 Colloid Science *Colloid Science* ed H R Kruyt (New York: Elsevier) ch 5 pp 194–244
[72] Dukhin S S 1974 Development of notions as to mechanism of electrokinetic phenomena and the structure of the colloid micelle *Surface and Colloid Science* vol 7 ed E Matijevic (New York: Wiley Interscience) ch 1 pp 1–49
[73] Dukhin S S and Derjaguin B V 1974 Equilibrium double layer and electrokinetic phenomena *Surface and Colloid Science* ed E Matijevic vol 7 (New York: Wiley Interscience) ch 2 pp 50–272
[74] Derjaguin B V and Dukhin S S 1974 Nonequilibrium double layer and electrokinetic phenomena *Surface and Colloid Science* vol 7 ed E Matijevic (New York: Wiley Interscience) ch 3 pp 273–335
[75] Hunter R J 1981 *Zeta Potential in Colloid Science* (New York: Academic)
[76] von Smoluchowski M 1903 Contribution a la theorie de l'endosmose electrique et de quelques phenomenes correlatifs *Bull. Intern. Acad. Sci. Cracovie Ser.* A **8** 182–200
[77] Huckel E 1924 Die kataphorese der kugel *Phys.* Z **25** 204–10
[78] Landau L D and Lifshitz E M 1987 *Fluid Mechanics* (Amsterdam: Pergamon)
[79] Henry D C 1931 The cataphoresis of suspended particles. i. the equation of cataphoresis *Proc. Roy. Soc. London* A **133** 106–29
[80] Ohshima H 2006 *Theory of Colloid and Interfacial Electric Phenomena* (Amsterdam: Elsevier)
[81] Overbeek J T G 1943 Theorie der elektrophorese *Kolloid-Beih.* **54** 287–364
[82] Booth F 1950 The cataphoresis of spherical, solid non-conducting particles in a symmetrical electrolyte *Proc. Roy. Soc. Ser.* A **203** 514–33
[83] Wiersema P H, Loeb A L and Overbeek J T G 1966 Calculation of the electrophoretic mobility of a spherical colloid particle *J. Colloid Interface Sci.* **22** 70–99
[84] Stern O 1924 Zur theorie der elektrolytischen doppelschicht *Z. Electrochem* **30** 508–16
[85] Grahame D C 1947 The electrical double layer and the theory of electrocapillarity *Chem. Rev.* **41** 441–501
[86] Borukhov I, Andelman D and Orland H 1997 Steric effects in electrolytes: A modified Poisson–Boltzmann equation *Phys. Rev. Lett.* **79** 435–8

[87] Ben-Yaakov D, Harries D, Andelman D and Podgornik R 2009 Beyond standard Poisson–Boltzmann theory: Ion–specific interactions in aqueous solutions *J. Phys.: Condens. Matter* **21** 424106
[88] Ben-Yaakov D, Andelman D, Podgornik R and Harries D 2011 Ion–specific hydration effects: Extending the Poisson–Boltzmann theory *Curr. Opin. Colloid Interface Sci.* **11** 542–50
[89] Stokes R H 1972 Debye model and the primitive model for eletrolyte solutions *J. Chem. Phys.* **56** 3382–3
[90] Fleharty M E, van Swol F and Petsev D N 2016 Solvent role in the formation of electric double layers with surface charge regulation: A bystander or a key participant? *Phys. Rev. Lett.* **116** 048301
[91] Vangara R, Brown D C R, van Swol F and Petsev D N 2017 Electrolyte solution structure and its effect on the properties of electric double layers with surface charge regulation *J. Colloid Interface Sci.* **488** 180–9
[92] Vangara R, van Swol F and Petsev D N 2017 Solvation effects on the potential and charge distributions in electric double layers *J. Chem. Phys.* **147** 214704
[93] Vangara R, van Swol F and Petsev D N 2018 Solvophilic and solvophobic surfaces and non-coulombic surface interactions in charge regulating electric double layers *J. Chem. Phys.* **148** 044702

Chapter 4

Integral equation theory

4.1 Background

Following the success of van der Waals in 1873 with his thesis on the continuity of the liquid and gaseous states [1] the modern theory of liquids had a relatively slow start. Partly, this was due to the success of other approaches. Given that the liquid state falls between the less dense gaseous phase and the more dense solid phase it stands to reason that there were two kinds of theories. One viewed the liquid as dense gas, while the other considered the liquid a disordered solid. The latter, it turns out, was the most actively traveled road before World War II. The researchers were more numerous and better organized as a group.

In addition, crucial fundamental work by Ornstein and Zernike [2] on light scattering and opalescence that contained major breakthroughs went mostly unnoticed because it was published in a Dutch journal in 1914. As we shall see, the concepts that they introduced ultimately helped Percus and Yevick [3], and Rushbrooke and Scoins [4] in the 1950s to formulate a modern theory of the liquid state. This was built onto the fundamental statistical mechanics work by Yvon, Bogolubov [5], Kirkwood [6], Born and Green [7], during the first half of the twentieth century. A wonderful and thorough historical account of these developments can be found in John Rowlinson's book *Cohesion* [8]. There are many liquid state texts that provide excellent and detailed introductions and into integral equations including [9–11]

Integral equation theories express relationships between fundamental correlation functions. In particular, for a single component simple fluid these functions include the radial distribution function $g(r)$, the total correlation function $h(r) = g(r) - 1$ and the direct correlation function $c(r)$. The latter two functions, $h(r)$ and $c(r)$, are related through the Ornstein–Zernike (OZ) equation [12]. Knowledge of these functions provides detailed insight into the liquid structure [9] and, through application of statistical mechanics, allows the calculation of the pressure equation of state. We will illustrate and discuss that issue below.

doi:10.1088/978-0-7503-2276-8ch4

The OZ equation is

$$h(\mathbf{r}_{12}) = c(\mathbf{r}_{12}) + \rho \int d\mathbf{r}_3\, c(\mathbf{r}_{13})h(\mathbf{r}_{32}). \tag{4.1}$$

The integral part of this equation embodies a convolution of c and h. It is instructive to use the OZ equation for $h(\mathbf{r}_{32})$, viz.,

$$h(\mathbf{r}_{32}) = c(\mathbf{r}_{32}) + \rho \int d\mathbf{r}_4\, c(\mathbf{r}_{34})h(\mathbf{r}_{42}). \tag{4.2}$$

Now, if we insert the right hand side back into equation (4.1) and repeat we obtain a density expansion of the total correlation function in terms of a sum of correlation chains that start at particle 1 and end at particle 2. The first chain involves all possible positions of a particle 3. Similarly, the next term contains all possible chains connecting 1 and 2 involving all possible positions of two intermediate particles, 3 and 4, and so on,

$$h(\mathbf{r}_{12}) = c(\mathbf{r}_{12}) + \rho \int d\mathbf{r}_3\, c(\mathbf{r}_{13})c(\mathbf{r}_{32}) + \tag{4.3}$$

$$\rho^2 \int\int d\mathbf{r}_3 d\mathbf{r}_4\, c(\mathbf{r}_{13})c(\mathbf{r}_{34})c(\mathbf{r}_{42}) + \ldots. \tag{4.4}$$

This expansions illustrates the terminology, demonstrating that the *total* correlation between particles 1 and 2 is the sum of the *direct* correlation between 1 and 2, and all contribution chains of 3, 4, 5, ... particles.

Naturally, a convolution like equation (4.1) makes it impossible to extract $h(\mathbf{r}_{12})$ from $c(\mathbf{r}_{12})$, but of course, a Fourier Transformation can unravel the two, thus,

$$\hat{h}(\mathbf{k}) = \hat{c}(\mathbf{k}) + \rho\hat{c}(\mathbf{k})\hat{h}(\mathbf{k}), \tag{4.5}$$

where the $\hat{h}(\mathbf{k})$ and $\hat{c}(\mathbf{k})$ denote the Fourier transforms of $h(r)$ and $c(r)$, respectively, cf,

$$\hat{h}(\mathbf{k}) = \int d\mathbf{r}\; e^{-i\mathbf{k}h(\mathbf{r})} \tag{4.6}$$

and similarly for $\hat{c}(\mathbf{k})$. For a spherically symmetric function, $h(\mathbf{r}) = h(r)$, the integral over angles can be performed and the final expression simplifies to

$$h(k) = 4\pi \int_0^\infty dr\; rh(r)\frac{\sin(kr)}{k}. \tag{4.7}$$

The introduction of $c(r)$ was inspired by light scattering experiments [12], and it was believed to be generally of a shorter range that $h(r)$.

4.2 Percus–Yevick closure

To solve the OZ equation requires additional information, as both $h(r)$ and $c(r)$ are unknown. Typically, integral equation theories start with an approximation for $c(r)$. This is referred to as a closure equation. Different integral equation theories correspond to different closures.

A famous example concerns a pair-potential fluid, where the Percus–Yevick (PY) closure becomes

$$c(r) = g(r) - g(r)e^{\beta\phi(r)}. \tag{4.8}$$

This can be rewritten in terms of the functions $f(r)$ and $y(r)$ as follows:

$$c(r) = g(r) - y(r) \tag{4.9}$$

$$= f(r)y(r). \tag{4.10}$$

Here $f(r)$ denotes the Mayer *f-function*. In terms of the pair potential, $\phi(r)$ this function is defined by $f(r) = \exp[-\beta\phi(r)] - 1$.

In addition, we have introduced the so-called *y-function* or *cavity function*, $y(r) = g(r)\exp[\beta\phi(r)]$. This turns out to be a very convenient function as it is a *continuous* function even when the potential is *not*!

Finally, we note that the low-density limit of the direct correlation function is $c(r) = f(r)$. The PY approximation is in agreement with that since the low density of $y(r)$ is unity, given that in that limit $g(r) \approx \exp[-\beta\phi(r)]$.

For hard spheres of diameter d the pair potential is a discontinuous function, i.e.,

$$\phi(r) = \begin{cases} \infty & r \leqslant d \\ 0 & r > d \end{cases}. \tag{4.11}$$

As a result the direct correlation function is also discontinuous at $r = d$. In fact, $c(r)$ is short- and finite ranged. It is only nonzero for $r \leqslant d$, and with a discontinuity at $r = d$, thus,

$$c(r) = \begin{cases} 0 & r > d \\ y(r) & r \leqslant d \end{cases}. \tag{4.12}$$

It turns out that the PY integral equation for hard spheres can be solved analytically, using Laplace transforms. This was done by Thiele and Wertheim in the early 1960s [13, 14]. It is worthwhile to quote the results in detail, as hard spheres form the reference system for c-DFT and, moreover, hard sphere mixtures feature in the Fundamental Measure Theory (FMT) of Y Rosenfeld [15, 16]

Adopting a dimensionless variable $x = r/d$ and defining the packing fraction (i.e., the volume occupied by spheres) as $\eta = (\pi/6)\rho d^3$ the analytic solution of the PY equation for the direct correlation function is

$$c(x) = \begin{cases} 0 & x > 1 \\ -\left[\frac{1}{2}a_1\eta x^3 + 6a_2\eta x + a_1\right] & x \leqslant 1 \end{cases}. \tag{4.13}$$

The coefficients a_1 and a_2 are simple functions of the packing fraction:

$$a_1 \equiv \frac{(1 + 2\eta)^2}{(1 - \eta)^4} \tag{4.14}$$

$$a_2 \equiv -\frac{(1 + \frac{1}{2}\eta)^2}{(1 - \eta)^4}. \tag{4.15}$$

Having found the direct correlation function, everything else is now known, structurally (i.e., $g(r) = h(r) + 1$) and thermodynamically. That is, $h(r)$ is obtained from the OZ equation. Expressions for the pressure can be obtained from in two ways. One route is to use $g(r)$ and the virial equation of state,

$$\frac{\beta p}{\rho} = 1 - \frac{2}{3}\beta\pi\rho \int_0^\infty dr\ r^3 g(r)\frac{d\phi(r)}{dr}. \tag{4.16}$$

For a hard sphere fluid this reduces to the familiar contact theorem

$$\frac{\beta p}{\rho} = 1 - \frac{2}{3}\pi\rho \int_0^\infty dr\ r^3 y(r)\frac{de^{-\beta\phi(r)}}{dr} \tag{4.17}$$

$$= 1 + 4\eta g(d) \tag{4.18}$$

where the only value of $g(r)$ that enters the pressure is the value at contact, $r = d$. The second route uses the compressibility equation:

$$\beta^{-1}\left(\frac{\partial\rho}{\partial p}\right)_T = 1 + 4\pi\rho \int_0^\infty dr\ r^2 h(r) \tag{4.19}$$

from which we obtain the pressure p as a function of ρ, just as experimentally, by integration with repeat to ρ. The two routes to the pressure are equivalent for an exact $g(r)$. But within the context of an approximate integral equation theory they do *not* necessarily provide the same answer. This is a reflection of the closure approximation we have made, see equation (4.8). Specifically, for the hard sphere fluid the virial route leads to a pressure p_v,

$$\frac{\beta p_v}{\rho} = \frac{1 + 2\eta + 3\eta^2}{(1 - \eta)^2} \tag{4.20}$$

whereas the compressibility route (HS–PY) produces

$$\frac{\beta p_c}{\rho} = \frac{1 + \eta + \eta^2}{(1 - \eta)^3}. \tag{4.21}$$

A virial expansion of both these pressures, and a comparison with the exact virial series, shows that both p_v and p_c produce the correct second and third virial coefficient. However, they differ in the fourth coefficient. The p_c expression is closer to the molecular simulation result, and that inspired Carnahan and Starling in 1969 [17] to propose a simple, but arbitrary, interpolation $p = 2/3p_c + 1/3p_v$ that proved to be remarkably accurate:

$$\frac{\beta p}{\rho} = \frac{1 + \eta + \eta^2 - \eta^3}{(1 - \eta)^3}. \tag{4.22}$$

A similar expression for *mixtures* followed later [18]. For completeness we also point out that, with the help of the OZ equation, we can rewrite the above equation as

$$\beta\left(\frac{\partial p}{\partial \rho}\right)_T = \frac{1}{1 + 4\pi\hat{h}(0)} = 1 - 4\pi\rho\hat{c}(0)\ . \tag{4.23}$$

Finally, it is worthwhile to point out that the earlier Scaled Particle Theory (SPT) developed by Reiss, Frisch and Lebowitz in 1959 [19] produces the same HS–PY compressibility equation of state. This is a curious fact as SPT is *not* an integral equation theory, but rather build on the concept of the reversible work that needs to be performed to insert a spherical cavity of diameter d_c into a collection of hard spheres of diameter d. SPT is discussed in the next chapter.

4.3 The hypernetted-chain closure

A different, and successful, closure is known as the hypernetted-chain (HNC) approximation, which, for a spherically symmetric fluids, reads

$$c(r) = g(r) - 1 - \ln y(r) \tag{4.24}$$

$$= f(r)y(r) + [y(r) - 1 - \ln y(r)] \tag{4.25}$$

where on the last line we have highlighted the term in square brackets that constitutes the difference with PY theory. Notice that, as was the case with the PY approximation, the direct correlation function has the correct behavior $c(r) \to f(r)$ in the low density limit.

For the HNC no analytical solution exists. Instead, as is the case for all closures, the solution can always be found numerically by iteration of the OZ and closure condition. To be specific, typically one makes an initial guess for $c(r)$. One then Fourier transforms this to obtain $\hat{c}(k)$. The OZ equation is then used to extract the corresponding $\hat{h}(k)$, from which one obtains $h(r)$ by taking the inverse Fourier transform. Inserting the latter into the closure equation (i.e., PY, HNC, etc) one then arrives at an improved guess for $c(r)$. This is essentially a description of Picard iteration. The new guess can entirely replace the old guess or, more commonly, construct a linear combination of the old and new guess. This is accomplished by the use of a mixing parameter $\in[0, 1]$. The stability of the iteration scheme can be controlled by carefully monitoring the convergence $(c_{\text{new}}(r) - c_{\text{old}}(r))^2$.

4.4 The mean spherical approximation (MSA)

The mean spherical approximation, or MSA, is yet another example of a particular closure. First introduced by Lebowitz and Percus in 1966 [20]. What makes the approximation attractive is that it has an analytical solution for certain fluids of interest to this text, including the restricted primitive model of electrolyte solutions and polar liquids. We consider a pair potential that has a hard core or diameter d and a soft tail potential $\phi_1(r)$ for separations larger than d. Typical examples include the square-well fluid and the Sutherland potential [8]. For such systems the MSA closure is,

$$g(r) = 0; \; r < d \tag{4.26}$$

$$c(r) = -\beta\phi_1(r); \; r \geqslant d. \tag{4.27}$$

Notice that for the hard sphere fluid $\phi_1(r) = 0$, and thus for that fluid the MSA closure reduces to the PY approximation.

4.4.1 Electrolytes

MSA is a particularly useful approximation when it comes to electrolytes in the restricted primitive model, where all the core are hard spheres of diameter d and the soft tail potential $\phi_1(r)$ is a Coulombic term, dropping the subscript '1',

$$\begin{aligned} \phi_{ij}(r) &= \frac{q_i q_j}{4\pi\varepsilon\varepsilon_0 r}; \; r > d \\ \mathrm{i}, j &= +, - \end{aligned} \tag{4.28}$$

where $q_i = z_i e$ with e the electron charge and z_i the charge number of ion i. With that the MSA approximation reads

$$g_{ij}(r) = 0; \; r < d \tag{4.29}$$

$$c_{ij}(r) = -\frac{\beta q_i q_j}{4\pi\varepsilon\varepsilon_0 r}; \; r \geqslant d. \tag{4.30}$$

We also need the mixture form of the OZ equation

$$h_{ij}(\mathbf{r}, \mathbf{r}') = c_{ij}(\mathbf{r}, \mathbf{r}') + \rho \sum_{\nu} x_{\nu} \int d\mathbf{x} \; c_{i\nu}(\mathbf{r}, \mathbf{x}) h_{\nu j}(\mathbf{x}, \mathbf{r}'). \tag{4.31}$$

Here ρ is the total number density of the mixture, and x_{nu} denotes the molefraction of species ν.

Waisman and Lebowitz [21, 22, 23] were able to solve this mixture problem analytically in 1970 and 1972, building on the results for a hard sphere mixture, Their result is

$$c_{ij}(r) = \begin{cases} c_{ij}^{hs}(r) - \dfrac{\beta q_i q_j}{4\pi\varepsilon\varepsilon_0 r}(2\tau - \tau^2 r/d) & r < d \\ -\dfrac{\beta q_i q_j}{4\pi\varepsilon\varepsilon_0 r} & r > d \end{cases} \tag{4.32}$$

where $c_{ij}^{hs}(r)$ is the correlation function for the corresponding hard sphere mixture. The parameter τ is defined as

$$\tau \equiv 1 + x^{-1}\left[1 - \sqrt{(1 + 2x)}\right] \tag{4.33}$$

where x is the inverse Debye screening length, i.e., $x = \kappa d = \sqrt{2\beta\rho^0 d^2 q^2 / \varepsilon\varepsilon_0 k_B T}$.

From the final result for $c_{ij}(r)$ (4.32) one can now obtain expressions for the radial distribution functions, for the like ions, $g_{++}(r) = g_{--}(r)$, and the unlike ions $g_{+-}(r)$. These expressions are provided in appendix B. Given the results for the radial distribution functions one can then obtain closed expressions for thermodynamic quantities. Appendix B also provides expressions for the internal energy, the osmotic coefficient and the Helmholtz free energy in terms of the inverse Debye screening length x.

Finally, when the hard cores shrink to zero, i.e. in $\lim d \to 0$, the MSA results reduce to Debye Hückel theory results.

4.5 Hard sphere mixtures

Hard sphere mixtures are a crucial reference fluid whose properties today are essential to perturbation theories of mixtures. In addition, the repulsive reference part of classical-DFT leans heavily on the HS mixture, which forms the basis of Rosenfeld's celebrated Fundamental Measure Theory (FMT) [15, 16]. In 1964 Lebowitz obtained an analytic solution for PY approximation for a hard sphere fluid binary mixture [24], by extending the Thiele–Wertheim [13,14] method. The binary mixture under consideration was an additive mixture where the cross diameter equals the arithmetic mean of the pure diameters, i.e.,

$$d_{12} = (d_{11} + d_{22})/2. \tag{4.34}$$

The PY solution for the direct correlation functions c_{11}, c_{22} and c_{12} are

$$c_{11}(r) = \begin{cases} -a_1 - b_1 r - cr^3 & r < d_{11} \\ 0 & r > d_{11} \end{cases} \tag{4.35}$$

$$c_{22}(r) = \begin{cases} -a_2 - b_2 r - cr^3 & r < d_{22} \\ 0 & r > d_{22} \end{cases} \tag{4.36}$$

$$c_{12}(r) = \begin{cases} -a_1 & r < \lambda \\ -a_1 - (bx^2 + 4\lambda cx^3 + cx^4)/r & \lambda \leqslant r \leqslant d_{12}. \\ 0 & r > d_{12} \end{cases} \tag{4.37}$$

Here $x = r - \lambda$ and $\lambda = (d_{22} - d_{11})/2$ and we choose to denote the largest diameter with d_{22}. We also introduced a 6 constants, these have the following values,

$$a_i = \frac{\partial \beta p_c(\rho_1, \rho_2)}{\partial \rho_i}; \; i = 1, 2 \tag{4.38}$$

$$b_1 = -\frac{6\eta_1}{d_{11}} g_{c11}{}^2 - \frac{6\eta_2 d_{12}^2}{d_{22}^3} g_{c12}{}^2 \tag{4.39}$$

$$b_2 = -\frac{6\eta_2}{d_{22}} g_{c22}{}^2 - \frac{6\eta_1 d_{12}^2}{d_{11}^3} g_{c12}{}^2 \tag{4.40}$$

$$b = d_{12}g_{c12}\left(-\frac{6\eta_1}{d_{11}^2}g_{c11} - \frac{6\eta_2}{d_{22}^2}g_{c22}\right) \tag{4.41}$$

$$c = -\left(\frac{a_1\eta_1}{d_{11}^3} + \frac{a_2\eta_2}{d_{22}^3}\right), \tag{4.42}$$

here we have introduced the following abbreviations for the contact values $g_{cij} \equiv g_{ij}(d_{ij})$ for $i, j = 1, 2$. Similarly the packing fractions of each species is $\eta_i \equiv (\pi/6)\rho_i d_i^3$. The PY compressibility pressure, $p_c(\rho_1, \rho_2)$, is given by

$$\beta p_c = \frac{\xi_0}{1 - \xi_3} + \frac{\xi_1\xi_2}{(1 - \xi_3)^2} + \frac{\xi_2^3}{12\pi(1 - \xi_3)^3} \tag{4.43}$$

where we use the SPT inspired variables that mimic the averages of volume, surface area and diameter of the spheres that make up the mixture,

$$\xi_0 = \sum_i^m \rho_i \tag{4.44}$$

$$\xi_1 = (1/2)\sum_i^m \rho_i d_i \tag{4.45}$$

$$\xi_2 = \pi \sum_i^m \rho_i d_i^2 \tag{4.46}$$

$$\xi_3 = (\pi/6)\sum_i^m \rho_i d_i^3. \tag{4.47}$$

Here our mixture is a binary, so the number of components is $m = 2$. Notice that independently of the number of components m, the number of variables that enter equation (4.43) of state is always 4. This will be of great importance in the classical-DFT implementation. In addition, two mixtures that differ in the number of components and/or the diameters d_i can have an identical equation of state as long as the variables ξ_0–ξ_3 are the same.

4.6 The Ornstein–Zernike equations approach to studying electric double layers

The Ornstein–Zernike (OZ) equations offer a much better approach for the theoretical analysis and modeling of EDLs in comparison with the point-charge Poisson–Boltzmann method (see chapter 3). The OZ equations (and the accompanying closure approximations) can be applied at different levels of structural complexity. The most comprehensive analysis requires an account for all solution components (ions and solvent), which allows assessing the contributions of all types of interactions, Coulombic and non-Coulombic. A simpler approach is to ignore the explicit solvent molecular structure and interactions, and limit the focus on the ions, their finite size and the Coulombic effects. Finally, if the ionic finite size is ignored, the OZ approach offers the same results as the Poisson–Boltzmann structureless model. An excellent review on the topic has been published by Attard [25] who offered a detailed presentation on the application of modern statistical–mechanical approaches to EDLs.

The selection of a closure approximation is an important part of the implementation of the OZ approach. The asymptotic behavior of direct correlation function between ions $i\,j$ is $c_{ij} \approx -\beta q_i q_j / \varepsilon\varepsilon_0$ (see also equation (4.30)). This means that the MSA approximation is a natural choice for moderately charged species. If that condition is not fulfilled, the typical closure choice is the HNC approximation, which provides better results (e.g., in comparison with the PY) for long-ranged interactions [9].

An advantage of the OZ integral equations approach is their ability provide structural information. Even in the primitive limit, the excluded volume contribution to the ion–ion correlation function have a noticeable effect. An example is the screening length parameter, which becomes a function of the ion diameter and overall salt concentration [25–28].

The OZ integral equation method allows to explicitly take into account the solvent contribution. A particularly interesting case is the ion–dipole mixture, which represents an aqueous electrolyte solution. Such systems are rather complex, and therefore the MSA is a reasonable choice for a closure [29–31]. The presence of the dipole molecules leads to two additional types of interactions, such as ion–dipole

$$\phi_{id}(r,\,\omega) = -\frac{q_i \mathbf{p}(\omega)\cdot\mathbf{n}}{4\pi\varepsilon_0 r^2}, \quad \text{for } r > \frac{d_i + d_d}{2} \tag{4.48}$$

and dipole–dipole

$$\phi_{dd}(r,\,\omega_1,\,\omega_2) = -\frac{\mathbf{p}_1(\omega_1)\cdot(3\mathbf{nn} - \mathbf{I})\cdot\mathbf{p}_2(\omega_2)}{4\pi\varepsilon_0 r^2}, \quad \text{for } r > d_d \tag{4.49}$$

where $\mathbf{p}(\omega)$ is the dipole moment, which is dependent on the solid angle ω. The diameters d_i and d_d are for the ions and dipoles, respectively. The remaining symbols are the unit vector $\mathbf{n} = \mathbf{r}/r$ and the unit tensor $\mathbf{I}$.

The surface–dipole interaction is

$$\phi_{wd}(\mathbf{r},\,\omega) = -\frac{Q_w \mathbf{p}(\omega)\cdot\mathbf{n}}{4\pi\varepsilon_0 r^2}, \quad \text{for } r > \frac{d_d}{2}. \tag{4.50}$$

The local relative dielectric permittivity ε depends on the ion–dipole mixture structure and is derived from the OZ solution.

The dipolar orientation requires an addition integration in the OZ equation over the angular variable, or [29, 30]

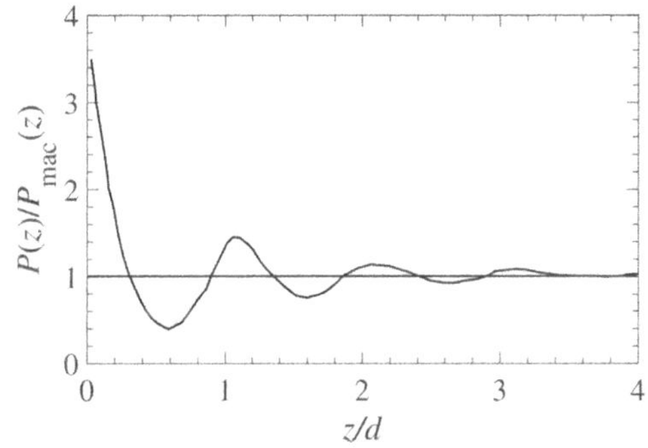

Figure 4.1. Relative structural effect on the polarization density near a charged planar surface. The dipole density is $\rho d_d^3 = 0.8$, $T = 298$ K, and $\varepsilon = 80$. (Reprinted from [30], with the permission of AIP Publishing.)

$$h_{jk}(\mathbf{r}_1, \omega_1, \mathbf{r}_2, \omega_2) = c_{jk}(\mathbf{r}_1, \omega_1, \mathbf{r}_2, \omega_2) + \sum_l \frac{\rho_l}{4\pi} \int \int d\mathbf{r}_3 d\omega_3 \, c_{jl}(\mathbf{r}_1, \omega_1, \mathbf{r}_3, \omega_3) \, h_{lk}(\mathbf{r}_3, \omega_3, \mathbf{r}_2, \omega_2). \tag{4.51}$$

A planar, charge wall can be modeled as a colloidal particle with a diameter $d_c \to \infty$. In this case, the total dipole–wall correlation function $h_{wd}(z, \omega)$ depends on the distance z and a single angle ω. It enables one to obtain the local polarization density $P(z)$ due to the arrangement of the dipole molecules, which is affected by the presence of the charged wall–solution interface and the dissolved ions [30, 31]

$$P(z) = \frac{\rho_d}{4\pi} \int d\omega h_{wd}(z, \omega)(\mathbf{p}(\omega) \cdot \mathbf{n}). \tag{4.52}$$

In contrast, the Debye and Huckel macroscopic continuum limit for the polarization density predicts that $P_{\text{mac}} \approx \sigma \exp(-\kappa z)$, where σ is the surface charge density.

Figure 4.1 demonstrates the effect of the precise solution structure on the local polarization density. The oscillations illustrate the ordering of the dipoles near the charged surface. This level of detail is completely lost in the continuum Debye and Huckel limit.

References

[1] van der Waals J D 1873 Over de continuities van den gas- en vloeistoftoestand. Thesis, Leiden

[2] Ornstein L S and Zernike F 1914 Accidental deviations of density and opalescence at the critical point of a sinqle substance *Proc. Acad. Sci. (Amsterdam)* **17** 793–806

[3] Parr R G and Yang W 1989 *Density-Functional Theory of Atoms and Molecules* (New York: Oxford University Press)

[4] Rushbrooke G S G S and Scoins H I 1953 On the Theory of Liquids *Proc Roy. Soc. Lond.* A **216** 203–18

[5] Bogolubov N 1946 *J. Phys (USSR)* **10** 257
Bogolubov N 1946 *J. Phys (USSR)* **10** 265

[6] Kirkwood J G 1935 Statistical mechanics of fluid mixtures *J. Chem. Phys.* **3** 300–13

[7] Born M and Green H S 1946 A General Kinetic Theory of Liquids. 1. The Molecular Distribution Functions *Proc Roy. Soc. Lond.* A **188** 10–8
[8] Rowlinson J S and Cohesion 2002 *A Scientific History of Intermolecular Forces* (Cambridge: Cambridge University Press)
[9] Hansen J P and McDonald I R 2006 *Theory of Simple Liquids* (New York: Academic)
[10] McQuarrie D A 2000 *Statistical Mechanics* (Mill Valley, CA: University Science Books)
[11] Goodstein D L 1985 *States of Matter* (New York: Dover Publications Inc)
[12] Ornstein L S and Zernike F 1914 Accidental deviations of density and opalescence at the critical point of a sinqle substance *Proc. Acad. Sci. (Amsterdam)* **17** 793–806
[13] Thiele E 1963 equation of state for hard spheres *J. Chem. Phys.* **39** 474–9
[14] Wertheim M S 1963 Exact solution of the percus-yevick integral equation for hard spheres *Phys. Rev. Lett.* **10** 321–3
[15] Rosenfeld Y 1988 Scaled Field Particle Theory of the Structure and the Thermodynamics of Isotropic Hard Particle Fluids *J. Chem. Phys.* **89** 4272–87
[16] Rosenfeld Y 1989 Free-energy model for the inhomogeneous hard-sphere fluid mixture and density functional theory of freezing *Phys. Rev. Lett.* **63** 980–3
[17] Carnahan N F and Starling K E 1969 *J. Chem. Phys.* **51** 635
[18] Mansoori G A, Carnahan N F, Starling K E and Leland T W Jr 1971 Equilibrium thermodynamic properties of the mixture of hard spheres *J. Chem. Phys.* **54** 1523–5
[19] Reiss H, Frisch H L and Lebowitz J L 1959 Statistical mechanics of rigid spheres *J. Chem. Phys.* **31** 369–80
[20] Lebowitz J L and Percus J K 1966 Mean spherical model for lattice gases with extended hard cores and continuum fluids *Phys. Rev.* **144** 251–8
[21] Waisman E and Lebowitz J L 1972 Mean spherical model integral equation for charged hard spheres. II. Results *J. Chem. Phys.* **56** 3093
[22] Waisman E and Lebowitz J L 1970 Exact solution of an integral equation for the structure of a primitive model of electrolytes *J. Chem. Phys.* **52** 4307
[23] Waisman E and Lebowitz J L 1972 Mean spherical model integral equation for charged hard spheres. I. Method of solution *J. Chem. Phys.* **56** 3086
[24] Lebowitz J L 1964 Exact solution of generalized percus-yevick equation for a mixture of hard spheres *Phys. Rev.* **133** A895–9
[25] Attard P 1996 Electrolytes and the electric double layer *Adv. Chem. Phys.* **92** 1–159
[26] Stell G and Lebowitz J L 1968 Equilibrium properties of a system of charged particles *J. Chem. Phys.* **48** 3706–17
[27] Kjellander R and Mitchell D J 1994 Dressed-ion theory for electrolyte solutions: A Debye-Hückel-like reformulation of the exact theory for the primitive model *J. Chem. Phys.* **101** 603–26
[28] Attard P 1993 Asymptotic analysis of primitive model electrolytes and the electrical double layer *Phys. Rev.* E **48** 3604–21
[29] Chan D Y C, Mitchell D J, Ninham B W and Pailthorpe B A 1978 On the theory of dipolar fluids and iondipole mixtures *J. Chem. Phys.* **69** 691–6
[30] Carnie S and Chan D Y C 1980 The structure of electrolytes at charged surfaces: ion-dipole mixtures *J. Chem. Phys.* **73** 2949–57
[31] Carnie S and Chan D Y C 1982 Ionic and dipolar adsorption from an ion-dipole mixture *J. Chem. Soc., Faraday Trans.* **2** 695–722

IOP Publishing

Molecular Theory of Electric Double Layers

Dimiter N Petsev, Frank van Swol and Laura J D Frink

Chapter 5

Perturbation and mean field theory

5.1 Background

In this chapter we take a look at exact and approximate theoretical approaches that are not integral equation theories but that are an integral part of liquid state theory [1–4].

We will start out with virial coefficients, B_n, that involve integrations over n particles. B_n feature in so-called exact virial expansions in density. For simple models, such as the hard sphere model, a number of the early terms of the series are exactly known. Subsequent terms have been evaluated numerically. Truncated virial series are convenient representations of dilute gases. In addition, they can be used for wall–fluid problems [5].

Next, we present perturbation theories, where a fluid can be considered to have a fluid structure that is similar to a so-called, known, reference fluid. The difference can be considered a perturbation: a fluid's free energy is considered the sum of the free energy of a so-called reference fluid and a perturbative contribution. This idea is a crucial part of c-DFT and discussed in section 5.3.

The mean field approach of, for example, van der Waals can be viewed as a crude form of perturbation theory. It considers the hard sphere fluid as the reference fluid, and sets $g(r) = 1$ for distances larger than the hard sphere diameter. The mean field approximation is a remarkably successful theory for a homogeneous fluid and lattice systems. But beyond that it is also an important part of the c-DFT approach to homogenous (lattice and off-lattice) models.

5.2 Virial expansions

In the limit of low density, ρ, the pressure p of all gases approaches the ideal gas law

$$\frac{\beta p}{\rho} = 1. \tag{5.1}$$

doi:10.1088/978-0-7503-2276-8ch5

Experimentally it was found that, along an isotherm, in the approach to $\rho \to 0$, The pressure would exhibit positive or negative deviations, and it became customary to perform a Taylor series expansion in ρ

$$\begin{aligned} \frac{\beta p}{\rho} &= 1 + B_2\rho + B_3\rho^2 + \ldots \\ &= \sum_{n=1}^{\infty} B_n\rho^{n-1} \end{aligned} \tag{5.2}$$

where $B_1 = 1$ and $B_n = B_n(T)$; $n \geqslant 2$. The coefficients B_n are known as *virial* coefficients. B_2 is known as the second virial coefficient. In the lim $\rho \to 0$, $\beta p/\rho$ will approach unity from above (or below) depending on whether B_2 is positive or negative, respectively. For many fluids there exists a temperature where $B_2(T) = 0$. At this particular temperature, T_b known as the Boyle temperature, the gas nearly behaves as an ideal gas, even though it is interacting.

The virial coefficients B_n can be calculated for a given pair potential, thus,

$$\begin{aligned} B_2(T) &= -\frac{1}{2V}\int\int d\mathbf{r}_1 d\mathbf{r}_2\,[e^{-\beta\phi(\mathbf{r}_{12})} - 1] \\ &= -\frac{1}{2V}\int d\mathbf{r}_1 \int d\mathbf{r}_{12}\,[e^{-\beta\phi(\mathbf{r}_{12})} - 1] \\ &= - \, -\frac{1}{2}\int d\mathbf{r}_{12}\,[e^{-\beta\phi(\mathbf{r}_{12})} - 1] \\ &= -2\pi\int_0^{\infty} dr\; r^2[e^{-\beta\phi(r)} - 1] \\ &= -2\pi\int_0^{\infty} dr\; r^2 f_{12}. \end{aligned} \tag{5.3}$$

On the first line, the integrations are performed over the entire volume, V. Since a reasonable classical potential approaches zero fairly quickly the integration variables can be replaced by a single relative coordinate $\mathbf{r}_{12} = \mathbf{r}_2 - \mathbf{r}_1$. Line three results from performing the integration over $\mathbf{r}_1$, while line four results from restricting the potential to be spherically symmetric: $\phi(\mathbf{r}_{12}) = \phi(r_{12})$ and performing the integration over all angles. Finally, the last line uses the Mayer f-function notation (e.g., see [2]), that is commonly used in theories of imperfect gases as the term $exp(-\beta\phi) - 1$ arises frequently.

Similarly, the third virial coefficient is given as,

$$\begin{aligned} B_3(T) &= -\frac{1}{3V}\int\int\int d\mathbf{r}_1 d\mathbf{r}_2 d\mathbf{r}_3 f_{12} f_{13} f_{23} \\ &= -\frac{1}{3}\int\int d\mathbf{r}_{12} d\mathbf{r}_{13} f_{12} f_{13} f_{23}. \end{aligned} \tag{5.4}$$

The virial coefficients can be calculated by numerical integration. For simple forms, such as the hard sphere and square-well potential the second virial coefficient can be calculated analytically. For hard spheres the potential is either ∞ for $r < d$ or 0 for $r > d$, and independent of T. Hence, for hard spheres the Mayer f-function is either −1 or 0, and thus

$$B_2^{hs} = \frac{2\pi d^3}{3}$$
$$B_3^{hs} = \frac{5\pi^2 d^6}{18} = \frac{5}{8}\left(B_2^{hs}\right)^2. \tag{5.5}$$

Considerably more coefficients are known up to B_7 thanks to Ree and Hoover [6], and more recently Kofala *et al* [7] reported coefficients up to B_{10}.

Another useful but simple potential is the square-well (sw) potential,

$$\phi(r) = \begin{cases} \infty & r > d \\ -\epsilon & d < r < \lambda d \\ 0 & d < r \end{cases} \tag{5.6}$$

here λ controls the width of the attractive well. Usually, $\lambda = 1.5$ because for that value the resulting $B_2(T)$ closely mimics the results of more sophisticated potentials (e.g., Lennard–Jones) as well as experiments for argon.

$$B_2^{sw}(T) = B_2^{hs}[1 - (\lambda^3 - 1)(e^{\beta\epsilon} - 1)]. \tag{5.7}$$

When $\lambda \to 1$ or $\epsilon \to 0$ the $B_2^{sw}(T) \to B_2^{hs}$ as expected. Although as mentioned $\lambda = 1.5$, Vega and co-workers [8] have explored the liquid–vapor phase diagrams for the entire family of square-well potentials.[1]

Although it is extremely useful to have the virial series for a potential, the question of convergence is a separate matter. For hard spheres a radius of convergence in density has been determined but is very small. For potentials with attractions (i.e., square well, Lennard–Jones) the truncated virial equation is convenient for the determination of liquid–vapor equilibrium. It can be used for the vapor branch. Sir John Rowlinson [9] observed that truncating the series at $B_3(T)$ is usually fine, because, in practice, when $B_3(T)$ is not enough the next few coefficients will not be sufficient either.

5.3 Zwanzig's perturbation theory

The similarity of liquid structures among simple liquids forms the inspiration for perturbation theory. It took several attempts to ultimately produce an excellent theory that was accurate in its prediction of thermodynamic properties. This was ultimately accomplished by Weeks, Chandler and Andersen (WCA) in 1971 [10]. Once we know the radial distribution function, and assuming that we have a pair potential fluid, the entire thermodynamics of the fluid are known. Thus, the internal energy (per particle) is

---

[1] Finally, we point out that negative values of ϵ turn the square-well potential into a square shoulder potential, which has other applications.

$$\begin{aligned}\frac{\beta\mathscr{U}}{N} &= \frac{\beta\mathscr{U}^{id}}{N} + \frac{\beta\mathscr{U}^{ex}}{N} \\ &= \frac{3}{2} + 2\beta\pi\rho\int_0^{\infty} dr\ r^2 g(r)\phi(r),\end{aligned} \tag{5.8}$$

where the first line is a definition of an excess property, here the excess internal energy. In addition, we have the virial equation (4.17) which expresses the pressure as an integral over the virial, $r^3 g(r) d\phi(r)/dr$. Thus,

$$\frac{\beta p}{\rho} = \frac{\beta p^{id}}{\rho} + \frac{\beta p^{ex}}{\rho} \tag{5.9}$$

$$= 1 - \frac{2}{3}\beta\pi\rho\int_0^{\infty} dr\ r^3 g(r)\frac{d\phi(r)}{dr}. \tag{5.10}$$

These expressions allow the calculation of Helmholtz free energies, along an isochore or an isotherm, by integrating the thermodynamic derivatives,

$$\left(\frac{\partial\mathscr{F}/T}{\partial 1/T}\right)_{\rho} = \mathscr{U} \tag{5.11}$$

$$\left(\frac{\partial\mathscr{F}}{\partial V}\right)_T = -\rho^2\left(\frac{\partial\mathscr{F}/N}{\partial\rho}\right)_T = -p. \tag{5.12}$$

The notation for the radial distribution function above, as is common, suppresses further dependences. But we stress that $g(r)$ also varies with density and temperature, i.e., formally

$$g(r) = g(r;\rho, T). \tag{5.13}$$

In 1954 it was R Zwanzig [11] that started the formal introduction of perturbation theory. This led to the WCA theory we know today, in the late sixties, before the advent of J A Barker and D Henderson (DH) [12], or WCA [10] had formulated the first successful perturbation theory of simple fluids. However, the subsequent WCA theory proved to be more accurate and equally easy to use and, over the years, WCA theory replaced DH theory (see for example [1]).

Zwanzig started with the assumption that the total potential energy of an $\hat{N}$-particle fluid is into two contributions [11],

$$\mathscr{U}_N = \mathscr{U}_N^{(0)} + \mathscr{U}_N^{(1)}, \tag{5.14}$$

where the first term on RHS is the energy of the unperturbed system or reference system. At this point the way in which the reference and perturbation terms are defined is immaterial. Later we shall see that in practice for a successful perturbation theory it is crucially important how the division is made. The second term on the RHS is the perturbation energy. In what follows we shall often assume that the reference system is a hard sphere fluid. Zwanzig showed that substitution of (5.14) into the canonical partition function Z_N leads to

$$Z_N = Z_N^{(0)} \langle e^{-\beta \mathscr{U}_N^{(1)}} \rangle_0. \tag{5.15}$$

Here the angle bracket factor (with subscript 0) indicates the ensemble average of the Boltzmann factor of the perturbation energy taken over the *reference* system! It is instructive to note how this expression can immediately be used in a molecular simulation (MC or MD). One simply generates a series of configurations by either Monte Carlo moves or MD time steps, using the *reference* interaction potential (e.g., hard spheres). And one calculates an ensemble (or time) average of $\exp(-\beta \mathscr{U}_N^{(1)})$ along the way.

Using the fact that $\mathscr{F}$, the Helmholtz free energy, is simply the negative of the logarithm of the partition function, $\mathscr{F} = -kT \ln Z_N$, we can rewrite Zwanzig's basic result as,

$$\beta \mathscr{F} = \beta \mathscr{F}_0 + \beta \mathscr{F}^{(1)} \tag{5.16}$$

$$= \beta \mathscr{F}_0 - \ln \langle e^{-\beta \mathscr{U}_N^{(1)}} \rangle_0 \tag{5.17}$$

$$\approx \beta F_0 + \beta \langle \mathscr{U}_N^{(1)} \rangle_0 - \frac{1}{2} \beta^2 [\langle (\mathscr{U}_N^{(1)})^2 \rangle_0 - \langle \mathscr{U}_N^{(1)} \rangle_0^2] + \ldots \tag{5.18}$$

where on the last line we have expanded the exponential in powers of β and the ellipses denote higher order terms. The latter all are made up of products and powers of $\langle \mathscr{U}_N^{(1)} \rangle_0$, that is, averages taken in the *reference* system.

If the perturbation potential is a pair potential, $\mathscr{U}_N^{(1)} - 1/2 \sum_{i,j} \phi^{(1)}(r_{ij})$, the expression simplifies. Keeping just the first term of the free energy expansion,

$$\beta \mathscr{F} \approx \beta \mathscr{F}_0 + \langle \mathscr{U}_N^{(1)} \rangle_0 \tag{5.19}$$

$$= \beta \mathscr{F}_0 + 2\beta \pi \rho^2 V \int_0^\infty dr\, r^2 \phi^{(1)}(r) g_0(r). \tag{5.20}$$

This is the final result that we anticipated: on the last line we arrived at an integral of the perturbation potential over the reference fluid structure (e.g., the hard sphere radial distribution function).

Notice also that in a molecular simulation we can readily compute $\langle \mathscr{U}_N^{(1)} \rangle_0$. And (see the paragraph above) we can directly compare this result to the full result $\langle \exp(-\beta \mathscr{U}_N^{(1)}) \rangle_0$. Doing so provides us with an exact measure of the consequence of the series expansion used in line three of (5.18).

5.3.1 Barker and Henderson perturbation theory [12]

As mentioned at the start of the Zwanzig derivation of formal perturbation theory [11], the division into reference and perturbation energy (see equation (5.15)) was still arbitrary. The success of practical implementation of these ideas rests entirely on the judicious choice of a division of the pair potential. As mentioned, this was accomplished by the late 1960s. At that time there had been several groundbreaking simulation studies of the Lennard–Jones (LJ) fluid that had provided information on a variety of aspects, the equation of state, the radial distribution function, the diffusion coefficient, liquid–vapor data and the critical parameters (density, temperature and pressure).

One very impressive study in 1964 was the first (non-hard sphere) MD study of LJ fluid by Aneesur Rahman of Argonne National Laboratory. This influential paper introduced the predictor–corrector algorithm to MD and calculated, among other thermodynamic quantities, distribution functions, $h(r)$ as well as functions related to scattering and integral equation theory, e.g., $\hat{h}(k)$ and $S(k)$. In addition, Rahman presented an analysis of the LJ transport. On a more curious note, Rahman used $N = 864$ particles, the largest number at the time. Typical values during that era for N were 108 and 216, and 256.[2]

Another seminal MD paper was that of Loup Verlet [49] published in 1967, where he introduced the efficient Störmer method to solve Newton's equations of motion in a way that preserved time reversibility. That symplectic algorithm is now known as the Verlet algorithm. In addition, Verlet introduced his famous neighbor lists to reduce the computational effort of calculating the forces on each particle. In a series of papers he (and sometimes with D Levesque or J-P Hansen) generated important information on the equation of state and liquid–vapor coexistence.

In was in this environment that researchers were very focused on bringing Zwanzig's perturbation theory [11] to fruition, that is, to turn it into a successful description of a simple fluid. Barker and Henderson were the first researchers to succeed in this endeavor in 1967. As mentioned the effectiveness of perturbation theory depends on a good choice for dividing the interaction potential into a reference and a perturbative part. This is precisely where the choices of BH [12] and WCA [10] differ. We will focus on the LJ potential. BH elected to divide the full potential, $\phi(r)$ into a repulsive reference $\phi^{(0)}(r)$ and attractive perturbative potential $\phi^{(1)}(r)$, namely,

$$\phi^{(0)}(r) = \begin{cases} \phi(r) & r \leqslant \sigma \\ 0 & r > \sigma \end{cases} \tag{5.21}$$

$$\phi^{(1)}(r) = \begin{cases} 0 & r \leqslant \sigma \\ \phi(r) & r > \sigma \end{cases}. \tag{5.22}$$

With that division one can immediately obtain an expression for the Helmholtz free energy, namely equation (5.20). However, we require a value for βF_0 and we need to evaluate $\beta F^{(1)}$ we need $g_0(r)$, and as Henderson [15] points out, those are no easier to accomplish than are βF and $g(r)$. This is where the choice of $\phi^{(0)}(r)$ is crucial, BH

[2] There are exceptions. For instance, Chapela *et al* [13] performed LJ liquid–vapor simulation studies with 255 atoms, after they had unfortunately 'lost' one particle [14]

showed that since the reference potential is so steep, it can be replaced by hard sphere values, i.e.,

$$F_0 \approx F_{hs} \tag{5.23}$$

$$g_0(r) \approx g_{hs}(r). \tag{5.24}$$

This still leaves one final choice to be made, namely the hard sphere diameter that best describes the steep repulsive reference potential $\phi^{(0)}(r)$. This is a problem that was first faced by Rowlinson in 1964, who considered the problem of selecting the best hard sphere diameter to represent a $\epsilon(\sigma/r)^n$ potential.[3]

BH showed, using an expansion of F, that the optimum choice for $d = d(T)$ is

$$\begin{aligned} d &= -\int_0^\sigma drf(r) \\ &= -\int_0^\sigma dr\ [e^{-\beta\phi^{(0)}(r)} - 1]. \end{aligned} \tag{5.25}$$

In practice, d is close to σ and is found to be only a weak function of T.

5.3.2 Weeks, Chandler and Andersen theory

Weeks, Chandler and Andersen (WCA) [10] decided on a different split of the fluid potential $\phi(r)$. Whereas BH [12] made the split where the potential changes from positive to negative, WCA argued that a better split would result if one split where the *force* changes from positive to negative. Thus, WCA split the full potential at the minimum, the distance $r_{\min}$ where the force is zero, i.e., at $r_{\min} = 2^{1/6}\sigma$

$$\phi^{(0)}(r) = \begin{cases} \phi(r) + \epsilon & r \leqslant r_{\min} \\ 0 & r > r_{\min} \end{cases} \tag{5.26}$$

$$\phi^{(1)}(r) = \begin{cases} -\epsilon & r \leqslant r_{\min} \\ \phi(r) & r > r_{\min} \end{cases}. \tag{5.27}$$

Just as in BH theory, to use equation (5.20), one is faced with the problem that the free energy of the reference system, F_0 (and other thermodynamic properties) are not well known. So, WCA also sought to replace F_0 by an equivalent hard sphere value. Like BH theory, this ultimately is a matter of assigning a hard sphere diameter, d_{hs}, to the reference fluid. In BH theory this leads to temperature dependent diameter (see equation (5.25)). In WCA theory we will find a diameter that depends on T *as well as* ρ.

First, WCA make the observation that the background (or cavity) correlation function, $y(r)$, of the hard sphere fluid is similar to that of the reference system, i.e., $y_0(r) \approx y_{hs}(r)$. Or, after rearranging, we have

$$g_0(r) \approx y_{hs}(r)e^{-\beta\phi^{(0)}(r)}. \tag{5.28}$$

[3] The Rowlinson solution was $d = -\int_0^\infty drf(r) = \sigma(\beta\epsilon)^{-n}(1 + 0.577\,215\,7/n)$. Note that the upper limit is larger than that of BH.

For the criterion for determining d_{hs}, WCA choose to stipulate that the compressibility (4.20) shall be the same for the reference fluid and the hard sphere fluid. Substituting approximation (5.28) into (4.20) one obtains the sought criterion for the hard sphere diameter:

$$\int_0^\infty dr\ r^2 y_{hs}(r)[e^{-\beta\phi^{(0)}(r)} - 1] = \int_0^\infty dr\ r^2 y_{hs}(r)[e^{-\beta\phi_{(hs)}(r)} - 1] \tag{5.29}$$

which we can rearrange to

$$\int_0^\infty dr\ r^2 y_{hs}(r)[e^{-\beta\phi^{(0)}(r)} - e^{-\beta\phi_{(hs)}(r)}] = 0. \tag{5.30}$$

The term between the square brackets, $\exp(-\beta\phi^{(0)}(r)) - \exp(-\beta\phi_{(hs)}(r))$, is a rapidly varying function that was dubbed the 'blip function'.

This choice of hard sphere diameter makes the free energy accurate to the fourth order [16].

So far, we have discussed two successful perturbation theories in detail. They both lead to accurate results for, say, the LJ fluid. There had been other attempts prior to that which differed in the choice of reference fluid and a perturbation. An example of that is the work by Frisch *et al* [17] and, independently, McQuarrie and Katz [18]. Both teams observed that the LJ potential is already formed as the sum of a repulsive term, r^{-12} and attractive r^{-6} contribution. This works well, especially at high temperatures ($kT/\epsilon \gtrsim 3$) when the reference fluid properties are calculated by molecular simulation. They used Rowlinson's criterion [19] to link the repulsive term to the hard sphere fluid. This works less well at high densities, and in that mode this perturbation theory comes out less successful than BH or WCA theory.

5.4 Mean field theory

The deceptive simplicity of the successful work by van der Waals [20] combined with the efforts of Weiss on ferromagnets [21], and Bragg and Williams [22] in lattice theory of alloys constitute a great starting point for mean field (MF) theory. In essence all three are examples of the mean field approximation as they arise in different scientific fields. From the two perturbation theories in the section above we can easily derive at a mean field theory by setting the hard sphere radial distribution function of the reference fluid to that of the low density hard spheres, i.e.,

$$g_0(r) = g_{hs}(r) = \begin{cases} 0 & r \leqslant d \\ 1 & r > d \end{cases}. \tag{5.31}$$

If we substitute this into equation (5.20), we obtain

$$\begin{aligned} \frac{\beta\mathscr{F}}{N} &\approx \frac{\beta\mathscr{F}_{hs}}{N} + 2\beta\pi\rho\int_d^\infty dr\ r^2\phi^{(1)}(r) \\ &= \frac{\beta\mathscr{F}_{hs}}{N} + \beta\rho a_{vdW} \end{aligned} \tag{5.32}$$

where the integral over the perturbing potential (together with the factor 2π) collapses into the van der Waals constant 'a'. We see that the latter constant is independent of temperature (and <0, for an attractive perturbation potential, $\phi^{(1)}(r)$).

By taking the thermodynamic derivative of F with respect to density (5.12), we arrive at the celebrated van der Waals equation [1, 20];

$$p = p_{hs} + a_{vdW}\rho^2. \tag{5.33}$$

It is easy to see where the terminology 'mean field' comes from. Starting from equation (5.17), we can expand and rewrite the perturbation factor as

$$\langle e^{-\beta \mathcal{U}_N^{(1)}} \rangle_0 \approx 1 - \beta \langle \mathcal{U}_N^{(1)} \rangle_0 + \ldots \approx e^{-\beta \langle \mathcal{U}_N^{(1)} \rangle_0} \tag{5.34}$$

upon neglecting terms β^2 and higher. In words, the average Boltzmann factor of the perturbation energy is approximated by the Boltzmann factor of the average (or mean) perturbation energy.

The van der Waals equation combines the hard sphere pressure p_{hs} with a MF contribution. At the time that van der Waals worked on this (ca, 1873) there was no good hard sphere equation of state. Van der Waals sought to improve upon the ideal gas equation of state by at least accounting for the excluded volume $V_{\text{excl}} = 4\pi d^3/3$ of a single sphere. Since the excluded volume is shared by two spheres, the free volume of a collection of N spheres is equal to $V - N2\pi d^3/3 = V - Nb_{vdW} = V - NB_2^{hs}$, using the expression for the second virial coefficient (5.5). With that, the van der Waals approximated the hard sphere pressure as

$$p_{hs} \approx \frac{N}{\beta(V - Nb_{vdW})} = \beta^{-1}\frac{\rho}{1 - B_2^{hs}\rho} \tag{5.35}$$

$$\approx \beta^{-1}\rho\Big(1 + B_2^{hs}\rho + (B_2^{hs})^2\rho^2 + \ldots\Big) \tag{5.36}$$

where on the last line we performed a virial expansion. This is correct up to the second coefficient. Comparing with the exact result (5.5) we see that the third coefficient is too large, and is out by a factor of 5/8. We point out that in practice, one often sticks with the first form of equation (5.35). That is, one does not make the connection with B^{hs}, but rather one leaves the parameter b_{vdW} as a free parameter that can be fitted to experimental data.

As an aside, the van der Waals pressure for the *one-dimensional* hard-rod system is, in fact, exact. Moreover, in one-dimension the mean field approximation is exact for a special system, a one-dimensional system of hard rods plus a very weak attraction of infinite range (such that the a_{vdW} integral exists and is finite) [23]. The impetus for van der Waals' work was the experimental work on liquefying gases. At the time, Thomas Andrews and Dewar in the UK and Kamerlingh–Onnes were involved a competitive effort to demonstrate that *all* gases could be liquified if the temperature was dropped sufficiently (e.g., see [24]).

The critical point temperature, below which is true, can be found with the help the condition that $(\partial p/\partial \rho)_T$ and $(\partial^2 p/\partial \rho^2)_T$ are simultaneously zero at that point. Simplifying the notation a little, dropping the 'vdW' subscript, one finds that

$$kT_c = -\frac{8a}{27b} \tag{5.37}$$

$$\rho_c = (3b)^{-1} \tag{5.38}$$

$$p_c = -\frac{a}{27b^2} \tag{5.39}$$

one can use the expressions for kT_c, ρ_c and p_c to rewrite the van der Waals equation in terms of reduced variables, $\tilde{T} = T/T_c$, $\tilde{\rho} = \rho/\rho_c$ and $\tilde{p} = p/p_c$

$$\tilde{p} = \frac{8\tilde{\rho}\tilde{T}}{3 - \tilde{\rho}} - 3\tilde{\rho}^2. \tag{5.40}$$

This form explicitly demonstrates the law of corresponding states, in that a *single* equation of state describes all fluids, once the equation of state is expressed in reduced variables. This highlights the basic notion that all fluids are similar.

The van der Waals contribution underlined the ability to describe first-order vapor-liquid phase transitions below T_c, and a higher order transition at T_c as well as continuity of the gaseous and liquid phases.

Along an isotherm $T < T_c$ the pressure is a continuous function (see figure 5.1), and as a function of volume, $V = 1/\rho$, exhibits one local maximum (point Q) and one local minimum (point S) connected by a stretch of isotherm (segment QRS) where the pressure derivative with respect to $1/\rho$ is positive. That is, a decrease in volume leads to a lowering in pressure and thus this part of the isotherm is *unstable*. Connecting the maxima of different isotherms produces a curve that ends at the critical point. This is shown in figure 5.2 for the van der Waals equation of state, and figure 5.3 for the hard-sphere improved mean field model. Similarly, connecting all

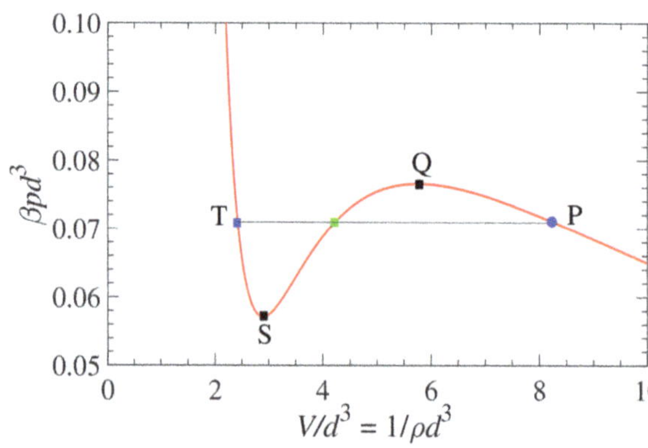

Figure 5.1. A subcritical isotherm for the Carnahan–Starling [25] plus mean field model. Here, $\tau \equiv kT\sigma^3/a = 017$.

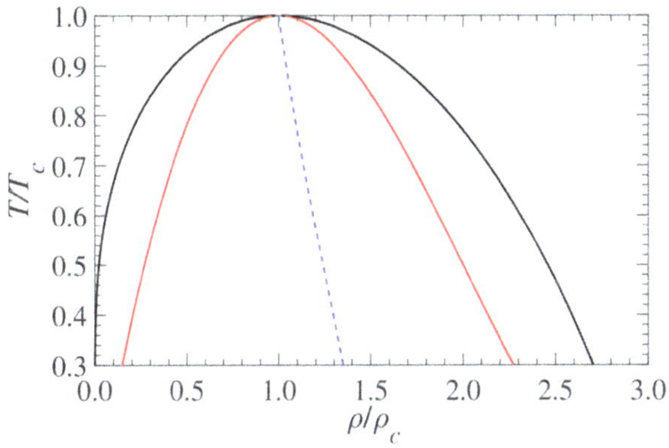

Figure 5.2. The phase diagram for the pure van der Waals equation of state (5.4). The outer solid black lines denote the binodals, the vapor and liquid coexistence densities. The inner solid red lines are the spinodals, i.e. the loci of the extrema (points Q and S in the p, V diagram (see figure 5.1)). The dashed line denotes the law of rectilinear diameters, i.e., $(\rho_v + \rho_l)/2$.

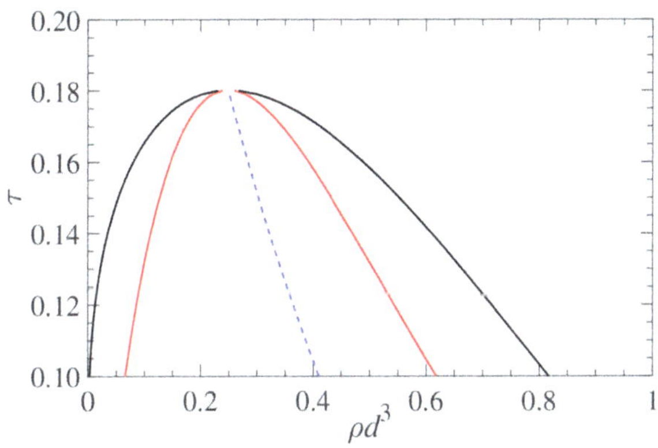

Figure 5.3. As figure 5.1 but now for the Carnahan–Starling [25] plus mean field equation of state. We plot the dimensionless temperature, $\tau \equiv kT\sigma^3/a$, versus the dimensionless density, $\rho\sigma^3$.

the minima produces another curve that smoothly joins the maxima loci curve at the critical point (figure 5.4). These two curves are known as the spinodal lines. Together they enclose a part of the V, T–diagram (or, alternatively (ρ, T)-diagram) that is unstable, and is referred to as the spinodal region. Although the spinodal region that we encounter here is entirely the consequence of employing a simple third degree equation of state, it does resemble reality. That is, a real fluid exhibits a similar region of instability, where an initially homogeneous fluid (or fluid mixture) is at a maximum in the free energy and spontaneously, and rapidly, phase separates following the tiniest of fluctuations. This process is referred to as spinodal decomposition, where the system breaks up into little domains that then undergo (a power law) growth with time. That is the characteristic size scales as $R(t) \sim t^{\alpha}$. For example, $\alpha = 1/3$ for diffusive growth.

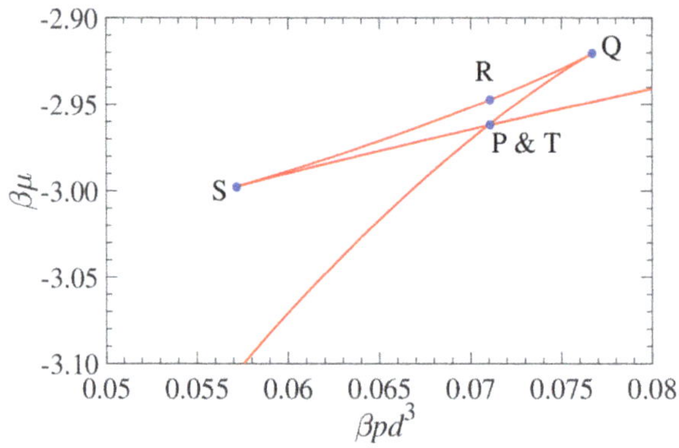

Figure 5.4. Chemical potential versus pressure, for a subcritical isotherm, $\tau = 0.17$. The labeled points are those of the isotherm shown in figure 5.1. There are two branches, gas and liquid, intersecting at the points labeled by P and T, which is the point of coexistence. Since $(\partial\mu/\partial p)_T = 1/\rho$ the steeper branch is the gas branch. For the same reason, the *unstable* section of the isotherm, QRS, meets the metastable sections, PQ (on the gas side), and ST (on the liquid), tangentially. This reflects the fact that the isotherm and its derivative are continuous at the spinodal points Q and S. Finally, we note that point R, is the intersection of the unstable section and the timeline $p = p_{\rm coex}$. Hence, in the μ, p - plane, point R lies directly above the coexisting points P and T.

Outside the spinodal region exists the so-called binodal regions, one on either side of the spinodal. Beyond the binodal lines the fluid is a thermodynamically stable one-phase homogeneous fluid (or fluid mixture). Within the binodal region the fluid is only metastable, it finds itself in free energy minimum that is, however, *not* a global minimum. As such the system is stable against small fluctuations, but given time will form a nucleus of the new incipient phase. Some nuclei will disappear but at some point one will grow to a critical nucleus size. After that the growth mechanism will take over and eventually a new phase appears. These effects are exploited in so-called cloud chambers and bubble chambers to detect the trajectory of energetic elementary particles, such as alpha and beta particles.

Van der Waals did not offer a construction to determine at which, for a given isotherm, densities coexistence would be reached. This omission was subsequently remedied by Maxwell who proposed in 1875 what is now known as the equal area rule [26]. In the p, $V = 1/\rho$ plane Maxwell placed the coexistence pressure, $p_{\rm coex}(T)$, a horizontal line, such that the area of the loop above the line equaled the area below the line, There are three intersects of this line with the van der Waals curve (cf, figure 5.1). On the low-volume side it intersects at $V_l = 1/\rho_l$ (point T of figure 5.1), while on the other high-volume side it intersects at $V_g = 1/\rho_g$ (point P of figure 5.1). that determines the sought two coexisting densities, $\rho_g(T)$ and $\rho_l(T)$. The third intersection (point R of figure 5.1) lies inside the spinodal region. The Maxwell construction for $p_{\rm coex}$ is therefore the solution to,

$$\int_{1/V_l}^{1/V_g} dV\ (p - p_{\rm coex}) = 0; \;\; \text{constant } T. \tag{5.41}$$

Unfortunately, no analytical expression exists for the coexistence properties.

Today, one finds the coexisting densities by stipulating equal pressures and equal chemical potentials for the gas and liquid state points. This chemical equilibrium condition was not formulated until a few years later, in 1877, by Willard Gibbs. In today's practice, the coexisting properties have to be found numerically, by solving simultaneous equations for the pressure and the chemical potential,

$$\begin{aligned} p(\rho_l, T) &= p(\rho_v, T) \\ \mu(\rho_l, T) &= \mu(\rho_v, T) \end{aligned} \tag{5.42}$$

where the chemical potential, as a function of density and temperature, is given by

$$\beta\mu - 3\ln\Lambda = \ln\rho - \ln(1 - b\rho) + \frac{b\rho}{1 - b\rho} + 2\beta a\rho; \tag{5.43}$$

where Λ is the De Broglie wavelength. This requires an iterative numerical procedure. Lekner [27] (and, he references others independently before him) has elegantly shown that using the Maxwell construction one can derive a parametric solution for the coexistence densities, that can be generated on a calculator. The binodal lines of figures 5.2 and 5.3 were determined by solving equations (5.42). It was verified that the solutions also satisfied the equal are constructions of equation (5.41).

At the time of its introduction, in the late 1800s, the van der Waals equation represented a major advance. As mentioned, the hard sphere pressure underpinning the equation was necessarily a crude approximation. Improvements on that front have been made in later years. Thus, Longuet-Higgins and Widom [28] substituted the Carnahan–Starling [25] equation of state for p_{hs}, into the equation. That improved the accuracy of the van der Waals equation of state, but they also extended it in the sense that the hard sphere system undergoes a fluid–solid transition at a density $\rho_f d^3 \approx 0.943$, and $\rho_s d^3 \approx 1.054$. By including this result, Longuet-Higgings and Widom extended the mean field phase diagram of van der Waals to include a liquid–solid and a gas–solid region including a triple point.

Given that the van der Waals equation is a mean field approximation we now know that some of its critical region properties are incorrect. As one example, the MF envelope of gas–liquid coexistence densities (i.e., the two binodals taken together) ends in a critical point as a parabola. That is, the shape is given by

$$\begin{aligned} \Delta\rho \equiv \rho_l - \rho_g \sim t^{\beta}; \quad \beta = \frac{1}{2} \\ t \equiv \frac{|T - T_c|}{T_c}. \end{aligned} \tag{5.44}$$

However, careful experiments have demonstrated that the envelope is considerably 'flatter' than a parabola near the critical point, see for instance [29]. The accepted value for all fluids is $\beta \approx 0.34$, tantalizingly close to $\frac{1}{3}$ but slightly different. Similarly, other critical exponents are off too [29].

Van der Waals equation for mixtures

Following the success of the van der Waals approach for pure fluids, van der Waals (1890) [20] and later van Laar (1904–1910) [30] expanded the mean field approach to binary mixtures. Van Konynenburg and Scott [31] have published an extensive paper (essentially the 1968 dissertation by van Konynenburg [32]) on the great variety of behaviors they were able to capture.

$$p_m = p_{hsm} + a_m\rho^2$$
$$p_{hsm} = \frac{N}{\beta(V - Nb_m)} \tag{5.45}$$

or, alternatively (5.46)

$$p_{hsm} = p_c \tag{5.47}$$

where the subscript m denotes a mixture property and hsm refers to a mixture of hard spheres. We have denoted two choices for the hard sphere mixture pressure: the mixture form of van der Waals (middle line) and the much improved PY compressibility pressure (see equation (4.43)). The van der Waals parameters are functions of composition, i.e., molefraction x_i. In particular, van der Waals proposed

$$b_m = \sum_I \sum_j x_i x_j b_{ij}$$
$$a_m = \sum_I \sum_j x_i x_j a_{ij}. \tag{5.48}$$

These rules are referred to as *mixing rules*. The mixed quantities b_{ij}, a_{ij}; $i \neq j$ still need to be specified. Traditionally they are specified in terms of the pure components, i.e., in terms of so-called *combining* rules. For example, for a binary mixture

$$b_{12} = b_{21} = \frac{1}{2}(b_{11} + b_{22})$$
$$a_{12} = a_{21} = \sqrt{a_{11}a_{22}}. \tag{5.49}$$

The volume term can be directly compared to equation (4.34).

The van Konynenburg and Scott (vKS) [31] work is focussed on binary mixtures. Restricting themselves mostly to the symmetric case of $b_{11} = b_{22}$,[4] the work shows the enormous richness of solution phase behavior. Thus, the work demonstrates both upper and lower critical solution temperatures (UCST and LCST), critical lines, azeotropes and double azeotropes. All the experimentally known binary mixture classes are covered by the van der Waals equation of state. That is, vKS identify Class I mixtures (with four sub classes) for mixtures of two components with similar gas–liquid critical temperatures. In such mixtures the critical points of the

[4] There are also some calculations for the non-symmetric case $b_{22} = 2b_{11}$.

pure components are continuously connected by a critical line. Class II (with five sub classes) mixtures of two components with very different gas–liquid critical temperatures. In these mixtures there is no continuous critical line joining the critical points of the pure components. vKS point out that it is convenient to describe the binary van der Waals mixtures in terms of three parameters, A, B and C,

$$\begin{aligned} A &= \frac{b_{22} - b_{11}}{b_{11} + b_{22}} \\ B &= \left(\frac{a_{22}}{b_{22}^2} - \frac{a_{11}}{b_{11}^2}\right) \Big/ \left(\frac{a_{11}}{b_{11}^2} + \frac{a_{22}}{b_{22}^2}\right) \\ C &= \left(\frac{a_{11}}{b_{11}^2} - \frac{2a_{12}}{b_{11}b_{22}} + \frac{a_{22}}{b_{22}^2}\right) \Big/ \left(\frac{a_{11}}{b_{11}^2} + \frac{a_{22}}{b_{22}^2}\right). \end{aligned} \tag{5.50}$$

As mentioned, vKS primarily considered the case $b_{11} = b_{22}$, i.e., $A = 0$. Figure 5.5 (which is figure 1 of vKS) is a summary diagram of phase behavior. For the case $A = 0$, it indicates where each of the nine mixture sub classes are located in the C, B plane.

5.4.1 The Weiss approximation [21]

Modern analysis of phase transitions and critical phenomena has uncovered that there are many similarities between certain lattice models and off-lattice fluids. For example, the Ising model [33][5] (or equivalently the lattice gas model, see the next section) belong to the same universality class, sharing critical exponents [29]. Similarly, their mean field approximations have much in common, including mean field values of critical exponents.

Pierre Weiss proposed the molecular field approximation for the Heisenberg models in 1907 [21]. We will consider a special simple example of those models, the so-called Ising model, a collection of N spins, $s_i = -1$ or $+1$. The Ising Hamiltonian is

$$\mathcal{H} = -J \sum_{<ij>} s_i s_j - h \sum_{1=1}^{N} s_i. \tag{5.51}$$

The summation is over all spins on site i, s_i, interacting with its nearest neighbors j only. The interaction energy, or coupling constant, is denoted J. The last term in the Hamiltonian represents the effect of a uniform external one-body magnetic field, h.

The Ising model can be solved exactly in dimensions 1 and 2. The latter was Onsager's tour de force [36]. In three dimensions, and well away from the critical point, the properties can be calculated with Monte Carlo simulations. However, to obtain the

[5] The Ising model was invented in 1920 by the physicist Wilhelm Lenz [34], and probably should be known as the Lenz model. Lenz gave it as a problem to his student Ernst Ising. The one-dimensional Ising model was solved by Ising in 1925 and published in his 1924 thesis. Ising found that it has no phase transition. It does in dimensions 2 and higher. The pronunciation of 'The Ising model' is very similar to 'de-icing model'. This can lead to serious confusion in daily life, as one Cornell statistical mechanics student of M E Fisher [35] found when on an interview trip to the Boeing Company.

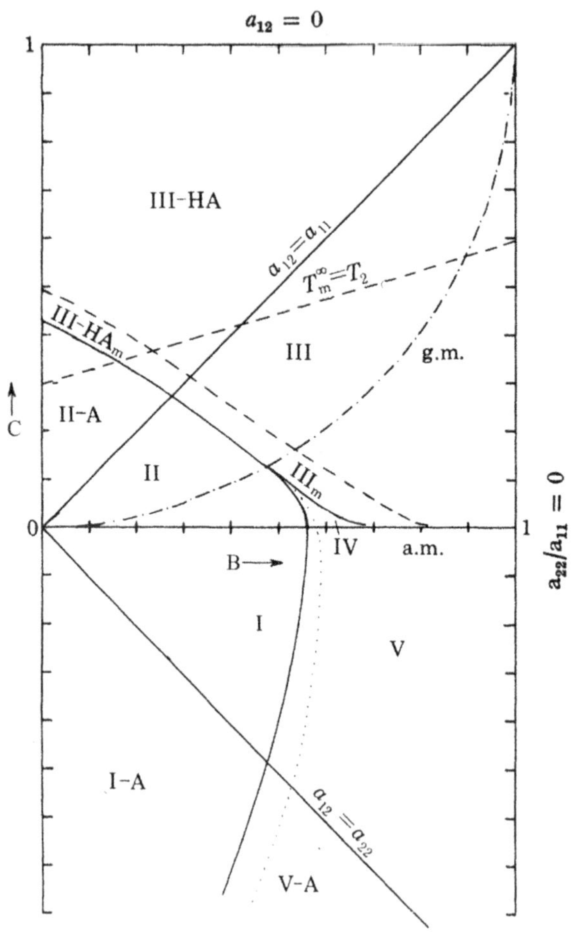

Figure 5.5. Diagram of phase behavior for equal size molecules (i.e., $A = 0$). Nine major regions of characteristic p, T-phase diagrams are separated by the boundaries (—). The quantities B and C are defined in the text (equation 5.50). (Reproduced from [31], with the permission of The Royal Society of Chemistry.)

mean field approximation of Hamiltonian (5.51) is straightforward. The key is to assume that fluctuations are small, such that a spin s_i interacts with an average of neighboring spins s_j. Thus we introduce the following notation for the average of a spin: $m \equiv <s_i> = N^{-1}\sum_{i=1}^{N} s_i$, and define the deviation from the average as:

$$\Delta s_i \equiv s_i - <s_i> . \tag{5.52}$$

Substituting this into equation (5.51) we have

$$\mathcal{H} = -J \sum_{<ij>} <s_i> + <s_i> \Delta s_j + <s_j> \Delta s_i + \Delta s_i \Delta s_j - h \sum_{1=1}^{N} s_i . \tag{5.53}$$

Now, the mean field approximation Hamiltonian is obtained at this point by neglecting the term $\Delta s_i \Delta s_j$. Thus, we obtain for the mean field Hamiltonian,

$$\begin{aligned}\mathcal{H}_{mf} &= -J\sum_{<ij>}\left[<s_i><s_j>+<s_i>\Delta s_j+<s_j>\Delta s_i\right]-h\sum_{1=1}^{N}s_i\\&= -J\sum_{<ij>}[s_i<s_j>+s_j<s_i>-<s_i><s_j>]-h\sum_{1=1}^{N}s_i\\&= -\frac{1}{2}Jqm\sum_{i-1}^{N}(2s_i-m)-h\sum_{1=1}^{N}s_i\\&= -\frac{1}{2}NJqm^2-(h+Jqm)\sum_{1=1}^{N}s_i.\end{aligned} \tag{5.54}$$

Here q denotes the number of nearest neighbors of a site, e.g., $q = 6$ for a three-dimensional cubic lattice, while $q = 4$ for a two-dimensional square lattice. The final result is that we have arrived at a solvable problem. That is, $\mathcal{H}_{mf}$ has the same form as a *noninteracting* Ising model in an *effective* external field, $h_e \equiv h + Jqm$.

Because the system has been simplified to that of a noninteracting spins in a field h_e, the calculation of the partition function also simplifies. Thus we find that,

$$\begin{aligned}Z_{mf} &= Tr(e^{-\beta\mathcal{H}_{mf}})\\&= e^{\beta NJqm^2/2}\,[2\cosh(\beta h_e)]^N\end{aligned} \tag{5.55}$$

and for the average magnetization, m,:

$$\begin{aligned}m &= \frac{1}{N}\sum_{i=1}^{N}<s_i>\\&= \frac{1}{NZ_{mf}}\sum_{i=1}^{N}Tr\left(s_i e^{-\beta\mathcal{H}_{mf}}\right)\\&= \frac{1}{\beta N}\frac{\partial\ln Z_{mf}}{\partial h_e}.\end{aligned} \tag{5.56}$$

Using equation (5.55) for Z_{mf} we can now express m in terms of the external field, h_e (and h), i.e.,

$$m = \tanh(\beta h_e) \tag{5.57}$$

$$= \tanh(\beta h + \beta Jqm). \tag{5.58}$$

Although a compact expression, this is in fact a transcendental equation that has to be solved graphically or numerically. To illustrate we set the external field to zero, i.e. $h = 0$. We then plot both the left hand and the right hand side of equation (5.58), and look for intersections. of the straight line and the tanh function. Three example curves are shown in figure 5.6, corresponding to $\beta Jq = 2,\ 1$, and 0.4. All cases have a solution at the origin, i.e. for $m = 0$.

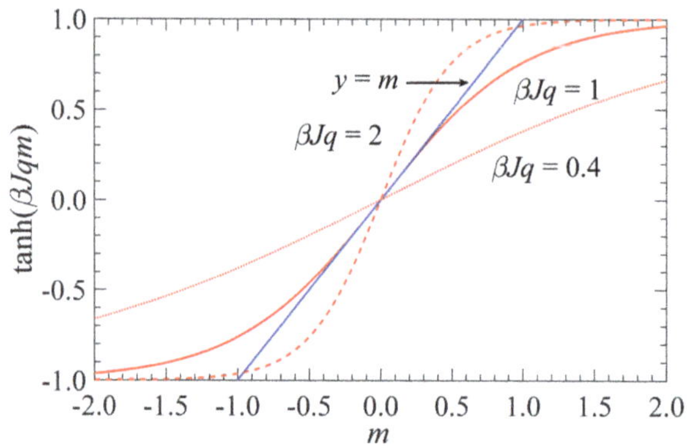

Figure 5.6. Graphical solution to equation (5.58). The straight blue line, $y = m$, is the left hand side of the equation. In red is the right hand side of the equation. We show three cases $\beta Jqm = 2$, 1, and 0.4. All cases have an intersection at $m = 0$. The case $\beta Jqm = 1$ corresponds to the critical case, where the blue line is tangent to the red tanh curve. The case $\beta Jqm = 0.4$ is a supercritical isotherm; it has no other intersections besides $m = 0$. Finally, $\beta Jqm = 2$ is a subcritical isotherm, it results in two additional intersections of the tanh curve: namely at roughly $m \pm 0.957\,50$.

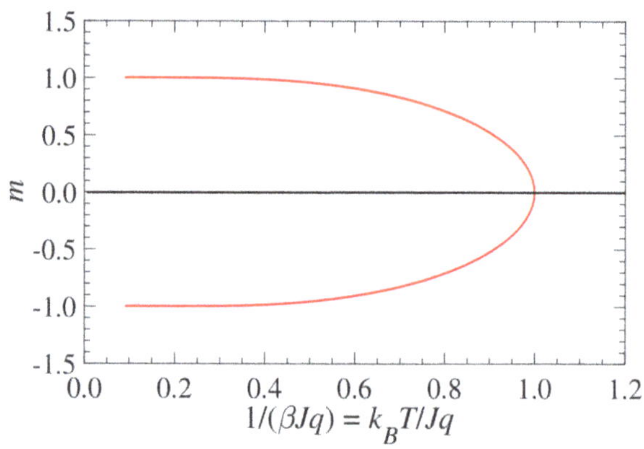

Figure 5.7. The resulting symmetric phase diagram, m versus temperature, resulting from a repeated numerical solution to equation (5.58).

The case $\beta Jq = 2$ has three intersections: $m = 0$ and $m = \pm|m_0|$ (where the subscript indicates the value of m at the intersection). This choice, $\beta Jq = 2$, corresponds to a temperature that is below the critical temperature, which occurs when $\beta Jq = 1$ or $kT_c = Jq$ and shows there are two anti-symmetric phases that coexist. The phase diagram is shown in figure 5.7.

When turned by 90 degrees, this phase diagram is equivalent to the fluid phase diagrams shown in figures 5.2 and 5.3.

It is instructive to inspect the accompanying free energy as a function of m for the phase transition case. We obtain the Helmholtz fee energy for zero field (i.e., $h = 0$) as the negative logarithm of the partition function (5.55), that is,

$$F_{mf} = -kT \ln Z_{mf} \tag{5.59}$$

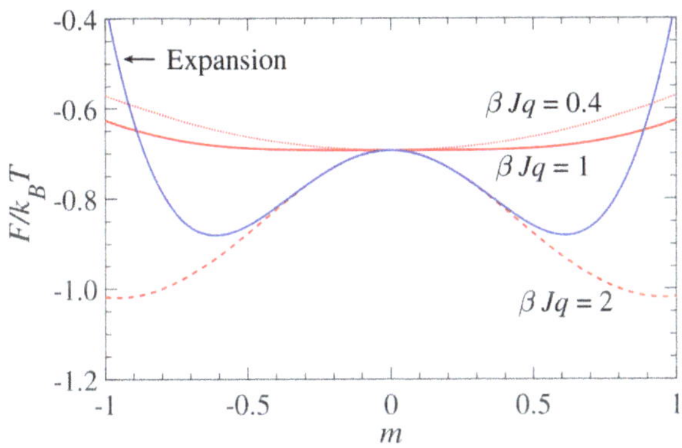

Figure 5.8. The mean field Helmholtz free energy per site, F_{mf}/NkT, versus the magnetization (see equation (5.61)). The same temperatures of figure 5.6 are shown here: $\beta Jqm = 1$ (critical isotherm, solid red curve), $\beta Jqm = 2$ (subcritical isotherm, dashed red curve), $\beta Jqm = 0.4$ (supercritical isotherm, dotted red curve). Note, the locations of minima of F_{mf}/NkT for $\beta Jqm = 2$ are identical to those found from the intersections: $m \pm 0.957\,50$. Also shown, in blue, is the result of the approximate F_{mf}/NkT (see equation (5.61)), based on the Taylor expansion of the $\ln(\cosh(\beta h_e))$ term. That curve is displayed as a solid blue curve. Notice that the approximate expression for F_{mf}/NkT captures the double minima feature of the subcritical free energy. For the critical and subcritical isotherms the approximation (not shown) is very close.

$$= \frac{1}{2}NJqm^2 - NkT\ln 2 - NkT\ln[\cosh(\beta h_e)]. \tag{5.60}$$

The results for the Helmholtz free energy are shown in figure 5.8, where the temperatures chosen are identical to those in figure 5.6. The coexistence values, $m \pm 0.957\,50$, found from the intersections of figure 5.6 correspond to the two minima of F_{mf}/NkT. The subcritical temperature curve displays two wells at $m \pm |m_o|$, while the supercritical isotherm has a single well (that gets to be quite flat for the critical isotherm). Those features are preserved when one performs a Taylor expansion of the free energy,

$$\frac{F_{mf}}{NkT} \approx \frac{1}{2}\beta Jqm^2 - \ln 2 - \ln\left[\frac{1}{2}(\beta h_e)^2 + \frac{1}{12}(\beta h_e)^4 + \mathcal{O}[(\beta h_e)^6]\right]. \tag{5.61}$$

We now specialize to the zero external field case $h = 0$, so $\beta h_e = \beta Jqm$. Rewriting and gathering terms in powers of m, we arrive at a quartic expansion of the free energy in m, with even powers of m

$$\frac{F_{mf}}{NkT} \approx -\ln 2 + \frac{1}{2}\left(\frac{T}{T_c}\right)^{-2}\frac{T - T_c}{T_c}m^2 + \frac{1}{12}\left(\frac{T}{T_c}\right)^{-4}m^4 + \mathcal{O}(m^6). \tag{5.62}$$

This approximate quartic curve has two minima below T_c. An example is shown (in blue) in figure 5.8.

The MFT generally overestimates the value of T_c [29]. For example, in two dimensions for a square lattice Onsager [36] found the exact solution to be

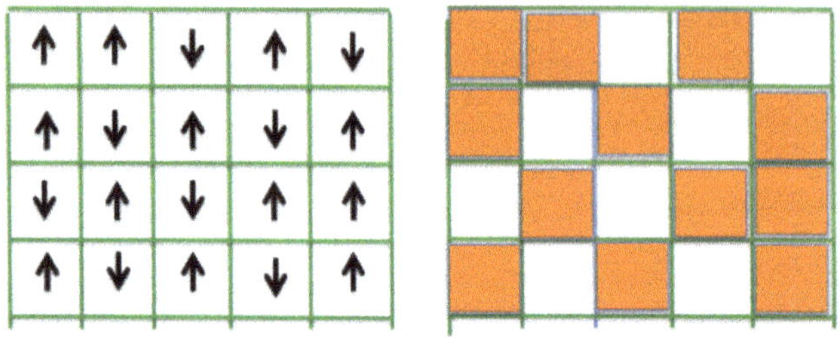

Figure 5.9. An illustration of the equivalence between the Ising model (on the left) and the lattice gas (on the right). Each up (down) arrow in the Ising model corresponds to a filled (empty) site in the lattice gas.

$kT_c = 2J(\ln(1 + \sqrt{2})^{-1}) \approx 2.27J$ while MFT finds a substantially higher value: $kT_c = 4J$. This is to be expected, as the neglected fluctuations destabilize the phase separation, pushing T_c down.

5.4.2 The lattice gas model

The lattice gas model was proposed by Cernuschi and Eyring [37] as a simplified liquid model.[6] The model consists of a simple lattice structure, e.g., square lattice in two dimensions of simple cubic in three. Each lattice site can be either filled (with a gas atom) or empty. Thus, the occupation number, n_i, of a site i is either 1 (filled) of 0 (empty).

$$\mathcal{H}_{lg} = -\epsilon \sum_{<ij>} n_i n_j - \mu \sum_i n_i \tag{5.63}$$

where μ denotes the chemical potential. Yang and Lee [38] proved in 1952 that the lattice gas model is isomorphic with the Ising model. The mapping can be made through

$$n_i = \frac{1}{2}(1 + s_i). \tag{5.64}$$

Notice that the total number of spins in the Ising model is constant (and equal to the number of lattice sites, N). In the lattice gas model on the other had the total number of atoms is *not* conserved, but equal to the number of filled sites. The larger the value of μ the more sites will end up filled. As a consequence, the Ising model is studied in the canonical ensemble, while the natural ensemble for the lattice gas is the Grand ensemble (figure 5.9).

To make the mapping between the two models more explicit we simply substitute mapping expression (5.64) into equation (5.63),

[6] Not to be confused with the lattice gas automata (LGA) that were the precursor to the lattice Boltzmann method (LBM) [39]. Both LGA and LBM are used for the simulation of fluid flow.

$$
\begin{aligned}
\mathcal{H}_{lg} &= -\frac{\epsilon}{4}\sum_{<ij>}(1+s_i)(1+s_j) - \frac{\mu}{2}\sum_i(1+s_i) \\
&= -\frac{\epsilon}{4}\left(\sum_{<ij>} s_i s_j + \sum_{<ij>} s_i + \sum_{<ij>} s_j + \sum_{<ij>} 1\right) - \frac{\mu}{2}\sum_i 1 - \frac{\mu}{2}\sum_i s_i \\
&= -\frac{\epsilon}{4}\left(\sum_{<ij>} s_i s_j + \frac{q}{2}\sum_i s_i + \frac{q}{2}\sum_j s_j + \frac{Nq}{2}\right) - \frac{N\mu}{2} - \frac{\mu}{2}\sum_i s_i \\
&= -\frac{\epsilon}{4}\sum_{<ij>} s_i s_j - \frac{\epsilon q}{4}\sum_i s_i - \frac{N\epsilon q}{8} - \frac{N\mu}{2} - \frac{\mu}{2}\sum_i s_i \\
&= -\frac{\epsilon}{4}\sum_{<ij>} s_i s_j - \left(\frac{\epsilon q}{4} + \frac{\mu}{2}\right)\sum_i s_i - \frac{N\epsilon q}{8} - \frac{N\mu}{2} \\
&= -J\sum_{<ij>} s_i s_j - h\sum_i s_i - C \\
&= \mathcal{H}_{\text{ising}} - C
\end{aligned}
\tag{5.65}
$$

here C denotes a constant term, *independent* of the spins s_i. On line 3 we made use of the identity $\sum_{<ij>} = (1/2)\sum_i \sum_{nn} = (q/2)\sum_i$.

Thus, the mapping of the isomorphism is now complete and has been obtained in explicit form. Specifically, the magnetic coupling constant $J = \epsilon/4$ and the external magnetic field is of strength $h = \epsilon q/4 + \mu/2$.

5.4.3 Scaled particle theory (SPT)

The introduction to scaled particle theory (SPT) appeared in 1959 in a paper by Reiss, Frisch, and Lebowitz (RFL) [40], although interestingly, the phrase 'scaled particle theory' does not appear in this paper. It was introduced as a statistical mechanical description for the fluid phase of the hard sphere model in three dimensions. It was found that the SPT equation of state results predicted agreed quite well with early computer simulations of that era.

We start by noting that the virial expression for the pressure of the hard sphere fluid is a simple expression,

$$
\begin{aligned}
\frac{\beta p}{\rho} &= 1 + \frac{2}{3}\pi\rho\sigma^3 g(\sigma) \\
&= 1 + 4\eta g(\sigma)
\end{aligned}
\tag{5.66}
$$

where $g(\sigma)$ denotes the contact value $g(\sigma+)$ of $g(r)$. So, to obtain the equation of state for hard spheres, we do not need the function $g(r)$ over its entire range, as integral equations seek to provide. SPT is a method to use geometrical information to improve the knowledge of $g(\sigma)$.

The theory proceeds by considering adding a hard sphere of diameter b, say, to a fluid of hard spheres of diameter σ and density ρ. We refer to the added sphere as a solute and the rest of the hard sphere fluid as solvent. The solute and solvent touch at a distance $(b + \sigma)/2$, which we will denote as λ, The value of the radial distribution function of solvent particles (of density ρ) around *and* in contact with the solute is introduced as $G(\lambda, \rho)$, where we have included ρ in the notation to stress that G is always a function of the fluid density. Similarly, the local density at the surface of the cavity is given by the product of G and the density, i.e., by $\rho G(\lambda, \rho)$.

Clearly, when the solute happens to have a diameter identical to the solvent, i.e. when $b = \sigma$ we have

$$G(\sigma, \rho) = g(\sigma, \rho) \tag{5.67}$$

hence,

SPT then aims to find the values of $G(\lambda, \rho)$ over the full range of distances from $\lambda = 0$ to $\lambda = \infty$.

Now, to successfully add a solute of diameter b is equivalent to finding a cavity of radius λ inside the fluid (figure 5.10).

RFL considered the reversible work, $W(\lambda)$, done when inserting a cavity of size λ. Knowing the reversible work allows us to write the probability, $p_0(\lambda)$, of observing a fluctuation of that size is given by the Boltzmann factor of the reversible work,

$$p_0(\lambda) = e^{-\beta W(\lambda)}. \tag{5.68}$$

The subscript 0 indicates that there are *no* centers of the fluid spheres whose center lies inside the radius λ.

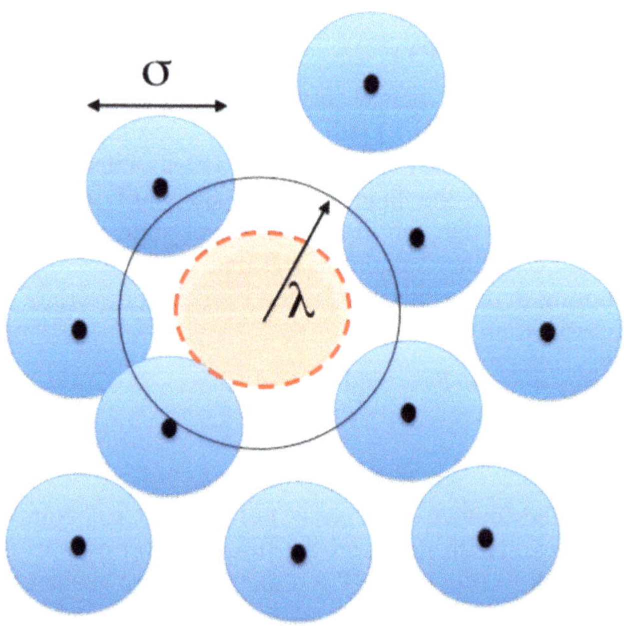

Figure 5.10. A cavity of radius λ inside a fluid of hard spheres of diameter σ. The cavity radius implies that no fluid (blue) particle centers fall inside it. It allows enough room to insert a solute particle (indicated by the pink sphere with a dashed-lined boundary) of diameter $2\lambda - \sigma$.

Now, the function G is related to the reversible work as follows. The differential of the work to change the cavity size by $d\lambda$ is equal to

$$\beta dW(\lambda) = \rho G(\lambda, \rho) 4\pi\lambda^2 d\lambda. \tag{5.69}$$

Hence,

$$\beta W(\lambda) = 4\pi \int_0^\lambda dr\ r^2 \rho G(r, \rho). \tag{5.70}$$

Exact results for $p_0(\lambda)$ help us to learn about the values $G(\lambda, \rho)$ assumes as λ is varied. We start with a cavity of zero radius $\lambda = 0$. This implies (cf, the above expression) that $W(0) = 0$, and from equation (5.68) we find $p_0(0) = 1$, as expected. In fact, p_0 is known for all sizes up to $\lambda \leqslant \sigma/2$, viz.,

$$p_0(\lambda) = 1 - \frac{4}{3}\pi\rho\lambda^3 \quad ; \lambda \leqslant \sigma/2. \tag{5.71}$$

For cavities with a radius in this size range there is at most *one* hard sphere of the fluid whose center lies inside the cavity. The probability of this occurring is $\frac{4}{3}\pi\rho\lambda^3$. Consequently, the probability of the cavity being free of such hard sphere centers is 1 minus that value, as stated in equation (5.71). From this result (using equation (5.69)) we obtain

$$G(\lambda, \rho) = \left(1 - \frac{4}{3}\pi\rho\lambda^3\right)^{-1} \quad ; \lambda \leqslant \sigma/2. \tag{5.72}$$

We can also consider derivatives of G with respect to λ. Taking the derivative of expression (5.72)

$$G'(\lambda, \rho) \equiv \frac{\partial G}{\partial \lambda} = 4\pi\rho\lambda^2\left(1 - \frac{4}{3}\pi\rho\lambda^3\right)^{-2} \quad ; \lambda \leqslant \sigma/2 \tag{5.73}$$

which as special cases for $\lambda = \sigma/2$ produce,

$$G(\sigma/2, \rho) = (1 - \eta)^{-1} \tag{5.74}$$

$$G'(\sigma/2, \rho) = 6\eta\sigma^{-1}(1 - \eta)^{-2} \tag{5.75}$$

where $\eta \equiv (\pi/6)\rho\sigma^3$.

Further, inserting expression (5.72) into (5.70), and setting the upper integration limit equal to $\lambda = \sigma/2$ produces:

$$\beta W(\sigma/2) = \ln(1 - \eta). \tag{5.76}$$

In the limit of large λ, $\lambda \to \infty$, thermodynamics suggest that the reversible work becomes the sum of two terms, a volume term and, for the surface created a surface term. That is for large radius λ we write

$$W(\lambda) = wv - \gamma v^{2/3} \tag{5.77}$$

where the volume $v = (4/3)\pi\lambda^3$, and w and γ are independent of v, but functions of ρ. The surface tension, γ, for a hard wall is negative, hence we included the minus sign in (5.77).

Differentiation of W with respect to the cavity volume v shows that the derivative is the kinetic energy times the density at the cavity's surface, i.e.,

$$\frac{\partial W}{\partial v} = kT\rho G(\lambda, \rho) = p_{\perp} \tag{5.78}$$

where we used (5.69). $kT\rho G$ is the (purely kinetic) normal pressure at the cavity wall, which is denoted by $p_{\perp}$.

In the limit that the cavity grows to infinite size, the cavity's surface becomes a planar hard wall, and the $\rho G(\infty, \rho$ becomes the local density at that wall, hence

$$\beta p_{\perp} = \rho G(\infty, \rho) \tag{5.79}$$

providing us with another constraint on the behavior of $G(\lambda, \rho)$ since:

$$G(\infty, \rho) = 1 + 4\eta G(\sigma, \rho) \tag{5.80}$$

using equations (5.66) and (5.67).

Bulk and surface equations of state

To simplify the notation we start be introducing a dimensionless variable for the distance

$$y \equiv \frac{r}{\sigma} \tag{5.81}$$

and continue to use $\eta = (\pi/6)\rho\sigma^3$. Rewriting equation (5.69) we have

$$G(y, \eta) = \frac{1}{24\eta y^2}\frac{\beta\partial W}{\partial y}. \tag{5.82}$$

Similarly, in terms of η and y, the reversible work of a *large* spherical cavity, the sum a volume and a surface contribution, can be made explicit as

$$\beta W = \frac{4\pi}{3}\beta p\sigma^3 y^3 + 4\pi\beta\gamma\sigma^2 y^2\left(1 - \frac{2\delta_T}{y} + \dots\right); \qquad \text{large } y \tag{5.83}$$

where the (dimensionless) Tolman's length, $\delta_T = \delta_T(\eta)$, measures the extent to which a curved body's surface tension deviates from its planar limit.

if we substitute equation (5.83) into (5.82) and use (5.66) to replace $\beta p/\rho$ by $1 + 4\eta G(1, \eta)$

$$G(y, \eta) = 1 + 4\eta G(1, \eta) + \frac{\pi\beta\gamma\sigma^2}{3\eta}\frac{1}{y} - \frac{\pi\beta\gamma\sigma^2\delta_T}{3\eta}\frac{1}{y^2}; \qquad \text{large } y. \tag{5.84}$$

We note that this has the correct form, as in the limit $y \to \infty$ it reduces to (5.80).

In order to obtain equations of state for the pressure, the surface tension as well as Tolman's length we now make a crucial approximation. We will assume that, contrary to the underlying condition of expression (5.83) that the cavity radius is large, we can apply it to small radii, such as $y = 1/2$. That means that we can set $y = 1/2$ in (5.84) and substitute that expression into the two exact relations (5.75). In addition we can simply set $y = 1$ in (5.84). Taken together this produces three equations for the three unknowns $G(1, \eta)$, $\gamma(\eta)$ and $\delta_T(\eta)$:

$$\frac{1}{1-\eta} = 1 + 4\eta G(1, \eta) + \frac{2\pi\beta\gamma\sigma^2}{3\eta} - \frac{4\pi\beta\gamma\sigma^2\delta_T}{3\eta} \tag{5.85}$$

$$\frac{6\eta}{(1-\eta)^2} = -\frac{4\pi\beta\gamma\sigma^2}{3\eta} + \frac{16\pi\beta\gamma\sigma^2\delta_T}{3\eta} \tag{5.86}$$

$$(1-4\eta)G(1, \eta) = 1 + \frac{\pi\beta\gamma\sigma^2}{3\eta} - \frac{\pi\beta\gamma\sigma^2\delta_T}{3\eta}. \tag{5.87}$$

Three equations and three unknowns. RFL [40] solve these simultaneous equations and produce the final result:

$$\begin{aligned} \frac{\beta p}{\rho} &= 1 + 4\eta G(1, \eta) \\ &= \frac{1+\eta+\eta^2}{1-\eta)^3} \end{aligned} \tag{5.88}$$

$$\beta\gamma\sigma^2 = -\frac{9}{2\pi}\frac{\eta^2(1+\eta)}{1-\eta)^3} \tag{5.89}$$

$$\delta_T = \frac{1}{4}\left(1 - \frac{1-\eta}{1+\eta}\right). \tag{5.90}$$

As has been pointed out elsewhere in this text, surprisingly, the SPT pressure equation of state is identical to the exact solution to the PY equation (specifically, the compressibility route), which was obtained four years later by Wertheim [41] as well as Thiele [42]. The surface equation of state was tested in 1984 [43] in a MD study of hard spheres against a hard wall.

5.4.4 Fluid Mixtures

The original SPT work on pure hard sphere fluids by RFL was quickly followed by work on general fluids (1960), Reiss *et al* [44]. In 1963, Lebowitz, Helfand, and Preastgaard (LHP) [45] published a paper on the scaled particle theory of fluid mixtures. It is this paper that many years later stimulated Rosenfeld's Fundamental Measure Theory (FMT) [46] that had such an impact on the development of classical-DFT.

We will summarize the main points of the LHP paper. We start by considering an m-component fluid mixture of particles that have a hard-core of radius R_i.[7]
For small R, we have the exact expression:

$$\beta W(R) = -\ln\left[1 - \frac{4\pi}{3}\sum_i^m \rho_i(R + R_i)^3\right]; \quad R \leqslant 0. \tag{5.91}$$

Based on the pure fluid approach LHP propose that for *all* R we can approximate the reversible work by

$$\beta W(R) = \beta W(0) + \beta W'(0)R + \frac{1}{2}\beta W''(0)R^2 + \frac{4\pi}{3}\beta p R^3. \tag{5.92}$$

Now, differentiating equation (5.91) to generate $W(0)$, $W'(0)$ and $W''(0)$, we find,

$$\beta W(0) = -\ln\left[1 - \frac{4\pi}{3}\sum_i^m \rho_i R_i^3\right] = -\ln(1 - \xi_3) \tag{5.93}$$

$$\beta W'(0) = \frac{6\xi_2}{1 - \xi_3} \tag{5.94}$$

$$\beta W''(0) = \frac{12\xi_1}{1 - \xi_3} + \frac{18\xi_2^2}{(1 - \xi_3)^2} \tag{5.95}$$

where LHP introduced the following short-hand notation[8]

$$\xi_n = \frac{\pi}{6}\sum_i^m \rho_i(2R_i)^n;\ n = 0, 1, 2, 3. \tag{5.96}$$

Using this notation one finds [45] the pressure,

$$\frac{\pi}{6}\beta p = \frac{\xi_0}{1 - \xi_3} + \frac{18\xi_1\xi_2}{(1 - \xi_3)^2} + \frac{18\xi_2^3}{(1 - \xi_3)^3} \tag{5.97}$$

and the surface tension at a hard wall,

$$\beta\gamma = \sum_i^m \beta p_i R_i + \frac{\frac{1}{2}\xi_1}{1 - \xi_3} + \frac{\frac{3}{4}\xi_2^2}{(1 - \xi_3)^2} \tag{5.98}$$

[7] LHP consider a fluid mixture where the particles also interact with an attractive potential ϕ_{ij} for distances larger than $R_i + R_j$. We will limit ourselves to hard sphere mixtures and set $\phi_{ij} = 0$.

[8] It is important to recognize the beauty and efficiency of the four quantities ξ_0, –, ξ_3. No matter how large the number of components m, the number of ξ_n is always four! This elegant feature is carried over into the fundamental measure theory (FMT) of Rosenfeld [46].

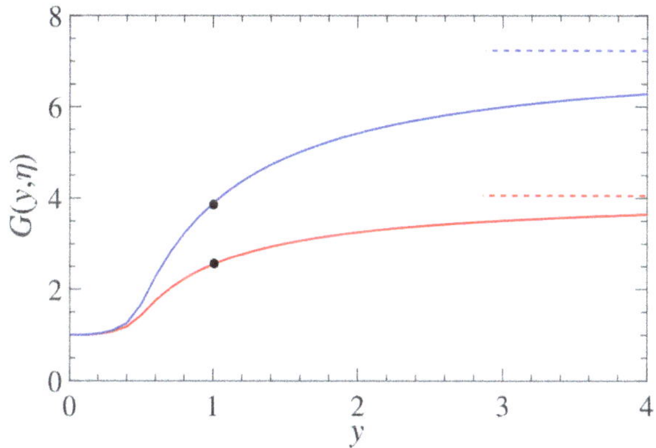

Figure 5.11. The SPT function G as given by expression (5.84) versus cavity radius, $y = r/\sigma$. We show two fluid packing fractions, $\eta = 0.3$ in red, and $\eta = 0.4$ in blue. The required values for $1 + 4\eta G(1, \eta) = \beta p/\rho$, $\beta\gamma\sigma^2$ and Tolman's length δ_T are taken from equation (5.87). As pointed out in the text, the approximation is accurate for large y (where it approaches $G(\infty, \eta) = \beta p/\rho$, indicated by dashed lines). RFL extended its use by assuming accuracy down to $y = 1/2$. At $y = 1$ the value of G equals the contact value of the radial distribution function, g, see equation (5.67). This is indicated by a solid black dot on each curve. Finally, for $y \leqslant 1/2$ the function $G(y, \eta)$ is exact, and given by equation (5.72).

where we can express the ith fractional pressure in terms of the density at the hard wall times kT, i.e., $p_i = kT\rho_i^w$ is the contribution to the pressure of the ith component (which for a hard wall is entirely kinetic), and $\sum_i^m p_i = p$, As has to be the case, both the expression for the pressure (5.97) and the surface tension (5.98) reduce to their pure counterparts (5.90).

5.4.5 One-dimensional hard sphere fluid

Following the RFL paper, in 1961 Helfand, Frisch and Lebowitz (HFL) published a paper addressing SPT for two and one-dimensional fluids [47]. Here, we will briefly consider the one-dimensional fluid, or Tonks gas,[9] as it has an exact solution. That original solution was due to Tonks in 1936 [48].

Following equation (5.71) the probability that a cavity of size is empty

$$p_0(\lambda) = 1 - 2\rho\sigma\lambda \quad ; \lambda \leqslant 1/2 \tag{5.99}$$

so that

$$G(\lambda, \rho) = (2\rho\sigma)^{-1} \frac{\beta \partial W}{\partial \lambda} \tag{5.100}$$

$$= \frac{1}{1 - 2\rho\sigma\lambda} \quad ; \lambda \leqslant 1/2. \tag{5.101}$$

[9] Only one phase for a one-dimensional fluid exists, as there are no phase transitions in one-dimension.

Now, for $\lambda > 1/2$ G is constant. Based on the continuity of G [40], we then have,

$$G(\lambda \geqslant \frac{1}{2}, \rho) = G(1, \rho) = G(\infty, \rho) = (1 - \rho\sigma)^{-1} \tag{5.102}$$

which, because $\beta p/\rho = G(\infty, \rho)$, leads us to the exact Tonks' result (figure 5.11).

In 1986 Robledo and Rowlinson reported the exact results (grand potential and distribution functions) for the system of hard rods on a line of *finite length*. The c-DFT solution of a 1D hard-rod fluid between two rigid walls is also exact.

References

[1] Hansen J P and McDonald I R 2006 *Theory of Simple Liquids* (New York: Academic)

[2] McQuarrie D A 2000 *Statistical Mechanics* (University Science Books)

[3] Goodstein D L 1985 *States of Matter* (New York: Dover Publications Inc)

[4] Eyring H, Henderson D, Stover B J and Eyring E M 1982 *Statistical Mechanics and Dynamics* 2nd edn (New York: Wiley)

[5] Bellemans A 1962 *Physica* **28** 493

[6] Ree F H and Hoover W G 1967 Hard sphere virial coefficients *J. Chem. Phys* **46** 4181

[7] Kolafa J, Labik S and Malijevsky A 2004 Accurate equation of state of the hard sphere fluid in stable and metastable regions *Phys. Chem. Chem. Phys.* **6** 2335–40

[8] de Miguel E, Rull L F, Jackson G, Vega L and McLure I A 1992 Phase equilibria and critical behavior of square-well fluids of variable width by Gibbs ensemble Monte Carlo simulation *J. Chem. Phys.* **96** 2296–305

[9] Rowlinson J S 1984 Private communication

[10] Weeks J D, Chandler D and Andersen H C 1971 Role of repulsive forces in determining the equilibrium structure of simple liquids *J. Chem. Phys.* **54** 5237–47

[11] Robert W 1954 Zwanzig. High temperature equation of state by a perturbation method. i. nonpolar gases *J. Chem. Phys.* **22** 1420–6

[12] Barker J A and Henderson D 1967 Perturbation theory and equation of state for fluids: The square-well potential *J. Chem. Phys.* **47** 2856–61

[13] Thompson S M, Chapela G A, Saville G and Rowlinson J S 1977 Computer-simukation of a gas–liquid interface *Faraday Trans. 2* **73** 1133–44

[14] Thompson S M 1986 Private communication

[15] Henderson D 2010 Rowlinson's concept of an effective hard sphere diameter *Journal of Chemical & Engineering Data* **55** 4507–8

[16] Andersen H C, Chandler D and Weeks J D 1972 Roles of repulsive and attractive forces in liquids: The optimized random phase approximation *J. Chem. Phys.* **56** 3812–23

[17] Frisch H L, Katz J L, Praestgaard E and Lebowitz J L 1966 High-temperature equation of state of argon *The Journal of Physical Chemistry* **70** 2016–20

[18] McQuarrie D A and Katz J L 1966 High-temperature equation of state *The Journal of Chemical Physics* **44** 2393–7

[19] Rowlinson J S 1964 The statistical mechanics of systems with steep intermolecular potentials *Molecular Physics* **8** 107–15

[20] van der Waals J D 1873 Over de continuities van den gas-en vloeistoftoestand *Thesis* Leiden

[21] Weiss P 1907 L'hypothèse du champ moléculaire et la propriété ferromagnétique *J. Phys. Theor. Appl* **6** 661–90

[22] Bragg W and Williams E 1935 title *Proc. Roy. Soc. Lond.* A 151 540–66

[23] Kac M, Uhlenbeck G E and Hemmer P C 1963 On the van der waals theory of the vapor-liquid equilibrium. i. discussion of a one-dimensional model *Journal of Mathematical Physics* **4** 216–28
[24] Rowlinson J S and Cohesion 2002 *A Scientific History of Intermolecular Forces* (Cambridge, UK: Cambridge University Press)
[25] Carnahan N F and Starling K E 1969 *J. Chem. Phys.* **51** 635
[26] Maxwell J C 1875 Xxii.—on the dynamical evidence of the molecular constitution of bodies *J. Chem. Soc.* **13** 493–508
[27] Lekner J 1982 Parametric solution of the van der waals liquid-vapor coexistence curve *Am. J. Phys.* **50** 161–3
[28] Longuet-Higgins H C and Widom B 1964 A rigid sphere model for the melting of argon *Mol. Phys.* **8** 549–56
[29] Stanley H E 1971 *Introduction to Phase Transitions and Critical Phenomena* (New York: Oxford University Press)
[30] van Laar J J 1910 The vapor pressure of binary mixtures *Z. Phys. Chem.* **72** 723–51
[31] van Konynenburg P H and Scott R L 1980 Critical lines and phase equilibria in binary van der waals mixtures *Phil. Trans. Roy. Soc. London* A **298** 495–540
[32] van Konynenburg P H 1968 Dissertation *PhD Thesis* UCLA, Los Angeles
[33] Ising E 1925 Beitrag zur theorie des ferromagnetismus *Physik* **31** 253–8
[34] Lenz W 1920 Beitrag zum verständnis der magnetischen erscheinungen in festen körpern *Phys. Z.* **21** 613–5
[35] Fisher M E 1986 Lectures on critical phenomena. Cornell
[36] Onsager L 1944 Crystal statistics. i. a two-dimensional model with an order-disorder transition *Phys. Rev.* **65** 117–49
[37] Cernuschi F and Eyring H 1939 An elementary theory of condensation *J. Chem. Phys.* **7** 547–51
[38] Yang N and Lee T D 1952 tstatistical theory of equations of state and phase transitions. i. theory of condensation *Phys. Rev.* **87** 404–9
[39] Huang H, Sukop M C and Lu X Y 2015 *Multiphase Lattice Boltzmann Methods: Theory and Application* (Oxford: Wiley)
[40] Reiss H, Frisch H L and Lebowitz J L 1959 Statistical mechanics of rigid spheres *J. Chem. Phys.* **31** 369–80
[41] Wertheim M 1963 *Phys. Rev. Lett.* **8** 321
[42] Thiele E 1963 Equation of state for hard spheres *J. Chem. Phys.* **39** 474–9
[43] Henderson J R and van Swol F 1984 *Mol. Phys.* **51** 991
[44] Reiss H, Frisch H L, Helfand E and Lebowitz J L 1960 Aspects of the statistical thermodynamics of real fluids *J. Chem. Phys.* **32** 119–24
[45] Lebowitz J L 1964 Exact solution of generalized percus-yevick equation for a mixture of hard spheres *Phys. Rev.* **133** A895–9
[46] Rosenfeld Y 1989 *Phys. Rev. Lett.* **63** 980
[47] Helfand E, Frisch H L and Lebowitz J L 1961 Scaled particle theory of fluid mixtures *J. Chem. Phys.* **34** 1037
[48] Tonks L 1936 The complete equation of state of one, two and three-dimensional gases of hard elastic spheres *Phys. Rev.* **50** 955–63
[49] Verlet L 1967 Computer "experiments" on classical fluids. I. Thermodynamical properties of Lennard–Jones molecules *Phys. Rev.* **159** 98–103

Chapter 6

Density functional theory

Density functional theory (DFT) has been extensively applied in investigation of the structure of both quantum mechanical systems and classical inhomogeneous fluids. The development of DFTs began with the work of Hohenberg and Kohn in 1964 in the context of electronic structure so we begin with a brief overview of quantum systems including the motivation for developing DFT, the fundamental theorems on which it is based, and a summary of the how density functionals are constructed in practice. More complete descriptions of the theoretical [1] and practical implementation [2] of DFT for electronic structure can be found elsewhere. Many of the themes that underpin DFT for quantum systems are also foundational to DFTs for inhomogeneous fluids at interfaces. Chapter 7 will discuss the application of DFT to electrolyte interfaces, while this chapter provides a more general introduction.

For clarity, the parts of this discussion that are specific to electronic structure will be denoted as q-DFT while c-DFT will be used for the discussion of inhomogeneous classical fluids. Where the text applies to both, there is no prefix. In the literature, the quantum community typically uses DFT without qualification. This is understandable since DFT was implemented first for quantum systems, and since the size of the field is larger. However, the lack of qualification does at times lead to confusion.

6.1 Density functional theory for electronic structure

6.1.1 History

The fundamental postulate of quantum mechanics is that there exists a wave function, Ψ, that fully describes the quantum state of the system. When the Hamiltonian operator, H, is applied to Ψ, a scalar energy observable, E, can be computed

$$H\Psi(\mathbf{x}_1, \dots, \mathbf{x}_N) = [T + V + V_{nn} + U]\Psi = E\Psi(\mathbf{x}_1, \dots, \mathbf{x}_N), \tag{6.1}$$

doi:10.1088/978-0-7503-2276-8ch6

where the Hamiltonian operator includes the kinetic energy, T, and potential energy contributions from the nuclei–electron interactions, V, the nuclei–nuclei interactions, V_{nn}, and the electron–electron interactions, U. The wave function depends on the coordinates $\mathbf{x}_i$ which include both position, $\mathbf{r}_i$, and spin coordinates, s_i, of the N electrons. The variational principle at play in quantum mechanical systems is that the correct wave function, Ψ, minimizes the energy at the ground state, $E = E_0$. Once the wave function is known, all properties of the system maybe be computed. For example, the electron density distribution, $\rho_e(\mathbf{r})$, is

$$\rho_e(\mathbf{r}_1) = N\int\cdots\int|\Psi(\mathbf{x}_1, \mathbf{x}_2, \ldots, \mathbf{x}_N)|^2\, ds_1 d\mathbf{x}_2\cdots d\mathbf{x}_N, \tag{6.2}$$

where the wave function is normalized such that $\int|\Psi|^2 d\mathbf{x}^N = 1$. This implies that given a $\rho_e(\mathbf{r})$ the total number of electrons in the system, N, will be recovered by integrating over space,

$$N = \int\rho_e(\mathbf{r})d\mathbf{r}. \tag{6.3}$$

In 1926, Erwin Schrödinger published the fundamental partial differential equation that describes the wave function for quantum mechanical systems within the construct above [3]. The time independent version of Schrödinger's equation, for N electrons at positions $\mathbf{r}$ and M nuclei with nuclear charge Z at positions $\mathbf{R}$, has

$$T = -\frac{\hbar^2}{2m}\sum_{i=1}^{N}\nabla_i^2 \tag{6.4}$$

$$V = \sum_{i=1}^{N}\left(\sum_{A=1}^{M}\frac{Z_A e^2}{|\mathbf{r}_i - \mathbf{R}_A|}\right) \tag{6.5}$$

$$V_{nn} = \sum_{A=1}^{M}\left(\sum_{B<A}\frac{Z_A Z_B e^2}{|\mathbf{R}_B - \mathbf{R}_A|}\right) \tag{6.6}$$

$$U = \sum_{i=1}^{N}\sum_{j<i}U(\mathbf{r}_i, \mathbf{r}_j) = \sum_{i=1}^{N}\sum_{j<i}\frac{e^2}{|\mathbf{r}_i - \mathbf{r}_j|}. \tag{6.7}$$

Nuclei movement (kinetic energy) can be eliminated by noting that there is a large difference in mass, and therefore timescales of motion, between the nuclei and the electrons in the system. This simplification is known as the Born–Oppenheimer approximation [4]. The V_{nn} term is often not included explicitly because for fixed $\mathbf{R}$, it only contributes a constant to E. We include it here both for completeness, and because those positions $\mathbf{R}$ may be optimized by computing the energy surface associated with perturbations of $\mathbf{R}$. For the case of molecules, this would be an optimization of molecular geometry. Simultaneous optimization of the nuclei positions, $\mathbf{R}$, and hence the one-body field, $v(\mathbf{r}, \{\mathbf{R}\})$, along with the electron state ($\Psi(\mathbf{x})$ or $\rho_e(\mathbf{r})$) is an example of a self-consistent field (SCF) calculation.[1]

While exact, Schrödinger's equation is not solvable for many systems. However, the 1-electron system does have an exact analytical solution for the wave function. Wave Function Theory (WFT) methods seek to find suitable wave functions as a linear combination of some suitable basis set functions. One natural choice for molecules uses the analytical 1-electron orbitals as the basis set, another approach might replace those orbitals with Gaussian functions. In solid periodic systems, plane waves basis sets are typically more convenient.

In addition to the mathematical approximations of the basis sets, physical approximations are also required to use WFT to compute the electronic properties of molecular or solid systems. For example, the Hartree–Fock approach replaces the pair interactions in U for effective 1-electron interaction with an average field due to other electrons in the system [5–7]. At the time the Hartree–Fock method was developed, its $O(n^4)$ scaling properties (where n is the number of basis sets) was prohibitive for large systems [8]. However, work on improving the numerical scaling of HF methods continues. For example, a recent parallel implementation that is asymptotically linear with molecular size has been applied to a 160 atom system of Elaiophylin [9].

The difficulties associated with using the wave function as the fundamental variable initially motivated the search for approximations that would simplify the problem. One important approximation was the Thomas–Fermi model developed in 1927 [10, 11], which describes the kinetic energy of the electronic ground state for an atom (single nuclei) as a functional of the electron density, $\rho_e(\mathbf{r})$,[2] as

$$T_{TF}[\rho_e] = \frac{3}{10}(3\pi^2)^{2/3}\int \rho_e^{5/3}(\mathbf{r})d\mathbf{r}. \tag{6.8}$$

The energy of this model is

$$E_{TF} = T_{TF}[\rho_e] - Z\int v(\mathbf{r})\rho_e(\mathbf{r})d\mathbf{r} + \frac{1}{2}\int\int \frac{\rho_e(\mathbf{r}_1)\rho_e(\mathbf{r}_2)}{|\mathbf{r}_1 - \mathbf{r}_2|}d\mathbf{r}_1 d\mathbf{r}_2, \tag{6.9}$$

where electron–electron interactions are treated classically (the last term is $U_{cl}[\rho_e]$), exchange and correlation of electrons is ignored, and the one-body external field, $v(\mathbf{r})$ is

$$v(\mathbf{r}) = -Z\int \frac{1}{|\mathbf{r} - \mathbf{R}|}dr. \tag{6.10}$$

[1] Note that the term self-consistent field theory is used in a variety of ways in the literature. Here, we will use the terms 'self-consistent calculation' to cases where two or more variables are simultaneously solved, 'effective field' theory or calculation will apply to any case where fields are in some way recast in order to make a problem tractable, and 'self-consistent field theory' will refer to problems where an external (usually one-body) field must be optimized along with the fluid structure (electronic or classical).

[2] Note that in the functional notation, the square brackets imply that the operand is a function, as in $E_{TF}[\rho_e] \equiv E_{TF}[\rho_e(\mathbf{r})]$. In what follows, this dependence is implied by all square brackets.

Important extensions to the Thomas–Fermi model included the addition of the Dirac exchange functional,

$$K_D[\rho_e] = \frac{3}{4}\left(\frac{3}{\pi}\right)^{1/3} \int \rho_e^{4/3}(\mathbf{r})d\mathbf{r}, \tag{6.11}$$

and the von Weizsacker correction to the kinetic energy,

$$T[\rho_e] = T_{TF}[\rho_e] + \lambda T_W[\rho_e], \tag{6.12}$$

where a squared gradient approximation is used,

$$T_W[\rho_e] = \frac{1}{8} \int \frac{|\nabla \rho_e(\mathbf{r})|^2}{\rho_e(\mathbf{r})} d\mathbf{r}. \tag{6.13}$$

Further improvements to the kinetic energy terms with expansions to higher order gradients have also been developed [12].

While the Thomas–Fermi model was not successful in describing the properties of molecule (for which it predicts dissociation of atoms in all cases) or even atoms to high accuracy, a model can be constructed that also includes Dirac exchange and gradient corrections to the kinetic energy to improve the prediction. Perhaps most importantly, these approximate models, which predate the development of DFT, suggested that $\rho_e(\mathbf{r})$ might replace the wave function as the central parameter needed to describe electronic structure.

6.1.2 Density functional theory formal proofs

Hohenberg and Kohn presented the two theorems that lay the foundation for density functional theory as a formally exact theory in 1964 [13]. The first theorem states that there is a one-to-one mapping between the ground state density, $\rho_e(\mathbf{r})$, and the external potential, $v(\mathbf{r})$. The proof by contradiction demonstrates that it is not possible to have two different Hamiltonians, H and H', with two different external fields, $v(\mathbf{r})$ and $v'(\mathbf{r})$, that would result in the same $\rho_e(\mathbf{r})$. This implies a unique mapping between $v(\mathbf{r})$ and $\rho_e(\mathbf{r})$. Furthermore, since the system is fully determined by $\rho_e(\mathbf{r})$ (nuclei positions as well as electron density), there must also be a one-to-one mapping between $\rho_e(\mathbf{r})$ and the ground state wave function, Ψ. Therefore, all the properties of the ground state can be found from $\rho_e(\mathbf{r})$.

The second theorem of Hohenberg and Kohn essentially shows that $\rho_e(\mathbf{r})$ obeys a variational principle. Specifically, given a trial system with $\tilde{\rho}_e(\mathbf{r})$, the energy of a system with a particular v will be minimized by the ground state

$$E_0 \leqslant E_v[\tilde{\rho}_e(\mathbf{r})]. \tag{6.14}$$

With these two theorems, the ground state energy of the many-electron system can be simply written as

$$E_v[\rho_e] = \int \rho_e(\mathbf{r}) v(\mathbf{r}) d\mathbf{r} + F[\rho_e], \tag{6.15}$$

where F includes all the kinetic energy and the electron–electron interactions. In electronic structure calculations, there is a set number of electrons. Therefore, to solve the system, this constraint is introduced, and the Lagrangian is

$$E_{\upsilon}[\rho_e] - \mu\left(\int \rho_e(\mathbf{r})d\mathbf{r} - N\right), \tag{6.16}$$

where μ is the Lagrange multiplier of the constraint. The Euler–Lagrange equation that must be solved in a q-DFT calculation is then

$$\frac{\delta E_{\upsilon}[\rho_e(\mathbf{r})]}{\delta \rho_e(\mathbf{r})} - \mu = 0. \tag{6.17}$$

Hohenberg and Kohn's formal proofs of the existence of the variational principle along with the mapping between $\rho_e(\mathbf{r})$ and $\upsilon(\mathbf{r})$ showed that the electronic structure problem can be cast in terms of $\rho_e(\mathbf{r})$ rather than Ψ without any loss of generality. Up to this point, q-DFT is a formally exact theory. From a computational point of view, this mapping reduces a system with $3N$ spatial coordinates to a problem with just 3 spatial coordinates so it has the potential for an enormous improvement in numerical performance.

Finally, note that the Hamiltonians and external fields of the Hohenberg–Kohn theorems are not restricted to quantum systems or Coulombic external fields. These proofs also apply to classical systems of various scales as described in more detail in the next section. Thus, these theorems provide a multiscale basis for DFT. This should perhaps not be surprising since the general variational principle and Hamiltonian mechanics apply across scales as well.

6.1.3 Quantum-DFT implementations

The approximations in q-DFT implementations come when defining F and its components. Once $\rho_e(\mathbf{r})$ is cast as the central variable of the system, F in equation (6.15) is typically split into

$$F[\rho_e] = T[\rho_e] + U_{cl}[\rho_e] + \Delta E_{xc}[\rho_e], \tag{6.18}$$

where T is the kinetic energy, U_{cl} is the classical part of the electron–electron interactions, and an exchange-correlation term, ΔE_{xc}, includes all of the non-classical contributions to the electron–electron potential energy and corrects for the self-interaction error associated with the classical Coulomb electron–electron interaction. $E_{xc}[\rho_e]$ may also include kinetic energy corrections (as in Kohn–Sham q-DFT [14]).

Returning briefly to early developments discussed in section 6.1.1, it is straightforward to see that the Thomas–Fermi, Dirac, and Weizsacker approaches are specific approximations to DFT. For example, in the Thomas–Fermi model, a local density approximation (LDA) is used for T, and $\Delta E_{xc} = 0$. More generally, any functional where the operator is related only to the local density, $L[\rho_e] = \int f(\rho_e(\mathbf{r}))d\mathbf{r}$ is an LDA. Gradient approximations are another class of functionals, and the von Weizsacker correction in equation (6.12) is one example. DFT functionals constructed solely based on $\rho_e(\mathbf{r})$, are categorized as orbital-free DFTs (OF-DFTs). OF-DFTs are appealing in the elimination of wave function and orbitals from the system of equations.

A recent review outlines further work in the area of OF-DFTs [15] with particular emphasis on nonlocal density functionals. We note that nonlocal extensions to the kinetic energy term date back to the 1980s with the CAT functional of Chacon, Alvarellos, and Tarazona [16]. It sets

$$T_{CAT}^{nl}[\rho_e] = c_0 \int \rho_e(\mathbf{r})\bar{\rho}_e^{2/3}(\mathbf{r})d\mathbf{r}, \tag{6.19}$$

where

$$\bar{\rho}_e(\mathbf{r}) = \int w(k_0(\mathbf{r}), |\mathbf{r} - \mathbf{r}'|)\rho_e(\mathbf{r}')d\mathbf{r}'. \tag{6.20}$$

The weight function, $w(k_0, |\mathbf{r} - \mathbf{r}'|)$ is defined as a spherical average, and $k_0(\mathbf{r})$ is the local Fermi-wave vector $k_0(\mathbf{r}) = (3\pi^2\rho_e(\mathbf{r}))^{1/3}$.

Wang and Teter proposed adding a correction to the Thomas–Fermi and von Weizsacker models using a nonlocal functional of the form [17],

$$T_{WT} = T_{TF}[\rho_e] + T_{vW}[\rho_e] + \int\int \rho_e^{5/6}(\mathbf{r})w(k_0, |\mathbf{r} - \mathbf{r}'|)\rho_e^{5/6}(\mathbf{r}')d\mathbf{r}d\mathbf{r}'. \tag{6.21}$$

This approach was furthered by others, and culminated in the Wang–Govind, Carter functional [18], where the correction term is generalized to

$$T_{WGC} = T_{TF}[\rho_e] + T_{vW}[\rho_e] + \int\int \rho_e^{\alpha}(\mathbf{r})w(k_0^{\gamma}(\mathbf{r}, \mathbf{r}'), |\mathbf{r} - \mathbf{r}'|)\rho_e^{\beta}(\mathbf{r}')d\mathbf{r}d\mathbf{r}', \tag{6.22}$$

where $\alpha, \beta = 5/6 \pm \sqrt{5}/6$ and k_0^{γ} is the two-point Fermi-wave vector.

Development of nonlocal OF-DFT methods continues to be an active area of research, but is beyond the scope of this text. However, there is one feature of these calculations that allows for the study of very large systems. Specifically, local pseudopotentials are often used to represent both the nuclear potential and the core electrons of the system [15]. This is an effective field approach where the one-body field acting on the valence electrons includes both the nuclei and the core electron interactions. This approach is in some sense a coarse graining of the electronic structure problem.

The power of OF-DFT methods is apparent in materials science applications where they have been applied to study systems of up to $O(10^3 - 10^6)$ atoms in order to study collective phenomena such as dislocations and defects in solids, and the properties of liquid metals in the context of bulk liquids and liquid–solid interfaces [15]. In order to apply OF-DFT to the largest systems, advanced numerical methods are needed. For example, recent OF-DFT calculations of 1 024 000 Li atoms were performed on a Petascale computing platform with 65 536 cores using a small-box fast Fourier transform implementation (SBFFT) [19].

OF-DFTs provide the most direct comparisons for c-DFT methods described below, but there has been significantly more work done with hybrid q-DFTs beginning with the work of Kohn and Sham (KS) in 1965 [14]. The primary driver for electronic structure calculations of molecules is usually to achieve chemical accuracy with respect to bond energies and molecular properties. Kohn and Sham were first to develop an approach with significantly improved chemical accuracy over the then available OF-DFT methods. KS-q-DFT is a hybrid method because it combines both orbital and density functional terms Specifically, the kinetic energy is written as a functional of molecular orbitals, $T[\psi_i]$, and modeled with a non-interacting electron gas that has the same ground state $\rho_e(\mathbf{r})$ as the true system. The kinetic energy computation is thus similar to a Hartree–Fock calculation, but is less costly due to the noninteracting nature of the model. Coulomb interactions for electron–electron interactions retain an explicit dependence on the electron density, $U_{cl}[\rho_e]$. As a result, the KS-q-DFT method requires a self-consistent iterative solution of trial densities and the orbitals. Since the development of KS-q-DFT, many functionals (which usually means new definitions for E_{xc}) have been developed for both general purpose use and for specific applications [20].

6.1.4 Molecular models from Quantum-DFT

Quantum-DFT has been widely applied in molecular systems, solids, and semiconductors. It has been used to investigate structural effects (as in molecular geometry predictions or elastic properties), energetic effects (such as transition states, reactions, or interactions between atoms or molecules), magnetic properties, polarization, collective effects, and other response functions including spectra. This section considers force calculations.

Calculating the force on a particular atomic nucleus directly from the system energy is straightforward in q-DFT calculations as

$$\mathbf{F}_A = -\frac{\partial E(\{\mathbf{R}\})}{\partial \mathbf{R}_A}. \tag{6.23}$$

To optimize molecular geometry, the nuclei positions would be adjusted until

$$\mathbf{F}_A = -\frac{\partial E(\{\mathbf{R}\})}{\partial \mathbf{R}_A} = 0 \tag{6.24}$$

for all A.

If the q-DFT calculation involves just two neutral atoms, and the energy, E, is computed for many atomic separations, $R_{ij} = |\mathbf{R}_i - \mathbf{R}_j|$, both a repulsive short range force due to electronic overlap and a longer range attractive force will be observed. The attractive force (called a dispersion or van der Waals force) is proportional to R_{ij}^{-7}. This attraction is due to the distortion of $\rho_e(\mathbf{r})$ such that there is an increased concentration of electrons between the atoms [21]. The attractive force is thus due to the nuclei interacting with the distorted charge distribution.

In molecular dynamics (MD) or Monte Carlo (MC) simulations of classical systems, $O(10^3 - 10^7)$ atoms may be required to determine equilibrium (or ensemble averaged) properties. These simulations require interaction potentials that do not explicitly include electrons. Therefore, ad-hoc potentials are defined to allow insights from quantum level calculations to be built into interaction potentials. One example is the Lennard–Jones (LJ) potential which is designed to model the short range repulsion and longer range dispersion effects observed between pairs of neutral atoms [22]. The LJ potential is

$$u_{ij}(r) = 4\epsilon_{ij}\left[\left(\frac{\sigma_{ij}}{r}\right)^{12} - \left(\frac{\sigma_{ij}}{r}\right)^{6}\right], \tag{6.25}$$

where r now denotes the distance between two atoms ($r \equiv R_{ij}$), σ is a reference length associated with the ij pair, and ϵ_{ij} sets the energy scale of the pair interaction. The parameters, ϵ_{ij} and σ_{ij} can be set with an optimal fit of the LJ function to the q-DFT predictions for $E(R_{ij})$.

Of course, if the atoms are charged, Coulombic effects (interactions proportional to $1/r$) will be important. Furthermore, many systems of interest are not atoms, but rather molecules. Those molecules not only have additional structure (bonds, bond angles, elasticity etc), but can also exhibit other effects that originate in their electronic structure including polarization, hydrogen bonding, and even magnetization. While q-DFT can probe all of these things, translating the results into molecular models that can be used in large classical fluid simulations at nonzero temperatures is often not straightforward.

For example, water can be modeled with molecular detail using the TIP4P model with four bonded sites [23]. The four sites include the three atoms and a site that models the center of the charge of the oxygen atom. However, water models have also been constructed with as few as one site to as many as six [24–26]. Note that when water or other molecules are reduced to a single spherical site, they are considered coarse-grained (CG) models. Reducing water interactions to a single site implicitly requires averaging over the orientation dependence of the interaction potential and potentially neglecting dipole and hydrogen bonding effects. Nevertheless, CG approaches such as the MARTINI force field [27, 28] are important tools for modeling large biomolecular systems where employing multi-site models would be prohibitive. A recent study of Polyethyleneimine (PEI) and PEI-DNA interactions by Beu, Ailenei, and Costinaş illustrates the design and application of CG models [29]. In their work the CG models are developed both with *ab initio* (from quantum mechanics) simulations and with reference to experimental data. The CG approach allows for calculations of 500 000 CG beads which is the equivalent of 2×10^6 atoms.

These examples are included to illustrate that there is a clear path for introducing specific knowledge from electronic structure (q-DFT or WFT) calculations into computations of many-atom (or molecule) systems where the length and time scales are far larger. In some cases, parameters of ad-hoc force fields (or pair potentials) may be fit from quantum (or ab-initio) calculations. In other cases limiting behaviors observed in electronic structure calculations are built into the pair potential (force field), while the parameters are tuned using experimental data. In either case, there is always an art to choosing or constructing a classical potential model appropriate for a given system. The corollary is also true, that all molecular models, no mater how detailed, are approximate representations of both the underlying electronic structure problem and the real (experimental) system.

6.1.5 DFT for finite temperature

The Hohenberg–Kohn theorems in section 6.1.2 were extended to a finite temperature electron gas in the grand canonical ensemble (fixed chemical potential, μ) by Mermin in 1965 [30]. Once again, a proof by contradiction was used to demonstrate $\rho(\mathbf{r})$ at finite temperature is uniquely determined by the external field, $V^{\text{ext}}(\mathbf{r})$. In this case, the associated variational problem requires minimization of the grand potential. Specifically

$$\left.\frac{\delta\Omega}{\delta\rho}\right|_{\mu,T} = 0, \tag{6.26}$$

where

$$\Omega[\rho] = \tilde{\Omega}[\rho] + \int V^{\text{ext}}(\mathbf{r})\rho(\mathbf{r})d\mathbf{r}, \tag{6.27}$$

and $\tilde{\Omega}$ is the intrinsic grand free energy. This development served as the bridge between q-DFT and c-DFT. The variational principle in equation (6.26) applies to both types of systems where $\rho(\mathbf{r})$ is a particle density in c-DFT and an electron density in q-DFT.

6.2 Density functional theory for classical fluids

Work on classical systems began in the 1970s [31] with significant foundational contributions in both theory and practical application of c-DFT over the next 20 years [32, 33]. Since then, c-DFT has found its way into several texts on molecular modeling and liquid state theory [34–40]. Classical-DFT has been the subject of both research collections [41, 42] and review summaries [32, 43–46] in addition to the many contributions in the wider literature. Here we begin by considering analogies between classical and quantum systems in order to emphasize the multiscale nature of DFT.

6.2.1 Comparing quantum-DFT and classical-DFT

There are a number of similarities (and a few differences) between quantum and classical systems that are helpful to understand in the context of DFT. They are summarized here:

- **The Hamiltonian**: Both quantum and classical systems have N-particle Hamiltonians with potential energy and kinetic energy contributions. The Hamiltonian for a classical fluid with N atoms of mass m in a one-body external field Hamiltonian is

$$H(\mathbf{r}^N, \mathbf{p}^N) = K(\mathbf{p}^N) + E(\mathbf{r}^N) + V(\mathbf{r}^N), \tag{6.28}$$

where the $K = \sum_{i=1}^{N} \mathbf{p}_i^2/2m$ is the kinetic energy, $V = \sum_{i=1}^{N} V^{\text{ext}}(\mathbf{r}_i)$ includes the particle interactions with the one-body field, and E is the potential energy due to particle interactions. The canonical partition function, Z, for a system with fixed number of particles, temperature, and volume is

$$Z = \frac{1}{h^{3N}N!} \int d\mathbf{r}^N \int d\mathbf{p}^N e^{-\beta H(r^N,\, \mathbf{p}^N)}. \tag{6.29}$$

In classical systems, the kinetic energy is only a function of the momenta, $\mathbf{p}_i$, and the Gaussian integrals over the momenta may be solved analytically to find that

$$Z = \left(\sqrt{\frac{2m\pi}{\beta h^2}}\right) Q = \Lambda^{-3N} Q, \tag{6.30}$$

where $\Lambda = (\beta h^2/2\pi m)^{1/2}$ is the thermal de Broglie wavelength, and Q is the configuration integral, $Q = (1/N!) \int e^{-\beta(E+V^{\text{ext}})} dr^N$. Several partition functions can be defined depending on the ensemble of interest. These are all fundamentally related to Q.

- **Motivation for DFT**: In both quantum and classical cases, there is an N-body function (Ψ and Q respectively) that defines the state and properties of the system. For example, the thermodynamic properties of a classical fluid in the canonical ensemble can be derived from the Helmholtz free energy,

$$F = -\beta^{-1} \ln Z = \beta(-3N \ln \Lambda + \ln Q), \tag{6.31}$$

which depends on Q. As discussed earlier, the properties of quantum systems can be obtained from the wave function Ψ. However, Ψ and Q are only solvable for very simple systems.[3] So, in both q-DFT and c-DFT, there is need for a defining variable with reduced dimensions.

- **The Born–Oppenheimer approximation**: Both electronic structure and fluid structure near surfaces arise from the interactions of a mobile phase with a static external field. The static nature of the field depends in both cases on the Born–Oppenheimer approximation which assumes the time scales associated with motion of the nuclei (in q-DFT) or surfaces (in c-DFT) are much longer than the time scale required for equilibration of the mobile phase. This principle allows separation of molecular geometry optimization from the electronic structure problem in quantum systems. In classical systems, it allows separation of surface motions (for example protein configuration changes) from the calculations of fluid (solvation) structure.
- **DFT formal proofs**: The formal proofs for q-DFT and c-DFT follow essentially the same lines establishing that there is a unique density profile, $\rho(\mathbf{r})$, that minimizes the energy (or free energy, Ω, for classical systems) for a given external field. In the case of point charge electrolytes, Reiner and Radke

[3] The case of one-dimensional hard rod fluids is one exactly solvable system [34, 47].

showed that the functional minimization central to c-DFT, $\delta\Omega/\delta\rho = 0$, can instead be written as a minimization of Ω with respect to the electrostatic field, $\psi(\mathbf{r})$, $\delta\Omega/\delta\psi = 0$ [48]. However, this is a special case, and a standard density-centric approach to c-DFT is more easily extended to more detailed molecular models.

- **DFT implementations**: Both the formally exact q-DFT and the formally exact c-DFT rely on functionals that are approximate. So, one goal is to develop functionals that are highly accurate (in comparison with experiments or molecular simulations). Local functionals, gradient based functionals, and nonlocal functionals have all been developed in both cases. However, while q-DFT development focuses significantly on the design of kinetic energy functionals, functional development for c-DFT is usually focused on the complexity in the molecular models for fluid–fluid interactions.
- **Interaction potentials**: Interactions in electronic systems are all captured by the $1/r$ Coulomb potential. In contrast, classical interactions are essentially coarse-grained representations of electronic structure as discussed in section 6.1.4. As such, they come in many flavors including Coulomb interactions, Yukawa ($\exp(-\alpha r)/r$) interactions, dispersion ($1/r^6$) interactions, dipolar interactions, etc. The potential is often split into contributions corresponding to fundamental physical effects (as in the perturbation theory discussed in chapter 5), and functionals are developed separately for the different bits.
- **Coarse graining**: Atomic pair interaction potentials can be determined from q-DFT calculations of atomic or molecular forces and then used in molecular simulations that include the atoms but not the electrons. Similarly, surface interaction potentials (also known as solvation potentials or potentials of mean force, PMF) can be calculated from c-DFTs. The PMF found in c-DFT calculations can be used to perform particle simulations with implicit solvent rather than including the small molecules explicitly [49–51]. This is the classical analog of the Car-Parrinello method [52]. Note that solvation potentials may exhibit complex oscillatory behavior [53].
- **Self-consistent fields**: In q-DFT for molecular systems, the external field depends on the positions of the nuclei. A self-consistent field calculation is required to optimize the molecular geometry. In classical systems, a similar calculation could be done to find the optimal positions of a collection of surfaces for a given state point of the mobile phase. However, there are other situations in classical systems where the external field becomes part of the free energy minimization. One example is a chemically active surface. When a surface contains ionizable groups, the surface charge (and hence external field) is not known apriori, but must be computed in the c-DFT calculation (see details in section 7.6.1). A second example is a surface with a fixed density of tethered polymer chains. The chains are part of the surface because they are bonded to the surface, but their structure (and hence the field they exert on the fluid) depends on their interactions with the mobile fluid phase.

- **Dimensionality and numerical methods**: Virtually all electronic structure calculations of interest require three-dimensional numerical calculation of the electronic density profile. In contrast, there are many classical systems that exhibit interesting behavior that do not require 3D calculations. Planar, cylindrical, and spherical surfaces all allow for analytic integration in two dimensions leaving only a one-dimensional numerical calculation to be performed. Even in these simple geometries, inhomogeneous classical systems exhibit rich phase behavior. As such, it is perhaps not surprising that there has been historically more emphasis on developing numerical methods in the q-DFT community than in the c-DFT community. Some efforts to develop c-DFT capable of large two- and three-dimensional calculations are summarized in chapter 16.

The similarities and differences between the quantum systems and classical systems help clarify why the q-DFT and c-DFT have developed sometimes along similar lines and other times in quite different directions. The next section focuses on connections between c-DFT and other liquid state theories.

6.2.2 Classical-DFT and liquid state theory

Chapter 4 discusses the central role of correlation functions to liquid state theory. Classical-DFT can also be expressed in terms of these correlation functions. To simplify the notation, the description here focuses on single component fluids. Further details about the formal development of c-DFT can be found in [31, 32, 44].

When studying interfacial fluids, the system is considered to be open and in chemical equilibrium with a bulk fluid. The state point of the fluid is then set by the chemical potential, μ and the temperature, T, and the grand canonical ensemble describes the system. The grand free energy, Ω, is a Legendre transform of the Helmholtz free energy, F. For a fluid in an external field, $V^{\text{ext}}(\mathbf{r})$, it is

$$\Omega[\rho(\mathbf{r});\ V^{\text{ext}}(\mathbf{r})] = \mathcal{F}[\rho(\mathbf{r});\ V^{\text{ext}}(\mathbf{r})] - \mu \int \rho(\mathbf{r})d\mathbf{r}, \tag{6.32}$$

where $\mathcal{F}$ is used to highlight that the Helmholtz free energy is now a functional of the singlet density, $\rho(\mathbf{r})$. It is helpful to split $\mathcal{F}$ into the external field contribution and an intrinsic Helmholtz free energy, $\mathcal{F}^{\text{int}}$, where

$$\mathcal{F}[\rho(\mathbf{r});\ V^{\text{ext}}(\mathbf{r})] = \mathcal{F}^{\text{int}}[\rho(\mathbf{r})] + \int \rho(\mathbf{r})V^{\text{ext}}(\mathbf{r})d\mathbf{r}. \tag{6.33}$$

The intrinsic $\mathcal{F}^{\text{int}}$ is further split into ideal and excess contributions, so the grand free energy becomes

$$\Omega[\rho] = \mathcal{F}_{id}[\rho] + \mathcal{F}_{ex}[\rho] - \int [\mu - V^{\text{ext}}(\mathbf{r})]\rho(\mathbf{r})d\mathbf{r}, \tag{6.34}$$

where the free energy of a nonuniform system of ideal (noninteracting) particles [44] is

$$\mathcal{F}_{id}[\rho] = kT \int \rho(\mathbf{r})[\ln(\Lambda^D \rho(\mathbf{r})) - 1]d\mathbf{r}. \tag{6.35}$$

The excess term, $\mathcal{F}_{ex}$, then includes all the free energy contributions due to particle–particle interactions for the molecular model of interest. It may include classical mean field terms and corrections that are analogous to the exchange-correlation term in q-DFT.

Minimizing Ω in order to determine the equilibrium profile, $\rho(\mathbf{r})$, is the central problem in c-DFT, and results in an Euler–Lagrange equation,

$$\begin{aligned} 0 &= \left.\frac{\delta\Omega}{\delta\rho}\right|_{\mu,T} \\ &= \frac{\delta}{\delta\rho}\left(\mathcal{F}_{id}[\rho] + \mathcal{F}_{ex}[\rho] - \int [\mu - V^{\text{ext}}(\mathbf{r})]\rho(\mathbf{r})d\mathbf{r}\right) \\ &= \ln[\Lambda^3\rho(\mathbf{r})] + \frac{\delta\mathcal{F}_{ex}}{\delta\rho(\mathbf{r})} - \beta[\mu - V^{\text{ext}}(\mathbf{r})]. \end{aligned} \tag{6.36}$$

Taking a second functional derivative of $\mathcal{F}_{\text{ex}}$ can be shown to recover the direct correlation function [33, 44]

$$\frac{\delta^2\mathcal{F}_{\text{ex}}}{\delta\rho(\mathbf{r}_1)\delta\rho(\mathbf{r}_2)} = -\frac{1}{\rho(\mathbf{r}_1)}\delta(\mathbf{r}_1 - \mathbf{r}_2) - \frac{\delta\beta V^{\text{ext}}(\mathbf{r}, [\rho])}{\delta\rho(\mathbf{r}_2)} \tag{6.37}$$

$$= -c^{(2)}(\mathbf{r}_1, \mathbf{r}_2; [\rho(\mathbf{r})]), \tag{6.38}$$

where $c^{(2)}(\mathbf{r}_1, \mathbf{r}_2; [\rho(\mathbf{r})])$ is the second order direct correlation function (DCF) for the inhomogeneous fluid. If $V^{\text{ext}} = 0$ it will be the DCF discussed in chapter 4 for a bulk fluid. A hierarchy of correlation functions can be connected to the functional derivatives of $\mathcal{F}_{ex}$ as

$$c^{(1)}(\mathbf{r}) = -\frac{\delta(\beta\mathcal{F}_{ex}[\rho])}{\delta\rho(\mathbf{r})}, \tag{6.39}$$

$$c^{(2)}(\mathbf{r}_1, \mathbf{r}_2) = -\frac{\delta^2(\beta\mathcal{F}_{ex}[\rho])}{\delta\rho(\mathbf{r}_1)\delta\rho(\mathbf{r}_2)}, \dots \tag{6.40}$$

$$c^{(n)}(\mathbf{r}_1, \dots, \mathbf{r}_n) = \frac{\delta(c^{(n-1)}(\mathbf{r}_1, \dots, \mathbf{r}_{n-1}))}{\delta\rho(\mathbf{r}_n)}, \tag{6.41}$$

and the excess free energy can be expressed as a Taylor series expansion about a reference state, $\rho_0(\mathbf{r})$. Truncating after the second term,

$$\begin{aligned} \beta\mathcal{F}_{ex}[\rho] &= \beta\mathcal{F}_{ex}[\rho_0] \\ &- \int \Delta\rho(\mathbf{r})c^{(1)}([\rho_0]; \mathbf{r})d\mathbf{r} \\ &- \int_0^1 (1 - \alpha)\left(\int\int \Delta\rho(\mathbf{r}_1)\Delta\rho(\mathbf{r}_2)c^{(2)}([\rho_\alpha]; \mathbf{r}_1; \mathbf{r}_2)d\mathbf{r}_1 d\mathbf{r}_2\right)d\alpha \end{aligned} \tag{6.42}$$

where a linear path, $\rho_\alpha = \rho_0(\mathbf{r}) + \alpha(\rho(\mathbf{r}) - \rho_0(\mathbf{r}))$, between the reference state and the final density was assumed.

Some c-DFT functionals are developed in part with this direct correlation function hierarchy in mind. The reference fluid may be chosen to be a uniform bulk, $\rho_0(\mathbf{r}) = \rho_b$, and if the correlations are assumed to be independent of the coupling parameter so that $c^{(2)}([\rho_\alpha]; \mathbf{r}_1; \mathbf{r}_2) \approxeq c^{(2)}(\rho_b; |\mathbf{r}_1 - \mathbf{r}_2|)$, then to second order, the excess grand potential becomes

$$\begin{aligned}\Omega^{ex} = \Omega[\rho] - \Omega[\rho_b] &= \int V^{\text{ext}}(\mathbf{r})\rho(\mathbf{r})d\mathbf{r} \\ &+\beta^{-1}\int \rho(\mathbf{r})[\ln\rho(\mathbf{r}) - 1]d\mathbf{r} - \beta^{-1}\int \rho_b\big[\ln\rho_b - 1\big]d\mathbf{r} \\ &+\beta^{-1}\int\int c^{(2)}(\rho_b; |\mathbf{r}_1 - \mathbf{r}_2|)\Delta\rho(\mathbf{r}_1)\Delta\rho(\mathbf{r}_2)d\mathbf{r}_1 d\mathbf{r}_2,\end{aligned} \tag{6.43}$$

where $\Delta\rho(\mathbf{r}) = \rho(\mathbf{r}) - \rho_b$. This bulk approximation for $c^{(2)}([\rho_\alpha]; \mathbf{r}_1; \mathbf{r}_2)$ is computationally attractive because a single $c^{(2)}(r)$ can be obtained, and then the Euler–Lagrange equation of the c-DFT can be iterated to find the density profile of the inhomogeneous system. The $c^{(2)}(r)$ might come from a molecular simulation or from the solution of the Ornstein–Zernike equation for a bulk fluid. Depending on the closure, an analytical result might be possible (as is the case for primitive model electrolytes within the mean spherical approximation [54, 55]).

It is important to note that since equation (6.43) only includes the quadratic term of the Taylor series, it is unable to describe fluid phase transitions at interfaces [32]. One pathway for improvement is to include higher order terms, $c^{(3)}(\rho; \mathbf{r}_1, \mathbf{r}_2, \mathbf{r}_3)$. Another alternative is to develop a model for $c^{(2)}$ that reflects the density variations in the inhomogeneous system. For example, Ebner *et al* [56] made an approximation that replaced ρ_b with a mean $\rho_m = [\rho(\mathbf{r}_1) - \rho(\mathbf{r}_2)]/2$. The implementation required tabulation of $c^{(2)}(r_{12})$ at many bulk densities along with suitable extrapolations into the two phase system.

An alternative approach to c-DFT functional design is based on a perturbation theory approach for pairwise additive potentials, $u(\mathbf{r}_1, \mathbf{r}_2) = u(|\mathbf{r}_1 - \mathbf{r}_2|)$ (as detailed for uniform fluids in chapter 5). The total fluid–fluid energy is

$$\begin{aligned}U(\mathbf{r}^N) &= \frac{1}{2}\sum_{i=1}^{N}\sum_{j=1}^{N} u(\mathbf{r}_i, \mathbf{r}_j) \\ &= \frac{1}{2}\int\int u(\mathbf{r}, \mathbf{r}')\rho(\mathbf{r})(\rho(\mathbf{r}') - \delta(\mathbf{r} - \mathbf{r}'))d\mathbf{r}d\mathbf{r}'\end{aligned} \tag{6.44}$$

where the factor of 1/2 is needed to prevent double counting of the interactions, and the last term eliminates the self-interactions that arise in translating the discrete interaction double sum into the double integral over density. To verify, simply consider any collection of N discrete particles in an arbitrary (frozen) configuration, and set $\rho(\mathbf{r}) = \hat{\rho}(\mathbf{r}) = \sum_{i=1}^{N}\delta(\mathbf{r} - \mathbf{r}_i)$.

In these cases the functional derivative of the free energy with respect to the pair potential (at fixed $\mu - V^{\text{ext}}(\mathbf{r})$) is [32]

$$\frac{\delta\Omega[\rho]}{\delta u} = \frac{\delta\mathcal{F}[\rho]}{\delta u} = \frac{1}{2}\rho^{(2)}(\mathbf{r}_1, \mathbf{r}_2) \tag{6.45}$$

where the two-body density distribution function, $\rho^{(2)}$, is

$$\rho^{(2)}(\mathbf{r}_1, \mathbf{r}_2) = \langle\hat{\rho}(\mathbf{r}_1)\hat{\rho}(\mathbf{r}_2)\rangle - \langle\hat{\rho}(\mathbf{r}_1)\delta(\mathbf{r}_1 - \mathbf{r}_2))\rangle. \tag{6.46}$$

Of course the one-body density distribution function is simply, $\rho^{(1)}(\mathbf{r}) = \langle\hat{\rho}(\mathbf{r})\rangle$, where the angle brackets denote the ensemble average of the particle system.

Perturbation theories split the pairwise interactions, u_{ij}, into a reference fluid and a perturbation. Introduction of a coupling parameter, α, allows tuning from the reference fluid to the system of interest,

$$u_\alpha(\mathbf{r}_1, \mathbf{r}_2) = u_r(\mathbf{r}_1, \mathbf{r}_2) + \alpha u_p(\mathbf{r}_1, \mathbf{r}_2) \tag{6.47}$$

where $u_\alpha = u_{ij}$ when $\alpha = 1$. Separating the free energy into reference and perturbation contributions, and integrating equation (6.45) yields

$$\begin{aligned}\mathcal{F}[\rho] &= \mathcal{F}_r[\rho] + \mathcal{F}_p[\rho] \\ &\mathcal{F}_r[\rho] + \frac{1}{2}\int_0^1 d\alpha \int\int \rho^{(2)}(u_\alpha; \mathbf{r}_1, \mathbf{r}_2)u_p(\mathbf{r}_1, \mathbf{r}_2)d\mathbf{r}_1 d\mathbf{r}_2.\end{aligned} \tag{6.48}$$

The goal of this strategy is to develop an approximation for $\mathcal{F}_r[\rho]$ that is very accurate, and then to approximate the distribution function $\rho^{(2)}(u_\alpha; \mathbf{r}_1, \mathbf{r}_2)$. The simplest approach is a strict mean field approximation where

$$\rho^{(2)}(u_\alpha; \mathbf{r}_1, \mathbf{r}_2) = \rho(\mathbf{r}_1)\rho(\mathbf{r}_2). \tag{6.49}$$

A more sophisticated approximation might set

$$\rho^{(2)}(u_\alpha; \mathbf{r}_1, \mathbf{r}_2) = \rho(\mathbf{r}_1)\rho(\mathbf{r}_2)g_r(\bar{\rho}; \mathbf{r}_{12}), \tag{6.50}$$

where g_r is a bulk fluid pair correlation function, and $\bar{\rho}$ is an averaged or weighted density. The strict mean field approach has the advantage of eliminating the correlation function from the functional, but it can be expected that this approximation will work best if $\mathcal{F}_r[\rho]$ is accurate and contains much of the important structural information for the system.

Perturbation theories have been successful for dense uniform fluids as discussed in chapter 5. The reference system is often taken to be a hard sphere fluid. In c-DFT studies, the definition of the excess free energy of hard sphere mixtures, $\mathcal{F}_r = \mathcal{F}_{HS}^{ex}[\{\rho_i\}]$, is crucially important.

Early c-DFT studies considered local density (LDAs) and gradient approximations where the excess free energy density, f, for the reference fluid was simply taken to be a function of the local density and its gradients

$$\mathcal{F}_{HS}^{ex}([\rho(\mathbf{r})]) = \int f(\rho(\mathbf{r}), \nabla\rho(\mathbf{r}), \ldots)d\mathbf{r}. \tag{6.51}$$

The free energy penalty associated with rapidly varying density distributions near surfaces turns out to be too great in the LDA approximation, and so the predicted profiles do not exhibit the correct structure [57]. As a result, weighted density approximations (WDAs) were developed that are generally defined [32, 36] as

$$\mathcal{F}^{ex}_{HS}([\rho(\mathbf{r})]) = \int \rho(\mathbf{r})[f(\bar{\rho}(\mathbf{r}))/\bar{\rho}(\mathbf{r})]d\mathbf{r} \tag{6.52}$$

where $\bar{\rho}(\mathbf{r})$ is a locally averaged density such as

$$\bar{\rho}(\mathbf{r}) = \int w(|\mathbf{r} - \mathbf{r}'|)\rho(\mathbf{r}')d\mathbf{r}'. \tag{6.53}$$

One intuitive weight function from the generalized van der Waals approach of Nordholm and coworkers [40, 58, 59] is

$$w(\mathbf{r}, \mathbf{r}') = \frac{3}{4\pi d^3}\theta(|\mathbf{r} - \mathbf{r}'| - d). \tag{6.54}$$

This is a straightforward short ranged spherically averaged weighted density where the average is taken over the particle diameter, d. A more intricate WDA might have

$$\bar{\rho}(\mathbf{r}) = \int w(\bar{\rho}(\mathbf{r}), |\mathbf{r} - \mathbf{r}'|)\rho(\mathbf{r}')d\mathbf{r}' \tag{6.55}$$

where the weight function depends on the density (or nonlocal density). Equation (6.55) is analogous to equation (6.20) in the q-DFT case. In fact, both of these WDAs were developed by Tarazona [16, 36, 60], and this is one clear example of the crossover that has occurred between the q-DFT and c-DFT communities.

The importance of establishing accurate functionals for the reference fluid cannot be understated. That functional captures most of the fluid structure both at interfaces and in freezing. Many modern reference functionals have been developed for hard sphere fluids based on the fundamental measures theory (FMT) developed by Rosenfeld in 1989 [45, 61]. Of course, the hard sphere systems are not the only choice of reference fluid. The square well fluid is another candidate that has been considered [62, 63].

6.2.3 Properties of inhomogeneous fluids from c-DFT

The density profile, $\rho(\mathbf{r})$, is the fundamental variable obtained from a c-DFT calculation. Electrostatic potential, $\psi(\mathbf{r})$, and molecular orientation profiles are also important in charged systems and polar or molecular models of the fluid. As such most studies using c-DFT (from simple fluids to electrolytes to self-assembled systems) will present results for one or more of those profiles (as in chapter 6 and part II of this text). This section details some of the other quantities that are often obtained and reported in c-DFT investigations.

Surface excess properties and thermodynamics

Given density profiles, the total adsorption for each species in the system is

$$\Gamma_i = \int \rho_i(\mathbf{r})d\mathbf{r}. \tag{6.56}$$

Of course, this is an extensive quantity increasing with the size of the computational domain. To isolate the effects of the surface, it is usually best to focus on surface excess properties. The surface excess adsorption is defined as

$$\Gamma_i^{ex} = \int \left[\rho_i(\mathbf{r}) - \rho_i^b\right]d\mathbf{r}, \tag{6.57}$$

where $\{\rho_i^b\}$ are the set of bulk fluid densities in equilibrium with the inhomogeneous fluid.

Taking a thermodynamic approach, the grand potential free energy of an interfacial fluid is $\Omega[\rho(\mathbf{r})]$. In a bulk fluid, it is simply $\Omega = -pV$ where p is the pressure. The excess surface free energy is defined as

$$\Omega^{ex}[\{\rho_i(\mathbf{r})\}] = \Omega[\{\rho_i(\mathbf{r})\}] - \Omega^b[\{\rho_i^b\}] = \Omega[\{\rho_i(\mathbf{r})\}] + pV \tag{6.58}$$

assuming some natural choice for a Gibbs dividing surface [33]. Note that the notation for the surface free energy is sometimes (and later in this text) written $\Omega_s \equiv \Omega^{ex}$.

The total differential of excess free energy may be written in terms of the state parameters, $\{\xi_i\}$, of the system as [33]

$$d\Omega^{ex} = \left(\frac{\Omega^{ex}}{A}\right)dA - \sum_i \frac{\partial\Omega^{ex}}{\partial\xi_i}d\xi_i, \tag{6.59}$$

where $\{\xi_i\}$ includes both bulk and surface state variables. The coefficients defined by $\partial\Omega^{ex}/\partial\xi_i$ are thermodynamic conjugates to each of the state (or field) parameters. For example, variations in the bulk state point (T, μ_i) lead to surface versions of familiar bulk extensive variables,

$$\frac{\partial(\Omega^{ex}/A)}{\partial T} = -\frac{S^{ex}}{A} \tag{6.60}$$

$$\frac{\partial(\Omega^{ex}/A)}{\partial\mu_i} = -\Gamma_i^{ex}, \tag{6.61}$$

where S denotes the entropy. Variations in the surface geometry or arrangement result in

$$\frac{\partial(\Omega^{ex}/A)}{\partial A} = \frac{\Omega^{ex}}{A} \tag{6.62}$$

$$\frac{\partial(\Omega^{ex}/A)}{\partial H} = -f_s/A, \tag{6.63}$$

where the surface tension is $\gamma = \Omega^{ex}/A$ for the case of hard walls [64], and f_s is the solvation force defined above for two surface separated by H. Variations in the surface free energy with state variables that are surface–fluid interaction parameters (such as ϵ_{wf} and d_{wf} in equation (7.82)) lead to less familiar but equally valid thermodynamic densities [33].

In ideal charged systems, the surface potential, ψ^s, and the surface charge, q^s, are conjugate variables so for electrolytes near charged interfaces,

$$\frac{\partial(\Omega^{ex}/A)}{\partial\psi_s} = -q_s. \tag{6.64}$$

Equation (6.64) is also known as the Lippmann equation in electrochemistry texts. The differential capacitance is an important electrode property. It can be computed with c-DFT [65–67], and is defined as

$$C^{\text{diff}} = \frac{\partial q_s}{\partial\psi_s} = -\frac{\partial^2\Omega^{ex}}{d\psi_s^2}, \tag{6.65}$$

The differential capacitance can be split into a diffuse layer capacitance and an inner layer capacitance for finite size models of electrolytes [65].

As with bulk systems, Maxwell relations for interfacial fluids can be derived from the identity [68]

$$\frac{\partial}{\delta\xi_j}\frac{\partial\Omega^{\text{ex}}}{\partial\xi_i} = \frac{\partial}{\delta\xi_i}\frac{\partial\Omega^{\text{ex}}}{\partial\xi_j}. \tag{6.66}$$

Phase coexistence in interfacial fluids

Surface mediated phase coexistence is one of the most important application domains of c-DFT. Early investigations considered wetting, pre-wetting, and capillary condensation (and their drying corollaries) for fluids at surfaces and in porous materials [32, 64, 69]. Due to the expansive number of state variables, ξ_i, the complexity of the phase space is immense (some have called it 'bewildering' [70]). This phase space complexity can be leveraged in microfluidic and nanofluidic devices where a phase transition acts as a switch between two types of phenomenological behavior. The switch is turned on or off with a change in state variable(s). Having a way to study these phase transitions both at simple and complex heterogeneous surfaces is of great value.

Fluid–fluid interfaces: Fluid–fluid coexistence occurs between two bulk phases when the pressure, p, and chemical potentials are identical between the fluid phases, but they have different order parameters (or densities, $\{\rho_i^b\}$). The condition $p_I(\{\mu_i\}) = p_{II}(\{\mu_i\}) = p^{coex}$ can also be expressed as

$$\Omega_I^b[\{\rho_i^{b,I}\}] = \Omega_{II}^b[\{\rho_i^{b,II}\}], \tag{6.67}$$

since $\Omega = -pV$ for a bulk fluid. In order to find the density profile across a fluid–fluid interface, there are two steps:

1. Find the coexisting bulk state points (as in figures 5.4 and 7.13).
2. Perform c-DFT calculations using the two bulk coexisting states as boundary conditions.

Many examples of c-DFT studies of fluid interfaces can be found in the literature including liquid metal interfaces [71], hydrocarbon interfaces [72], electrolyte interfaces [73], and lipid bilayers interfaces [74, 75]. Fluid–fluid interfaces can be finicky from a numerical point of view. There are no mechanical constraints so the interface can travel during an iterative solution making it difficult to apply convergence criteria. It is often necessary to pin one (but only one in a multi-component system) density at a specific (but arbitrary) location.

Fluid–surface interfaces: Introducing a surface allows investigation of the wetting properties of the fluid. Layering transitions related to the addition of adsorbed layers can be observed when $p/p_{\text{coex}} < 1$ (the pre-wetting regime) [32]. In porous media, capillary condensation is a similar phenomena [76, 77]. At fluid–surface interfaces, the order parameters are $\{\Gamma_i^{\text{ex}}\}$, and coexisting phases I and II are identified when

$$\Omega_I^{\text{ex}}[\{\rho_i^I(\mathbf{r})\}] = \Omega_{II}^{\text{ex}}[\{\rho_i^{II}(\mathbf{r})\}]. \tag{6.68}$$

An example of pre-wetting layering transitions can be found in section 7.5.

Wetting of surfaces at $p/p_{coex} = 1$ is even more complex with the possibility of thick liquid films, long range correlations parallel to the surfaces, and interface fluctuations [70, 78, 79]. Detailed studies of wetting of surfaces by electrolytes have been presented by Oleksy and Hansen [73, 80–82].

From a macroscopic perspective the contact angle, θ, is used as a measure of the wetting properties of surfaces. It ranges between $-1 \leqslant \cos\theta \leqslant 1$ from complete drying to complete wetting. The Young's contact angle can be directly computed from the surface free energy of wall–liquid (WL), wall–vapor (WV), and liquid–vapor (LV) interfaces as [70, 83]

$$\cos\theta = \frac{\Omega_{WV}^{\text{ex}} - \Omega_{WL}^{\text{ex}}}{\Omega_{LV}^{\text{ex}}} \tag{6.69}$$

When attractive surface–fluid molecular interactions become stronger, $\cos\theta$ increases. This is one example of tuning a macroscopic observable using a molecular state parameter.

The solvation force and potential of mean force

The solvation force, f_s was defined in equation (6.63) using thermodynamic principles and will be discussed further below from a mechanical point of view. The solvation force is the fluid mediated interaction force between any two surfaces. There is direct experimental access to the solvation force using surface forces apparatus (SFA) or atomic force microscope (AFM) measurements [84]. The solvation force is related to a solvation potential (or potential of mean force, PMF) as mentioned in section 6.2.1. The PMF, $W(R)$, could be found numerically from the solvation force as

$$W(R) = -\int_{R=\infty}^{R=R} f_s(R)dR \tag{6.70}$$

where R is the distance between of two surfaces. However, considering equation (6.63), it is easy to see that the PMF can also be written,

$$W(R) = \Omega^{\text{ex}}(R) - \Omega^{\text{ex}}(R \to \infty). \quad (6.71)$$

The PMF is sensitive to all of the state variables associated with the fluid, and $W(R) \to 0$ as $R \to \infty$ as is needed for an interaction potential in a particle simulation with implicit solvation. If finite size solvent is included in a c-DFT calculation, the solvation potentials may be oscillatory [53].

Potentials of mean force, like correlation functions, are intimately tied to fluid structure. Percus showed that if a fluid particle is made explicit, placed at the origin, and treated as though it is a surface (or one-body field), the density profile of the remaining fluid particles will be [85, 86]

$$\rho(\mathbf{r})/\rho^b = g^{(2)}(\mathbf{r}), \quad (6.72)$$

where $g^{(2)}(r)$ is the second order pair correlation function of the isotropic fluid. Furthermore, the potential of mean force acting between that explicit fluid particle and a particle a distance r away from the origin is

$$W(r) = -kT \ln g(r). \quad (6.73)$$

The Percus particle insertion approach provides a way to test the definition of the PMF in equation (6.71) by applying the machinery of c-DFT in two ways. First, set up a system with one explicit fluid particle at the origin. Specifically, this will be a spherical 'surface' the same size as the fluid particles, with a 1-body external field identical to the pair potential of the fluid, $V^{\text{ext}}(\mathbf{r}) = u(|\mathbf{r}_{ij}|)$. Then compute $\rho(\mathbf{r})$ with c-DFT, and the PMF from equation (6.73). This requires either a single one-dimensional calculation in spherical coordinates or a single three-dimensional calculation in Cartesian coordinates.

A second approach sets up a series of systems with two explicit fluid particles separated by a distance R. The density profile is determined for various R in order to compute $\Omega^{ex}(R)$. This approach requires many three-dimensional calculations. Figure 6.1 compares these two routes to the PMF for a hard sphere fluid. The calculations were all done using the real space Cartesian algorithms outlined in section 16.1.5.

As discussed previously, $W(r)$ can be used as an interaction potential for an implicit solvent simulation. If the PMF is only available for flat planar surfaces, the Derjaguin approximation can be used to estimate the solvation potential for spherical particles [84]. In addition, figure 6.1 demonstrates that the density profile, (or $-kT \ln \rho(\mathbf{r})/\rho^b$) is always linked to the PMF experienced by the fluid particles in whatever 1-body external field they experience. Hence, the density profile provides direct access to understanding free energy barriers encountered by the fluid particles in inhomogeneous environments. In one study, solvation potentials have been applied in Brownian motion simulations of diffusion near surfaces where fluid layering creates barriers to the diffusion. In that case, the use of solvation potentials reduced a many body simulation of interacting particles to

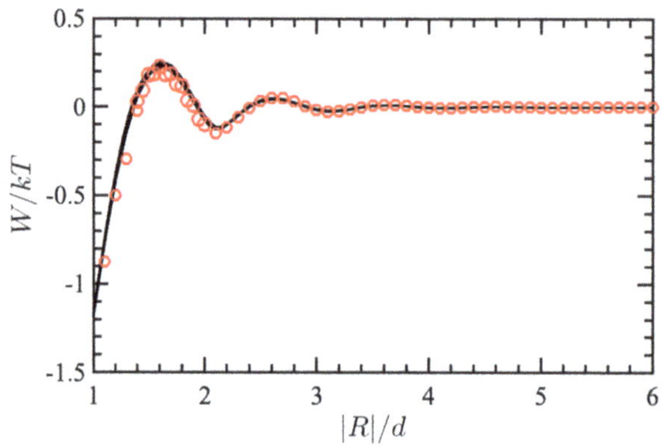

Figure 6.1. The potential of mean force between two hard sphere particles at a bulk fluid density of $\rho^b d^3 = 0.7$ as a function of the center–center distance, $|R|$. The black line is from a calculation of $W(r) = -kT \ln g(r)$ using the Percus test particle method. The red symbols are from calculations of $W(r) = \Omega^{\rm ex}(R_{ij}) - \Omega^{\rm ex}(\infty)$ where there are two explicit solvent particles. The slight thickening of the black line around the first peak reflects surface staircasing on the Cartesian mesh (which had a total of 4.17×10^6 nodes). The real space 3D Cartesian grid had a mesh spacing of $d/10$ in all dimensions for all cases except for the red symbols in the range $1.4 \leqslant |R|/d \leqslant 2.0$ where $h = d/20$.

a simulation of independent particles in an external field [87]. This connection between the density profiles and the energy landscape is also helpful in understanding electrokinetic systems where there are energy barriers to transport such as in ion channel proteins (see section 7.8) [88, 89].

Fluid structure

While many c-DFT investigations center on the singlet or one particle density profile, $\rho(\mathbf{r})$, it is also possible to compute second order correlation functions which are related to two-particle densities, $\rho_{ij}^{(2)}$ [66]. These two-particle densities near a surface can be found directly from c-DFT using the same Percus test particle route described above. In this case, both the wall and an explicit particle are needed to define the total external field. The explicit particle would be moved perpendicular to the wall, to generate $\rho_i(\mathbf{r} = z_p, \mathbf{r}')$ where the only important coordinate for the test particle is the perpendicular distance to the wall, z_p. It is clear that the density profile (due to the remaining fluid particles) will vary in plane as well as perpendicular to the wall in the vicinity of the test particle so in plane correlations may be studied.

Early studies on in plane correlations focused on wetting films [64, 78], while more recent investigations have considered both hard sphere systems [90] and electrolytes at capacitive interfaces [66]. In-plane fluid structure in capacitive systems has implications for 'blue energy' applications where the mixing of rivers and oceans along with the change in salinity may used to harvest energy [67]. These investigations have pursued a compressibility route (rather the Percus particle insertion approach). The compressibility route leverages the direct correlation function hierarchy in equation (6.41). Specifically the direct correlation function can be found from

$$c_{ij}^{(2)}(\mathbf{r}, \mathbf{r}') = -\frac{\delta^2(\beta \mathcal{F}_{ex}[\rho])}{\delta \rho_i(\mathbf{r}) \delta \rho_j(\mathbf{r}')}. \tag{6.74}$$

The pair total correlation function can then be found by solving the Ornstein–Zernike equation,

$$h_{ij}^{(2)}(\mathbf{r}, \mathbf{r}') = c_{ij}^{(2)}(\mathbf{r}, \mathbf{r}') + \sum_{\alpha=1}^{N} \int h_{i\alpha}^{(2)}(\mathbf{r}, \mathbf{r}'') \rho_\alpha(\mathbf{r}'') c_{\alpha j}^{(2)}(\mathbf{r}'', \mathbf{r}') d\mathbf{r}'', \tag{6.75}$$

which in turn can be used to compute the structure factor, $S(k)$, the pair distribution function, $g^{(2)}$ and the two-particle density, $\rho^{(2)}$.

In c-DFT calculations, the free energy derivatives in equation (6.74) can be computed although there are some complexities associated with finding the intersections of weight functions in hard sphere systems. There is some discussion of those issues in section 16.1.5 and [91] in the context of developing Newton's method solvers for c-DFT, and a very detailed discussion by Härtel, Kohl, and Schmiedeberg in their work on pair correlations in hard sphere mixtures near surfaces [90].

Sum rules

In addition to the thermodynamic route discussed above, there is a second route to interfacial system properties based on the requirements of equilibrium that is explicitly linked to the microscopic structure of the system. The resulting expressions are generally known as sum rules [33, 64, 92, 93].

Formally, Ω is related to the singlet density, by the functional derivative

$$\left(\frac{\delta \Omega}{\delta [\mu_i - V_i^{\text{ext}}(\mathbf{r})]} \right)_T = -\rho_i(\mathbf{r}), \tag{6.76}$$

and to the pair density, $\rho^{(2)}$, by the second functional derivative

$$\left(\frac{\delta^2 \Omega}{\delta [\mu_i - V_i^{\text{ext}}(\mathbf{r})] \delta [\mu_j - V_j^{\text{ext}}(\mathbf{r}')]} \right)_T = -\frac{1}{kT} \left[\rho_{ij}^{(2)}(\mathbf{r}, \mathbf{r}') - \rho_i(\mathbf{r}) \rho_j(\mathbf{r}') + \rho_i(\mathbf{r}) \delta_{ij}(\mathbf{r}, \mathbf{r}') \right]. \tag{6.77}$$

equation (6.76) can be used to generate sum rules as by applying [79]

$$\left(\frac{\partial \Omega}{\partial \xi} \right)_T = -\sum_j \int \rho_j(\mathbf{r}) \frac{\partial}{\partial \xi} \left[\mu_j - V_j^{\text{ext}}(\mathbf{r}) \right] d\mathbf{r} \tag{6.78}$$

using any of the state variables, ξ, mentioned above with the exceptions of temperature, T, and surface area, A. Additional compressibility sum rules are obtained beginning with equation (6.77) and taking derivatives, $\partial^2 \Omega / \partial \xi_i \partial \xi_j$ [33].

The adsorption sum rule

The simplest sum rule is found when equation (6.78) is applied to $\xi = \mu_i$. In that case, $\partial\mu_j/\partial\mu_i = \delta_{ij}$ and $\partial V^{\text{ext}}/\partial\mu_i = 0$. The sum rule results in a microscopic route to the excess adsorption that was anticipated in the definition in equation (6.57),

$$-\frac{\partial\Omega^{\text{ex}}/A}{\partial\mu_i} = \Gamma_i^{\text{ex}} = \int\left[\rho_i(\mathbf{r}) - \rho_i^{b}\right]d\mathbf{r}. \tag{6.79}$$

Note that it is straightforward to calculate the adsorption of species i from $\rho_i(r)$ for any system (or surface) geometry. Computing the solvation force is a bit more complicated as is discussed below.

The force sum rule

For a system with two planar surfaces, located at x_{s1} and x_{s2} the solvation force on each surface is the difference between the force on its interior face due to the confined fluid and the force on its exterior face due to the bulk fluid. By convention, a repulsive solvation force is positive and an attractive force is negative. However, since forces are directional, two identical surfaces would actually experience forces of equal magnitude, but opposite sign. The magnitudes of the forces can be written as pressures (force per unit area) for planar surfaces. For a planar system, the interior pressure is sometimes labeled $\tilde{p}$ and the pressure in the surrounding bulk fluid is p_b. Note that while the focus here is the normal force or pressure on the surfaces, the pressure is more properly a tensor as is discussed further in section 14.4.

Taking a slightly more general approach, consider two planar surfaces that exert different external fields on the fluid. The total external field at any fluid point is

$$V^{\text{ext}}(x) = V_1^{\text{ext}}(|x - x_{s1}|) + V_2^{\text{ext}}(|x - x_{s2}|). \tag{6.80}$$

The thermodynamic route to the solvation force acting on a given surface, ν, is

$$f_{s,\nu}/A = -\frac{\partial\Omega^{ex}/A}{\partial x_{s\nu}}. \tag{6.81}$$

Defining the interior face of either surface to have a unit normal $\mathbf{n}_{\text{in}}$ and the exterior face to have unit normal $\mathbf{n}_{\text{out}}$, the solvation force (per unit area) on the ν surface based on microscopic properties is

$$\begin{aligned} f_{s,\nu}/A &= -\mathbf{n}_{\text{in}}\tilde{p} - \mathbf{n}_{\text{out}}p_b \\ &= -\mathbf{n}_{\text{in}}\int_{x_{s1}}^{x_{s2}}\rho(x)\frac{dV_\nu^{\text{ext}}(|x - x_{s,i}|)}{d|x - x_{s,i}|}dx - \mathbf{n}_{\text{out}}p_b \end{aligned} \tag{6.82}$$

for a single component fluid. The case of planar surfaces is straightforward with $\mathbf{n} = [n_x, 0, 0]$.

In a system with N_S surfaces of arbitrary geometry, the net force acting on a particular surface, A, with all the surfaces located at $\{\mathbf{R}_S\}$ is

$$\mathbf{f}_{s,A} = \frac{\partial\Omega_s[\rho(\mathbf{r})]}{\partial\mathbf{R}_A}. \tag{6.83}$$

A numerical (finite difference) estimate of the solvation force via the thermodynamic route requires density profiles, $\rho_i(\mathbf{r})$, and Ω_s for various $\mathbf{R}_A$ while holding other surfaces in place. This works well when there are only two surfaces (planar or otherwise), but becomes tedious when there are 3 or more surfaces.

Alternatively, by applying the sum rule, it is possible to compute the solvation force on an arbitrary number of surfaces simultaneously from the density profile, $\rho(\mathbf{r})$. Assume that the external field at a position, $\mathbf{r}$, in the fluid is now

$$V^{\text{ext}}(\mathbf{r}) = \sum_{\nu=1}^{N_S} V_\nu^{\text{ext}}(|\mathbf{r} - \mathbf{R}_\nu|). \tag{6.84}$$

For this more general system, the net force on surface A is now written as a vector,

$$\mathbf{f}_{s,A}(\{\mathbf{R}_S\}) = -\int_V \mathbf{n}(\mathbf{r})\rho(\mathbf{r})\frac{\partial V_A^{\text{ext}}(|\mathbf{r} - \mathbf{R}_A|)}{\partial|\mathbf{r} - \mathbf{R}_A|}d\mathbf{r}, \tag{6.85}$$

where $\mathbf{n}(\mathbf{r}) = [n_x, n_y, n_z]$ is the unit vector in the direction from $\mathbf{R}_s$ to $\mathbf{r}$, and the integral is over the entire fluid volume.

Forces at hard surfaces

Finally, consider a surface that is a hard body. It is impenetrable, but otherwise exerts no potential on the fluid. The center of the body is at $\mathbf{r}_0$ (the origin), and its surface is defined by $\mathbf{r}_S$. The external field for this hard surface is defined by a step (or θ) function as

$$V_i^{\text{ext}}(\mathbf{r}) = \begin{cases} \infty \text{ if } |\mathbf{r} - \mathbf{r}_0| < \mathbf{r}_S, \\ 0 \text{ otherwise.} \end{cases} \tag{6.86}$$

Ultimately, this step function will transform the volume integral in equation (6.85) into a surface integral. To make this clear, consider a unit step function in three dimensions defined by some surface, $\mathbf{r}_S$, such that

$$u(\mathbf{r}) = \begin{cases} 1 \text{ if } |\mathbf{r} - \mathbf{r}_0| < \mathbf{r}_S, \\ 0 \text{ otherwise.} \end{cases} \tag{6.87}$$

The external field can be written as

$$V_i^{\text{ext}}(\mathbf{r}) = A - Au(\mathbf{r}), \tag{6.88}$$

where A is a constant (with the understanding that $A \to \infty$). Taking $\nabla V_i^{\text{ext}}(\mathbf{r}) = \nabla(A - Au(\mathbf{r}))$, the first term is clearly zero. The derivative of the unit step function is

$$\nabla u(\mathbf{r}) = n\delta(\mathbf{r} - \mathbf{r}_S), \tag{6.89}$$

where $\mathbf{n}$ is the normal to the surface, and $\delta(\mathbf{r} - \mathbf{r}_S)$ is the the Dirac delta function,

$$\delta(\mathbf{r} - \mathbf{r}_S) = \begin{cases} \infty \text{ if } (\mathbf{r} - \mathbf{r}_S) = 0, \\ 0 \text{ otherwise.} \end{cases} \tag{6.90}$$

A δ function that is defined on a manifold, has the property [94]

$$\int \delta(\mathbf{r} - \mathbf{r}_S)d\mathbf{r} = \int_S dS. \tag{6.91}$$

Given the definition of the δ function, the constant A must be irrelevant since $A\infty = \infty$. So the solvation force becomes (for a multicomponent system),

$$\begin{aligned} \mathbf{f}_s(\mathbf{R}) &= -\sum_i \int \mathbf{n}\delta(\mathbf{r} - \mathbf{r}_S)\rho_i(\mathbf{r})d\mathbf{r}. \\ &= -\sum_i \int_S \mathbf{n}\rho_i(\mathbf{r}_s)dS. \end{aligned} \tag{6.92}$$

Equation (6.92) shows that for hard surfaces of arbitrary geometry, the solvation force depends only on contact densities. If the body is a uniform planar surface, there is only one component to the solvation force (perpendicular to the surface), the density is uniform on r_S, and the step function discontinuity in V^{ext} occurs at x_S. In addition, the pressure on the exterior of the surfaces is the bulk pressure. So, the solvation force (with direction normal to the planar surface) is given simply by the contact values at x_S,

$$f_s(H)/A = \tilde{p} - p_b = \sum_i \rho_i(x_S) - p_b. \tag{6.93}$$

Equation (6.93) is then a familiar limiting case to the more arbitrary surface geometries discussed above [33]. For an external field that has a tail function in addition to a hard core, the two parts of the potential can be treated separately applying equation (6.85) for the tail and equation (6.92) for the hard core.

Figure 6.2 compares the force calculated from the thermodynamic and microscopic routes for the same case as in figure 6.1 where in both cases, two explicit hard spheres surfaces with diameter d are used. The figure also includes the Percus test particle route to the force where there is only one explicit test particle. In that case, the force is

$$f_s = -\frac{\partial W}{\partial \mathbf{r}} = \frac{\partial(\ln[\rho(\mathbf{r}))/\rho^b])}{\partial \mathbf{r}}. \tag{6.94}$$

In all cases, the Cartesian domain is discretized with $h = d/10$. The spherical surfaces are staircased (see section 16.1.5), so considerable discretization errors are possible on this rather coarse mesh. While some errors are observed in the vicinity of the first peak, all routes to the solvation force are otherwise in good agreement.

Finally, it is possible to express solvation forces calculated using contact densities for planar surfaces as either

$$f_s(H) = \sum_i \rho_i(\mathbf{r}_s, H) - p_b \tag{6.95}$$

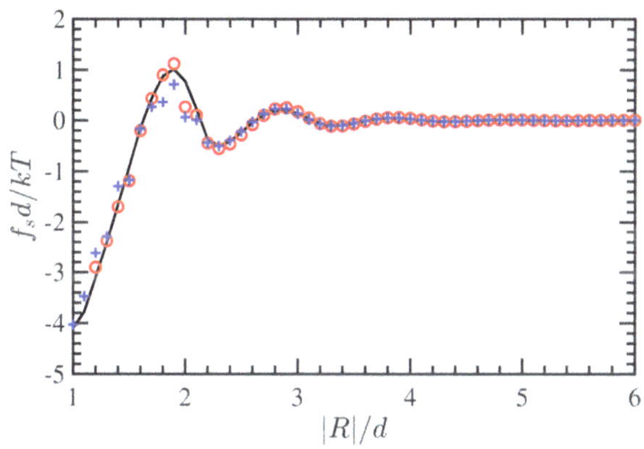

Figure 6.2. The solvation force acting on explicit hard sphere solvent atoms immersed at a bulk fluid density of $\rho^b d^3 = 0.7$ as a function of their center–center distance, $|R|$. The black line is from the Percus test particle route (computed from densities on a line along the x Cartesian axis), the red circles are computed via the thermodynamic route, and the blue (+) symbols are computed from contact values. In all cases, the real space Cartesian 3D grid had a mesh spacing of $h = d/10$.

or

$$f_s = \sum_i \rho_i(\mathbf{r}_s, H) - \sum_i \rho_i(\mathbf{r}_s, H = \infty) \tag{6.96}$$

where $H = \infty$ indicates a distance separation where the surfaces become independent. The two approaches should be identical; however, there will be small errors in contact densities due to mesh discretization. As a result, computing the force with equation (6.96) sometimes results in a helpful cancellation of errors.

References

[1] Parr R G and Yang W 1989 *Density-Functional Theory of Atoms and Molecules* (New York: Oxford University Press)

[2] Sholl D S and Steckel J A 2009 *Density Functional Theory, A Practical Introduction* (New York: Wiley)

[3] Schrödinger E 1926 An undulatory theory of the mechanics of atoms and molecules *Phys. Rev.* **28** 1049–70

[4] Born M and Oppenheimer R 1927 Zur quantentheorie der molekeln *Annal. der Physik (in German)* **389** 457–84

[5] Hartree D R 1928 The wave mechanics of an atom with a non-coulomb central field *Math Proc. Camb. Philos. Soc.* **24** 89–110

[6] Slater J C 1928 The self-consistent field and the structure of atoms *Phys. Rev.* **32** 339

[7] Fock V A 1930 Näherungsmethode zur lösung des quantenmechanischen mehrkörperproblems *Z. Phys. (in German)* **61** 795

[8] Froese Fischer C 1987 General hartree-fock program *Comp. Phys. Comm.* **43** 355–65

[9] Koppl C and Werner H-J 2016 Parallel and low-order scaling implementation of hartree-fock exchange using local density fitting *J. Chem. Theory Comput.* **12** 3122–34

[10] Thomas L H 1927 The calculation of atomic fields *Proc. Camb. Phil. Soc.* **23** 542–8

[11] Fermi E 1927 Un metodo statistico per la determinazione di alcune proprietà dell'atomo (in italian) *Rend. Accad. Naz. Lincei (in Italian)* **6** 602–7

[12] Hodges C H 1972 Quantum corrections to the Thomas–Fermi approximation-the kirzhnits method *Can. J. Phys.* **51** 1428–37

[13] Hohenberg P and Kohn W 1964 Inhomogeneous electron gas *Phys. Rev.* **136** B864–71

[14] Kohn W and Sham L J 1965 Self-consistent equations including exchange and correlation effects *Phys. Rev.* **140** A1133–8

[15] Witt W C, del Rio B G, Dieterich J M and Carter E A 2017 Orbital-free density functional theory for materials research *J. Mater. Res.* **33** 777–95

[16] Alvarellos J E, Chacon E and Tarazona P 1985 Nonlocal kinetic energy functional for nonhomogeneous electron systems *Phys. Rev.* B **32** 7868

[17] Wang L-W and Teter M P 1992 Kinetic-energy functional of the electron density *Phys. Rev.* B **45** 13196

[18] Govind N, Wang Y A and Carter E A 1999 Orbital-free kinetic-energy density functionals with a density-dependent kernel *Phys. Rev.* B **60** 16350

[19] Zhuang H, Wang L-W, Chen M, Jiang X-W and Carter E A 2016 Petascale orbital-free density functional theory enabled by small-box algorithms *J. Chem. Theory Comput.* **12** 2950–63

[20] Verma P and Truhlar D G 2020 Status and challenges of density functional theory *Trends in Chem.* **2** 302–18

[21] Feynman R P 1939 Forces in molecules *Phys. Rev.* **56** 340–3

[22] Lennard-Jones J E 1930 Perturbation problems in quantum mechanics *Proc. Roy. Soc. Lond. A.* **129** 598–615

[23] Madura J D, Impey R W, Jorgensen W L, Chandrasekhar J and Klein M L 1983 Comparison of simple potential functions for simulating liquid water *J. Chem. Phys.* **79** 926

[24] Nada H 2016 Anisotropy in geometrically rough structure of ice prismatic plane interface during growth: Development of a modified six-site model of h2o and a molecular dynamics simulation *J. Chem. Phys.* **145** 244706

[25] Izvekov S and Voth G A 2005 Multiscale coarse graining of liquid-state systems *J. Chem. Phys.* **123** 134105

[26] Stell G, Dyer K M, Perkyns J S and Montgomery Pettitt B 2009 Site-renormalised molecular fluid theory: on the utility of a two site model of water *Mol. Phys.* **107** 423–31

[27] Marrink S J, de Vries A H and Mark A E 2004 Coarse grained model for semiquantitative lipid simulations *J. Phys. Chem.* B **108** 750–60

[28] Yefimov S, Peter Tieleman D, de Vries A H, Marrink S J and Jelger Risselada H 2007 The martini force field: Coarse grained model for biomolecular simluations *J. Phys. Chem.* B **111** 7812–24

[29] Beu T A, Ailenei A-E and Constinaş R-I 2019 Martini force field for protonated polyethyleneimine *J. Comp. Chem.* **41** 349–61

[30] Mermin D 1965 Thermal properties of the inhomogeneous electron gas *Phys. Rev.* **137** A1441–3

[31] Evans R 1979 The nature of the liquid-vapour interface and other topics in the statistical mechanics of non-uniform, classical fluids *Adv. Phys.* **28** 143–200

[32] Evans R 1992 Density functionals in the theory of nonuniform fluids *Fundamentals of Inhomogeneous Fluids* ed D Henderson (New York: Dekker) ch 3 pp 85–175

[33] Henderson J R 1992 Statistical mechanical sum rules *Fundamentals of Inhomogeneous Fluids* ed D Henderson (New York: Dekker) ch 2 pp 23–84
[34] Davis H T 1996 *Statistical Mechanics of Phases, Interfaces, and Thin Films* (New York: Wiley-VCH)
[35] Barrat J-L and Hansen J-P 2003 *Basic Concepts for Simple and Complex Liquids* (Cambridge: Cambridge University Press)
[36] Tarazona P, Cuesta J A and Marinez-Ratón Y 2008 Density functional theories of hard particle systems *Theory and Simulation of Hard-Sphere Fluids and Related Systems* ed A Mulero (Berlin: Springer) ch 7 pp 247–342
[37] Hansen J-P and McDonald I R 2013 *Theory of Simple Liquids with Applications to Soft Matter* 4th ed. (Oxford: Academic)
[38] Forsman J, Woodward C E and Szparaga R 2015 Classical density functional theory of ionic solutions *Computational Electrostatics for Biological Applications* ed W Rocchia and M Spagnuolo (Switzerland: Springer) ch 2 pp 17–38
[39] Wu J (ed) 2017 *Variational Methods in Molecular Modeling* (Singapore: Springer) Molecular Modeling and Simulation Applications and Perspectives
[40] Nordholm S, Forsman J, Woodward C, Freasier B and Abbas Z 2019 *Generalized van der Waals Theory of Molecular Fluids* (Amsterdam: Elsevier)
[41] Special Issue on Density Functional Theory of Liquids *J. Phys.: Cond. Matter* volume 14 (Bristol: IOP Publishing)
[42] Evans R, Oettel M, Roth R and Kahl G (ed) 2016 New developments in classical density functional theory *J. Phys.: Cond. Matter* 28 240401 (Bristol: IOP Publishing)
[43] Wu J and Li Z 2007 Density-functional theory for complex fluids *Ann. Rev. Phys. Chem.* **58** 85–112
[44] Lutsko J F 2010 Recent developments in classical density functional theory *Adv. Chem. Phys.* **144** 1–92
[45] Roth R 2010 Fundamental measure theory for hard-sphere mixtures: a review *J. Phys.: Cond. Matter* **22** 063102
[46] te Vrugt M, Löwen H and Wittkowski R 2020 Classical dynamical density functional theory: from fundamentals to applications *Adv. Phys.* **69** 121–247
[47] Vanderlick T K, Scriven L E and Davis H T 1986 Solution of percus's equation for the density of hard rods in an external field *Phys. Rev.* A **34** 5130–1
[48] Reiner E S and Radke C J 1990 Variational approach to the electrostatic free energy in charged colloidal suspensions : General theory for open systems *J. Chem. Soc. Faraday Trans.* **86** 3901–12
[49] Freasier B C and Nordholm S 1983 A generalized van der waals model for solvation forces between particles in a colloidal suspension *J. Chem. Phys.* **79** 4431–8
[50] Löwen H, Madden P A and Hansen J-P 1992 Ab initio description of counterion screening in colloidal suspensions *Phys. Rev. Lett.* **68** 1081–4
[51] Löwen H, Hansen J-P and Madden P A 1993 Nonlinear counterion screening in colloidal suspensions *J. Chem. Phys.* **98** 3275–89
[52] Car R and Parrinello M 1985 Unified approach for molecular dynamics and density-functional theory *Phys. Rev. Lett.* **55** 2471–4
[53] Frink L J D and van Swol F 1994 Solvation forces and colloidal stability: A combined monte carlo and density functional theory approach *J. Chem. Phys.* **100** 9106–16

[54] Blum L 1975 Mean spherical model for asymmetric electrolytes i. method of solution *Mol. Phys.* **30** 1529–35
[55] Blum L and Hoeye J S 1977 Mean spherical model for asymmetric electrolytes. 2. thermodynamic properties and the pair correlation function *J. Phys. Chem.* **81** 1311–6
[56] Ebner C, Saam W F and Stroud D 1976 Density-functional theory of simple classical fluids. i. surfaces *Phys. Rev.* A **14** 2264–72
[57] Tarazona P, Marini Bettolo Marconi U and Evans R 1987 Phase equilibria of fluid interfaces and confined fluids: non-local versus local density functionals *Mol. Phys.* **60** 573–95
[58] Nordholm S and Haymet A D J 1980 Generalized van der Waals theory. I. Basic formulation and applications to uniform fluids *Aust. J. Chem.* **33** 2013–27
[59] Nordholm S, Johnson M and Freasier B C 1980 Generalized van der waals theory. iii. The prediction of hard sphere structure *Aust. J. Chem.* **33** 2139–50
[60] Tarazona P 1985 Free-energy density functional for hard spheres *Phys. Rev.* A **31** 2672–9
[61] Rosenfeld Y 1989 Free-energy model for the inhomogeneous hard-sphere fluid mixture and density functional theory of freezing *Phys. Rev. Lett.* **63** 980–3
[62] Bernet T, Piñeiro M M, Plantier F and Miqueu C 2017 Generalization of the fundamental-measure theory beyond hard potentials: the square-well fluid case *J. Phys. Chem.* C **121** 6184–90
[63] Bernet T, Piñeiro M M, Plantier F and Miqueu C 2018 Effect of structural considerations on the development of free energy functionals for the square-well fluid *Mol. Phys.* **116** 1977–89
[64] Henderson J R and van Swol F 1985 On the approach to complete wetting by gas at a liquid-wall interface *Mol. Phys.* **56** 1313–56
[65] Tang Z, Scriven L E and Davis H T 1992 A three-component model of the electrical double layer *J. Chem. Phys.* **97** 494–503
[66] Härtel A, Samin S and van Roij R 2016 Dense ionic fluids confined in planar capacitors: in and out-of-plane structure from classical density functional theory *J. Phys.: Cond. Matter* **28** 2344007
[67] Härtel A 2017 Structure of electric double layers in capacitive systems and to what extent (classical) density function theory describes it *J. Phys.: Cond. Matter* **29** 423003
[68] Evans R and Marini Bettolo Marconi U 1987 Phase equilibria and solvation forces for fluids confined between parallel walls *J. Chem. Phys.* **86** 7138–48
[69] Evans R, Marini Bettolo Marconi U and Tarazona P 1986 Fluids in narrow pores: Adsorption, capillary condensation, and critical points *J. Chem. Phys.* **84** 2376–99
[70] van Swol F and Henderson J R 1989 Wetting and drying transitions at a fluid-wall interface: Density functionl theory versus computer simulation *Phys. Rev.* A **40** 2567–78
[71] Chacón E, Reinaldo-Falagán M, Velasco E and Tarazona P 2001 Layering at free liquid surfaces *Phys. Rev. Lett.* **87** 133101
[72] Camaco Vergara E L, Kontogeorgis G M and Liang X 2020 On the study of the vapor-liquid interface of associating fluids with classical density functional theory *Fluid Phase Equilibria* **522** 112744
[73] Oleksy A and Hansen J-P 2009 Microscopic density functional theory of wetting and drying of a solid substrate by an explicit solvent model of ionic solutions *Mol. Phys.* **107** 2609–24
[74] Frink L J D and Frischknecht A L 2005 Density functional theory approach for coarse-grained lipid bilayers *Phys. Rev.* E **72** 041923
[75] Frink L J D, Frischknecht A L, Heroux M A, Parks M L and Salinger A G 2012 Toward quantitative coarse-grained models of lipids with fluids density functional theory *J. Chem. Theory Comput.* **8** 1393–408

[76] Evans R and Tarazona P 1984 Theory of condensation in narrow capillaries *Phys. Rev. Lett.* **52** 557–60
[77] Evans R, Marini Bettolo Marconi U and Tarazona P 1986 Capillary condensation and adsorption in cylindrical and slit-like pores *J. Chem. Soc., Faraday Trans. 2* **82** 1763–87
[78] Tarazona P, Evans R and Marini Bettolo Marconi U 1985 Pairwise correlations at a fluid-fluid interface. the influence of a wetting film *Mol. Phys.* **54** 1357–92
[79] van Swol F and Henderson J R 1986 Wetting at a fluid-wall interface. computer simulation and exact statistical sum rules *J. Chem. Soc. Faraday Trans. 2* **82** 1685–99
[80] Oleksy A and Hansen J-P 2006 Towards a microscopic theory of wetting by ionic solutions. i. surface properties of the semi-primitive model *Mol. Phys.* **104** 2871–83
[81] Oleksy A and Hansen J-P 2010 Wetting of a solid substrate by a 'civilized' model of ionic solutions *J. Chem. Phys.* **132** 204702
[82] Oleksy A and Hansen J-P 2011 Wetting and drying scenarios of ionic solutions *Mol. Phys.* **109** 1275–88
[83] Myhal V and Derzhko O 2017 Wetting in the presence of the electric field: The classical density functional theory study for a model system *Physica A: Stat. Mech. and its Applications* **474** 293–300
[84] Derjaguin B V, Churaev N V and Muller V M 1987 *Surface Forces* (New York: Plenum)
[85] Percus J K 1962 Approximation methods in classical statistical mechanics *Phys. Rev. Lett.* **8** 462–3
[86] Percus J K 1964 The pair distribution function in classical statistical mechanics *The Equilibrium Theory of Classical Fluids* ed H L Frisch and J L Lebowitz (New York: Benjamin. Inc.) ch II-3 pp II–33–II170
[87] Carmer J, van Swol F and Truskett T 2014 Note: Position-dependent and pair diffusivity profiles from steady-state solutions of color reaction-counterdiffusion problems *J. Chem. Phys.* **141** 046101
[88] Frink L J D, Salinger A G, Sears M P, Weinhold J D and Frishknecht A L 2002 Numerical challenges in the application of density functional theory to biology and nanotechnology *J. Phys.: Cond. Matter* **14** 12167–87
[89] Roux B, Allen T, Bernèche S and Im W 2004 Theoretical and computational models of biological ion channels *Quarterly Rev. of Biophys.* **37** 15–03
[90] Kohl M, Härtel A and Schmiedeberg M 2015 Anisotropic pair correlations in binary and multicomponent hard-sphere mixtures in the vicinity of a hard-wall: A combined density functional theory and simulation study *Phys. Rev.* E **92** 042310
[91] Frink L J D and Salinger A G 2000 Two- and three-dimensional nonlocal density functional theory for inhomogeneous fluids: I. Algorithms and parallelization *J. Comp. Phys.* **159** 407–24
[92] Martin P A 1988 Sum rules in charged fluids *Rev. Modern Phys.* **60** 1075–127
[93] Šamaj L 2013 Counter-ions at single charged wall: Sum rules *Eur. Phys. J.* E **36** 100
[94] Onural L 2006 Impulse functions over curves and surfaces *J. Math. Anal. Appl.* **322** 18–27

Chapter 7

Classical-DFT for electrolyte interfaces

In this chapter, the classical density functional theory (c-DFT) approach is presented for several molecular models and functional approximations for electrolyte interfaces. The intent is to provide a practical introduction to c-DFT that includes both theory and detailed examples of electrolyte interfaces as a general orientation to the field. Detailed recent research results are discussed in Part II of this text. The electric double layer has been extensively studied since the pioneering work of Gouy [1, 2] and Chapman [3] around 1910. Some of the review articles that are relevant to the discussion here include reviews by Hanson and Löwen [4] and Forsman, Woodward, and Szparaga [5] that provide theoretical developments from a c-DFT perspective for the electric double layer. Reviews of c-DFT for complex fluid applications by Wu and Li [6] and Messina [7] discuss cases where the charged surfaces are polyions (colloids, rods, etc) found in environments that may contain polyelectrolytes as well as small ions. A recent synopsis of the state of the art in c-DFTs has been presented by Evans, Oettel, Roth, and Kahl [8], and a comprehensive review focused on dynamic-DFTs has been recently contributed by Vrugt, Löwen and Wittkowski [9].

7.1 Molecular models of electrolytes

Every molecular theory calculation or molecular dynamics (MD) simulation of an electrolyte interface has at its heart a molecular model that describes the system. Figure 7.1 provides sketches of six common models for electrolytes that will be discussed in this chapter. Briefly, these molecular models are:

- **Point Charge Electrolytes**: In this model, the solvent (water) is a featureless dielectric continuum, and the ions are point particles. Continuum (see chapter 3) and c-DFT treatments of this model both result in the Poisson–Boltzmann (PB) equation. This model is widely used in situations where fast three-dimensional calculations are required.

doi:10.1088/978-0-7503-2276-8ch7

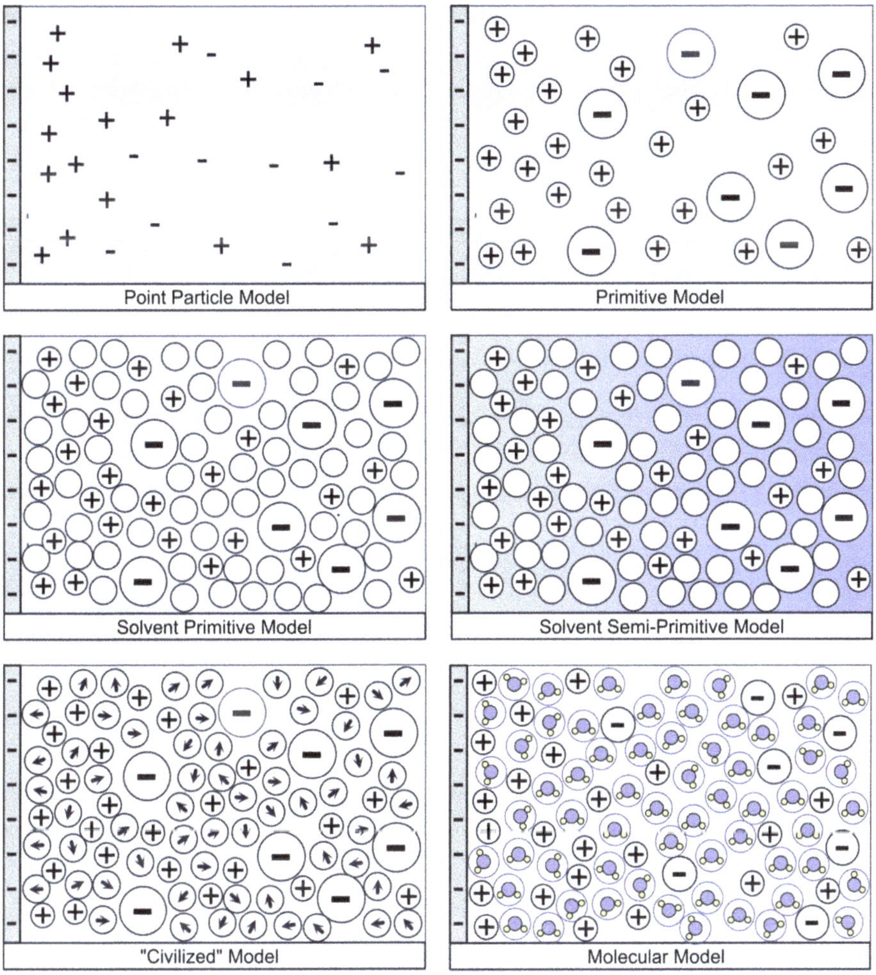

Figure 7.1. Sketches for a variety of model electrolytes near a charged planar interfaces. The blue background indicates a uniform permittivity assumed to describe the dielectric response of the solvent in the point particle, primitive, and solvent primitive models. The solvent semi-primitive model has color variations in the background to reflect a spatially varying permittivity. Explicit dipoles (drawn as vectors) are included in the water model in the civilized model case. The water in the molecular model panel shows the oxygen atoms in blue and the hydrogen atoms in yellow drawn using their covalent radii. The van der Waals radii of the oxygen atoms is also included as the larger circles that surround the water groups to illustrate that spherical water models are not unreasonable.

- **Primitive Model (PM) Electrolytes**: This model treats the ions as finite size charged particles while the solvent (water) remains a featureless dielectric continuum. When all of the ions are the same size, the model is called the restricted primitive model (RPM). Early applications of the RPM to electrical double layers can be found in the work of Grimson and Rickayzen [10], Mier-Y-Teran and coworkers [11] and Tang and coworkers [12, 13].

- **Solvent Primitive Model (SPM) Electrolytes**: This model includes the finite size of the dense solvent (water) while still assuming a constant dielectric constant throughout the fluid region. Solvent particles are generally spherical, but may include cohesive interactions in addition to volume exclusion effects. The SPM nomenclature was used by Pizio and Sokołlowski in a 2014 study of adsorption and capacitance in electric double layers [14]. Earlier application of the same type of model can be found in the work of Tang and coworkers from the 1990s [15, 16].
- **(Solvent) Semi-Primitive Model (SSPM) Electrolytes**: This model treats the solvent explicitly (as in the SPM), but also allows for a spatially varying permittivity, $\epsilon_r(\mathbf{r})$. Note that the term semi-primitive model was coined by Oleksy and Hanson in their studies of wetting of electrolyte interfaces [17, 18]. The current text uses SSPM to avoid an unfortunate collision of acronyms with the solvent primitive model (SPM).
- **"Civilized" Model Electrolytes**: The civilized models introduce a molecular interpretation of dielectric permittivity with an explicit dipole embedded in a spherical water model. Those dipoles represent the permanent dipoles on water molecules. The orientation of the dipole is sensitive to the local charge conditions in the electrolyte interface. Carnie and Chan solved the Ornstein–Zernike equation for an SPM model where the solvent had a dipole in 1980 [19]. Augousti, Rickayzen, and Moradi [20, 21] began applying c-DFT to dipolar systems in the late 1980s. Biben, Hansen, and Rosenfeld presented a general approach for electrolytes with dipolar solvent (and molecular solvent) in 1998 [22]. More recent applications have considered the structure of electrolytes near interfaces [23, 24].
- **Molecular Models of Water**: Detailed molecular models seek to explicitly include all of the atomic species. For water, a variety of 3-site models have been developed (for example, TIPS, SPC, TIP3P, SPC/E) where the partial charges on the oxygen and hydrogen atoms are $\delta_O \approx -0.82$ and $\delta_H \approx +0.41$. Additional sites are sometimes introduced to improve predictions for water in molecular simulations. c-DFT functionals for detailed multi-site water models have been presented by Levesque, Borgis, and their coworkers [25–27]. c-DFT has also been used to model water based on association sites that represent hydrogen bonding. This approach was developed by Chapman and coworkers in the late 1980s [28], and has seen broad application to water based systems [29–31].

7.2 Classical-DFT for point-charge electrolytes

7.2.1 Theory

The simplest molecular model of an electrolyte has point charge ions that interact via Coulomb forces in a featureless solvent with a relative permittivity of ϵ_r. The development here most closely follows the notation of van Roij [32]; however earlier variational treatments of point-charge electrolytes were presented by Sharp and Honig [33] as well as Reiner and Radke [34]. In both of the earlier studies, the focus was on developing exact expressions for the free energies of interacting double layers. For convenience in this development, a dielectric permittivity parameter $\epsilon = 4\pi\epsilon_r\epsilon_0$ will be used to simplify notation. To further simplify matters, a system with uniform dielectric properties (including both the fluid and the interiors of any surfaces) is considered first.

The molecular model for charged point particles is given by Coulomb interactions between the ions, u_{ij}. The interaction between the ions and the surface $V_i^{\text{ext}}(\mathbf{r})$ includes both Coulomb interactions and a volume exclusion term. The pair ion–ion Coulomb interaction is

$$u_{ij}^{c}(r_{ij}) = \frac{z_i z_j e^2}{\epsilon|\mathbf{r}_i - \mathbf{r}_j|} \tag{7.1}$$

where $r_{ij} = |\mathbf{r}_i - \mathbf{r}_j|$, z_i is the valence of species i, and $e = 1.602 \times 10^{-16}$C is the elementary charge. If the system contains a density distribution of fixed charges, $Q_s(\mathbf{r})$, the electrolyte will be inhomogeneous with a net charge number density of mobile ions given by

$$q(\mathbf{r}) = \sum_i z_i \rho_i(\mathbf{r}). \tag{7.2}$$

where $\rho_i(\mathbf{r})$ is the number density and z_i is the valence of each species i. The charge density is $q_c(\mathbf{r}) = eq(\mathbf{r})$. Charge neutrality requires that

$$\int q(\mathbf{r})d\mathbf{r} = \int Q_s(\mathbf{r})d\mathbf{r}. \tag{7.3}$$

The total electrostatic potential energy of the system can be written as

$$U_e = \frac{\lambda_B kT}{2} \int\int \frac{(Q_s(\mathbf{r}) + q(\mathbf{r}))(Q_s(\mathbf{r}') + q(\mathbf{r}'))}{|\mathbf{r} - \mathbf{r}'|} d\mathbf{r}d\mathbf{r}', \tag{7.4}$$

where $\lambda_B = e^2/\epsilon kT$ is the Bjerrum length, and Coulomb interactions between mobile ions, between fixed charges, and between the fixed charges and mobile ions are all included. From the point of view of the mobile species, the fixed charges exert an external field, V^{ext} that includes both an electrostatic contribution, $V_c^{\text{ext}}(\mathbf{r})$, and a neutral volume exclusion contribution, $V_n^{\text{ext}}(\mathbf{r})$. The electrostatic contributions are included in equation (7.4). The neutral term, $V_n^{\text{ext}}(\mathbf{r})$ makes the regions of fixed charge (often surfaces or fixed molecules) impenetrable to the mobile ions. For an ion of species i located at $\mathbf{r}$, the external field is

$$V_i^{\text{ext}}(\mathbf{r}) = V_{n,i}^{\text{ext}}(\mathbf{r}) + z_i e \int \frac{Q_s(\mathbf{r}')}{\epsilon|\mathbf{r} - \mathbf{r}'|} d\mathbf{r}'. \tag{7.5}$$

If the neutral part of the external field captures only volume exclusion, it is

$$\begin{aligned} V_{n,i}^{\text{ext}}(\mathbf{r}) &= \infty \ \text{ for } |\mathbf{r} - \mathbf{r}_s| < 0 \\ V_{n,i}^{\text{ext}}(\mathbf{r}) &= 0 \ \text{ otherwise.} \end{aligned} \tag{7.6}$$

where $\mathbf{r}_s$ defines the location of the fixed surface, and $|\mathbf{r} - \mathbf{r}_s| < 0$ indicates the region interior to a surface using a suitably chosen origin.

The c-DFT proceeds with a minimization of the grand potential with respect to the density distributions of the mobile ions,

$$\frac{\delta\Omega}{\delta\rho_i(\mathbf{r})} = 0 \text{ for all } i. \tag{7.7}$$

The grand potential in the mean field approximation (where correlations are neglected) is

$$\begin{aligned}\Omega[\{\rho_i\}] &= \mathcal{F}_{id} + \mathcal{F}_c - \sum_i \int \left[\mu_i - V_{n,i}^{\text{ext}}(\mathbf{r})\right]\rho_i(\mathbf{r})d\mathbf{r} \\ &= kT\sum_i \int \rho_i\left\{\ln[\Lambda_i^3\rho_i(\mathbf{r})] - 1\right\}d\mathbf{r} \\ &+ \frac{kT\lambda_B}{2}\int\int \frac{(Q_s(\mathbf{r}) + q(\mathbf{r}))(Q_s(\mathbf{r}') + q(\mathbf{r}'))}{|\mathbf{r} - \mathbf{r}'|}d\mathbf{r}d\mathbf{r}' \\ &- \sum_i \int \left[\mu_i - V_{n,i}^{\text{ext}}(\mathbf{r})\right]\rho_i(\mathbf{r})d\mathbf{r}.\end{aligned} \tag{7.8}$$

The chemical potentials in equation (7.8) are known because the inhomogeneous system is presumed to be in osmotic equilibrium with a bulk neutral electrolyte that has densities ρ_i^b and chemical potentials,

$$\mu_i = kT\ln\left(\rho_i^b\Lambda_i^3\right). \tag{7.9}$$

Note that in diffusive systems, the electrochemical potential,

$$\tilde{\mu}(\mathbf{r}) = \mu(\mathbf{r}) + z_i e\psi(\mathbf{r}). \tag{7.10}$$

will be spatially varying due to gradients in either chemical potential, μ or electrostatic potential, ψ. Electrokinetic systems are considered in section 7.8, but the remainder of this chapter considers equilibrium systems $\psi = 0$ and $\tilde{\mu}_i = \mu_i$ in the bulk fluid.

The dimensionless electrostatic potential, $\tilde{\psi} = e\psi/kT$, is formally defined by Coulomb's Law,

$$\tilde{\psi}(\mathbf{r}) = \lambda_B\int \frac{\left(Q_s(\mathbf{r}') + q(\mathbf{r}')\right)}{|\mathbf{r} - \mathbf{r}'|}d\mathbf{r}'. \tag{7.11}$$

Given this definition of electrostatic potential, equation (7.8) becomes

$$\begin{aligned}\Omega[\{\rho_i\}] &= kT\sum_i \int \rho_i(\mathbf{r})\left\{\ln\left[\frac{\rho_i(\mathbf{r})}{\rho_i^b}\right] - 1\right\}d\mathbf{r} \\ &+ \frac{kT}{2}\int (Q_s(\mathbf{r}) + q(\mathbf{r}))\tilde{\psi}(\mathbf{r})d\mathbf{r} + \sum_i \int \rho_i(\mathbf{r})V_{n,i}^{\text{ext}}(\mathbf{r})d\mathbf{r}.\end{aligned} \tag{7.12}$$

The free energy as expressed in equation (7.12) is a functional that can be written as $\Omega[\tilde{\psi}[\rho_i], q(\rho_i), \rho_i]$, and the total functional differentiation of Ω with respect to ρ will be

$$\frac{\delta\Omega}{\delta\rho_i} + \frac{\delta\Omega}{\delta q(\rho_i)}\frac{\partial q(\rho_i)}{\partial\rho_i} + \int d\mathbf{r}'\frac{\delta\Omega}{\delta\tilde{\psi}(\mathbf{r}')}\frac{\delta\tilde{\psi}[\rho_i](\mathbf{r}')}{\delta\rho_i(\mathbf{r})}. \tag{7.13}$$

The free energy is more often written simply as $\Omega[\{\rho_i\}]$, and the minimization problem is easily found (with the help of equation (7.13)) to be

$$\begin{aligned} 0 = \frac{\delta\Omega}{\delta\rho_i} &= kT\ln\left[\frac{\rho_i(\mathbf{r})}{\rho_i^b}\right] + \frac{kT}{2}z_i\tilde{\psi}(\mathbf{r}) \\ &+ \frac{kT}{2}\int\left(Q_s(\mathbf{r}') + q(\mathbf{r}')\right)\left(\frac{\lambda_B z_i}{|\mathbf{r}-\mathbf{r}'|}\right)d\mathbf{r}' + V_{n,i}^{\text{ext}}(\mathbf{r}) \\ &= \ln\left[\frac{\rho_i(\mathbf{r})}{\rho_i^b}\right] + z_i\tilde{\psi}(\mathbf{r}) + V_{n,i}^{\text{ext}}(\mathbf{r})/kT. \end{aligned} \tag{7.14}$$

Setting $\delta\Omega/\delta\rho_i = 0$, and recognizing that $V_{n,i}^{\text{ext}} = 0$ except where $\rho_i = 0$ (interiors of surfaces), the resulting Euler–Lagrange equation becomes the familiar Boltzmann distribution,

$$\rho_i(\mathbf{r})/\rho_i^b = \exp(-z_i\tilde{\psi}(\mathbf{r})). \tag{7.15}$$

Poisson's equation provides a necessary second relation between the electrostatic and density profiles,

$$\nabla^2\tilde{\psi}(\mathbf{r}) = -4\pi\lambda_B\left(q(\mathbf{r}) + Q_s(\mathbf{r})\right). \tag{7.16}$$

The coupled system of equations, equations (7.15) and (7.16), can be used to compute the structure of an electrolyte given an arbitrary fixed charge distribution, $Q_s(\mathbf{r})$.

Constant charge boundary

Frequently, $Q_s(\mathbf{r})$ is confined to a surface of known geometry represented by a surface charge density so $eQ_s(\mathbf{r}) = q_s(\mathbf{r}_s)$. For a uniform surface charge, $q_s(\mathbf{r}_s) = q_s\delta(|\mathbf{r} - \mathbf{r}_s|)$, and we note that often the notation for q_s is σ. The constant surface charge distribution may be applied as a boundary condition to Poisson's equation using Gauss' law,

$$q_s(\mathbf{r}_s) = -\frac{1}{4\pi\lambda_B}\mathbf{n}\cdot\nabla\tilde{\psi}(\mathbf{r})|_{\text{surface}}. \tag{7.17}$$

The free energy functional may be rewritten to reflect that the fixed charges are confined to surfaces as

$$\Omega[\{\rho_i\}]/kT = \sum_i \int \rho_i \left\{ \ln\left[\frac{\rho_i(\mathbf{r})}{\rho_i^b}\right] - 1 \right\} d\mathbf{r} + \frac{1}{kT}\sum_i \int \rho_i(\mathbf{r}) V_{n,i}^{\text{ext}}(\mathbf{r}) dr$$
$$+ \frac{1}{2}\int q(\mathbf{r})\tilde{\psi}(\mathbf{r}) d\mathbf{r} - \frac{1}{8\pi\lambda_B}\int_S (\mathbf{n}\cdot\nabla\tilde{\psi}(\mathbf{r}_s))\tilde{\psi}(\mathbf{r}_s) d\mathbf{r}_s. \quad (7.18)$$

System of equations

There are a few options for formulating the system of equations depending on whether the free energy is posed as a functional of the density fields, the electrostatic potential, or both. Here we summarize the tradeoffs of the three approaches:

- $\Omega[\{\rho_i(\mathbf{r})\}]$: Here, $\psi(\mathbf{r})$ is never introduced, and Poisson's equation need not be solved. Only the Euler–Lagrange equation that results from the minimization of equation (7.8) is solved. However, it is problematic to include an explicit representation of the long range $1/r$ Coulomb interactions in the Euler–Lagrange equation because no finite cutoff (or computational domain of finite size) is sufficient to guarantee charge neutrality. Furthermore, Coulomb interactions result in a dense matrix problem (for the real space implementation described in section 16.1.5).
- $\Omega[\psi(\mathbf{r})]$: This approach eliminates $\rho(r)$ as a variable (and the Euler–Lagrange equation from the system of equations) by substituting the Boltzmann distribution (equation (7.15)) into Poisson's equation (equation (7.16)) to recover the well known nonlinear Poisson–Boltzmann (NLPB) equation,

$$\nabla^2\tilde{\psi}(\mathbf{r}) = -\frac{4\pi}{\epsilon}\sum_i z_i e \rho_i^b \exp(-z_i\tilde{\psi}(\mathbf{r})). \quad (7.19)$$

Reiner and Radke discuss the stationary condition that is associated with the NLPB equation (see equation (22) of [34]),

$$0 = \delta I(\psi, \psi_s) = \delta\Bigg(\int_S Q_s^{RR}(\psi_s, \mathbf{r}_s) d\mathbf{r}_s - \frac{\epsilon}{8\pi}\int (\nabla\psi(\mathbf{r}))^2 d\mathbf{r}$$
$$- \sum_i z_i e \rho_i^b \int \exp(-z_i\tilde{\psi}(\mathbf{r})) d\mathbf{r}\Bigg) \quad (7.20)$$

where $I(\psi, \psi_s) \equiv \Omega - \Omega_0$ and Ω_0 is the free energy of a reference system with uniform bulk densities ($\psi = 0$) and neutral surfaces. The notation Q_s^{RR} is used here to distinguish this general surface term from the source term, Q_s in equation (7.16).

Reiner and Radke showed that the variational principle $\delta\Omega/\delta\rho_i(\mathbf{r}) = 0$ is equivalent to $\delta\Omega/\delta\psi(\mathbf{r}) = 0$ for point-charge electrolytes [34]. They further demonstrate that the surface term $\int_S Q_s^{RR}(\psi_s, \mathbf{r}_s) d\mathbf{r}_s$ is zero for constant potential surfaces, and is linked to surface chemistry for charge regulating surfaces [35]. Charge regulating surfaces are discussed further in section 7.6.1. A constant charge boundary has $\int_S Q_s^{RR}(\psi_s, \mathbf{r}_s)d\mathbf{r}_s = \int q_s(\mathbf{r}_s)\psi_s(\mathbf{r}_s)d\mathbf{r}_s$ consistent with equation (7.18). A proper treatment of the boundaries is critical in calculating potentials of mean force.

Framing the variational problem with $\psi(\mathbf{r})$ as the central variable is a natural approach for point charge models, and it can be extended to include correlations [36]. However, a density-centric approach is more easily extended to molecular model functionals not easily cast in terms of ψ.

- $\Omega[\{\rho_i(\mathbf{r})\}, \psi(\mathbf{r})]$: Finally, a coupled system of equations can be constructed that treats $\rho(\mathbf{r})$ and $\psi(\mathbf{r})$ as independent variables. Increased detail in the molecular model will lead to more complex Euler–Lagrange expressions for the particle distribution (thus generalizing the Boltzmann distribution in equation (7.15)). Poisson's equation is solved simultaneously with the Euler–Lagrange equation to obtain $\psi(\mathbf{r})$ and $\rho_i(\mathbf{r})$. This approach eliminates the explicit treatment of the long range pair potentials, transforms what could be a dense matrix problem in to a sparse matrix problem, guarantees charge neutrality, and provides a clear route to incorporating new molecular models (and their associated functionals). This is the approach taken in the remainder of this chapter with a focus on systematic improvement of molecular models and functionals for electrolytes.

7.2.2 Results

Density and electrostatic potential profiles

It is important that c-DFT developed above for the point charge electrolyte model recovers the well known NLPB description of electrostatic systems because it provides context for c-DFT investigations with more sophisticated molecular models. To demonstrate why improved approaches are needed beyond the NLPB model, consider a charged plane with surface charge per unit area of q_s. In this case, $Q_s(\mathbf{r}) = q_s\delta(|x - x_s|)$, so the fixed charge is all located at the interface, x_s, between an impenetrable surface and the electrolyte. Figure 7.2 shows the counterion density and electrostatic potential profiles for the NLPB model. As expected, there is an accumulation of counter ions at the charged surface and the electrostatic potential increases with decreasing electrolyte concentration or increasing surface charge. The decay length of the double layer is set by the Debye length, $\lambda_D = (4\pi\sum_i(z_i^2\rho_i^b e)/\epsilon kT)^{-1/2}$, and so the double layer increases in range with decreasing electrolyte concentration.

In any real system, the ions will have finite size, and the density of the ions closest to the surfaces will be limited by a close packing restriction. Figure 7.3 illustrates how contact densities and surface potentials vary in the NLPB model. The contact densities increase in an unbounded way as $y = 4\pi\lambda_B\lambda_D|q_s|$ increases which is an unphysical consequence of the underlying point charge model. One way to compensate for this deficiency of the NLPB approach is to treat that first layer of counterions as part of the surface allowing for a smaller effective surface charge in the NLPB calculation (thus introducing a Stern layer [37]). Alternatively, the c-DFT

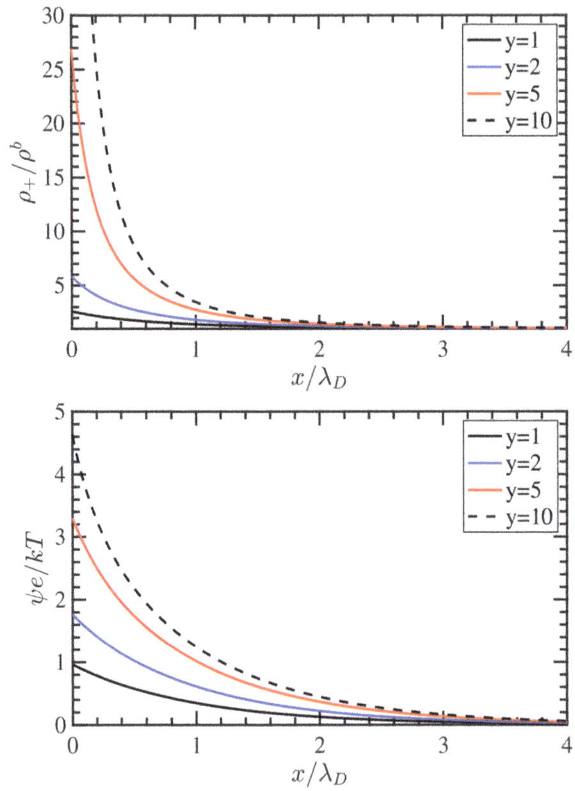

Figure 7.2. Counterion profiles near a charged wall NLPB calculations of a 1:1 electrolyte (upper), and the corresponding electrostatic potential profiles (lower). The curves indicate different values for a convenient dimensionless parameter, $y = 4\pi\lambda_B\lambda_D|q_s|$ where $\lambda_B = e^2/\epsilon kT$ is the Bjerrum length and $\lambda_D = \kappa^{-1} = (4\pi \sum_i (z_i^2\rho_i^b e)/\epsilon kT)^{-1/2}$ is the Debye length.

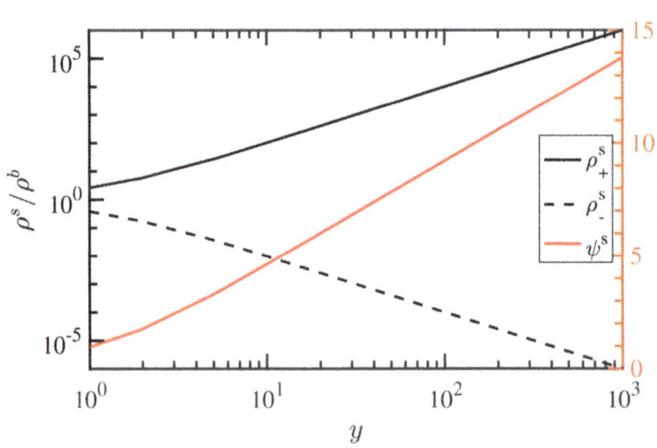

Figure 7.3. Contact densities of ions (left axis) and surface potentials (right axis) at a charged planar surface with known surface charge density, q_s treated with in the NLPB approximation. Again, $y = 4\pi\lambda_B\lambda_D|q_s|$.

approach can be extended to include the energy costs associated with excessive packing of particles near surfaces. Section 7.3 will describe how finite size fluid particles are included in the c-DFT functionals.

Length scales in electrolytes

There are several length scales that come into play when electrolytes are studied. While the point charge electrolyte does not have a natural molecular scale, any real system will have ions and solvent of finite size. As such, a characteristic length d may be introduced where typically $0.2 < d < 0.5$ nm to represent the size of an ion, a water molecule, or a solvated group. The Debye screening length, λ_D is the characteristic length associated with electrostatic decay, and λ_D increases as the bulk ionic strength of the electrolyte decreases. For example, a 1:1 electrolyte at a concentration of 0.1M has $\lambda_D \approx 1$ nm while the same electrolyte at a bulk concentration of 10^{-7}M has $\lambda_D \approx 1000$ nm. These electrostatic interactions are extremely long range compared to the scale of an atom or molecule. Setting $d = 0.2$ nm, the 10^{-7}M case has $\lambda_D/d = 5000$.

There are two additional length scales in electrolytes that depend on the properties of the solvent. The Bjerrum length, $\lambda_B = e^2/\epsilon kT$, is the distance over which two interacting charges will have a Coulomb potential $(e^2/\epsilon r)$ equal to kT. For water at room temperature, $\lambda_B \approx 0.72$ nm Similarly, when a charged wall is present, the Gouy–Chapman length, $\lambda_{GC} = 1/2\pi(z\lambda_B)|q_s|$ gives the length where the interaction between an ion (with valence z) and the charged wall is kT. The mean-field NLPB model is most applicable to a weak coupling regime where $\lambda_B/\lambda_{GC} < 1$ (for a 1:1 electrolyte in water at room temperature, this corresponds to $q_s < 0.3\text{nm}^{-2}$), and will break down for strong coupling when $\lambda_B/\lambda_{GC} > 10$.

Surfaces forces

Finally, consider forces between surfaces immersed in a point charge electrolyte. Forces are central to questions of the stability of colloidal dispersions for example where the theories on stability date back to the 1940s with the work of Derjaguin and Landau [38] and Verwey and Overbeek [39]. For an isolated charged wall, the normal pressure on the surface due to the inhomogeneous fluid must be equal and opposite the bulk pressure in the fluid for mechanical stability. The contact theorem for the normal pressure on an isolated surface is [40, 41]

$$p_b = kT\sum_i \rho_i(x_s) - \frac{2\pi q_s^2}{\epsilon} \tag{7.21}$$

Between two surfaces separated by H, the normal pressure must be constant at any x, and specifically [4],

$$\tilde{p}(H) = kT\sum_i \rho_i(x) - \frac{\epsilon}{8\pi}\left[-\frac{d\psi(x)}{dx}\right]^2. \tag{7.22}$$

At the midplane between two surfaces, the second term is zero by symmetry. The disjoining pressure, $\Pi(H)$, introduced in section 3.2.2 (or the solvation force f_s discussed in section 6.2.3) is

$$\begin{aligned} f_s(H) = \Pi(H) &= \tilde{p}(H) - p_b \\ &= \tilde{p}(H) - \tilde{p}(H = \infty). \end{aligned} \tag{7.23}$$

The solvation force can be computed either from the local microscopic properties as defined in equation (7.22) or using the thermodynamic route described in section 6.2.3

$$f_s = -\frac{\partial \Omega_s}{\partial H}, \tag{7.24}$$

where the surface excess grand potential is now Ω_s. Comparing the two routes to the force is useful for validating both the c-DFT functional of interest and the numerical implementation.

A recent analysis of the NLPB equation using Jacobi elliptic functions to consider the small separation limits of the disjoining pressure has been presented by Markovich, Andelman, and Podgornik [42]. They considered a variety of surface boundary conditions. For now, consider the predictions for constant charge (CC) and constant potential (CP) boundaries. They show that the disjoining pressures at small separations for CC boundaries is

$$\Pi_{CC} = \frac{kT}{\pi \lambda_B \lambda_{GC}} \frac{1}{d}, \tag{7.25}$$

while CP boundaries have

$$\Pi_{CP} = kT\rho_b \left(4\sinh^2(\psi_s/2) - \sinh^2(\psi_s)\frac{(d/\lambda_D)^2}{8} \right). \tag{7.26}$$

Since the leading term dominates for $d/\lambda_D < 1$, the disjoining pressure (or solvation force) is essentially constant at small separations for the point charge model with CP boundaries. Note that in a constant potential system, the surface charge varies with separation, and $q_s(H) \to 0$ as $H \to 0$.

Figure 7.4 shows the solvation force as a function of surface separations, H, for different electrolytes using both CC and CP boundary conditions. The axes presentation is chosen to highlight the short range predictions from equation (7.25) in the upper figure and the expected exponential decay of the force ($f_s \propto \exp(-H/\lambda_D)$) in the lower figure. Most of the curves were calculated using the contact route by evaluating equation (7.22) at the surface. The thermodynamic route to the force is included for one CC case and one CP case in the lower figure to demonstrate that there is good agreement between the two methods for all separations over many orders of magnitude in the force. All of the results were generated using a mesh spacing of $h = d/100$ where d is a characteristic molecular diameter set to 0.425 nm. On a discrete mesh, the calculation of $d\psi/dx$ improves

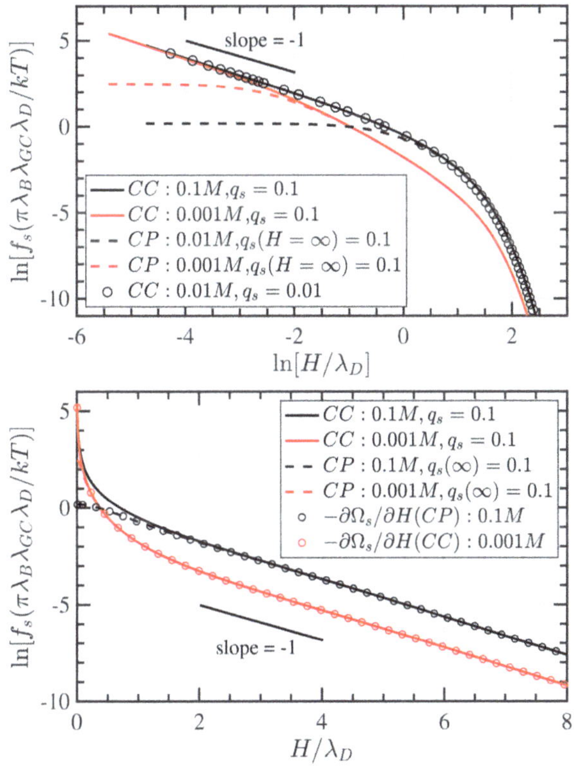

Figure 7.4. The solvation force between charged planar surfaces with the surface charge per area ($q_s = \sigma d^2/e$) and electrolyte concentration indicated in the legends. In the CP cases, the surface potentials were set to the value found in CC calculations of isolated plates ($H = \infty$). Specifically, $\psi_s e/kT = 3.211$ (0.1M) and $\psi_s e/kT = 7.735$ (0.001M). In all cases, $T = 298$ K and $\epsilon_r = 78.5$. The various constants were $\lambda_B/d = 1.680\,7$, $\lambda_D/d = 2.263$ (0.1 M), $\lambda_D/d = 7.156$ (0.01 M), $\lambda_D/d = 22.630$ (0.001 M), $\lambda_{GC}/d = 0.947$ (for $q_s d^2 = 0.1$), and $\lambda_{GC}/d = 9.47$ (for $q_s d^2 = 0.01$). The molecular diameter was set to $d = 0.425$ nm. Symbols in the lower figure show the thermodynamic route to the force.

with mesh refinement affecting the calculation of the pressure in equation (7.22). Note that in a finite element weak variational approach, the surface charge boundary condition is exact; it does not depend on a finite difference approximation for $\nabla\psi(\mathbf{r}_s)$.

The point-charge model of the electrolyte has been used for over a century to describe electrolytes, and it continues to be used in a wide range of applications to where solvation of molecules or macromolecules by electrolytes requires fast 3-dimenisonal calculations. However, the remainder of this chapter will focus on modifications that allow for more molecular detail. In some sense, all of these modifications can be considered to be modified Poisson–Boltzmann methods [43].

7.3 Classical-DFT for finite-size electrolytes

An obvious improvement to a point-particle model of an electrolyte would be a model with finite-size ions. In electrolytes (as opposed to plasmas or ionic liquids), it is also important to consider the ubiquitous high density solvent (most often water). It is well known from experiments and studies of neutral liquid systems at interfaces, that ordering is significant in high density fluids near surfaces [44, 45].

Several models with finite-size are sketched in figure 7.1. Primitive models (PM) consider only finite size ions; while solvent primitive models (SPM) include finite-size solvents [46, 47]. A PM electrolyte where all ions are the same size is called the restricted primitive model (RPM). The 3-component model (3CM) of Tang, Scriven, and Davis was one of the earliest treatments of finite-size electrolytes with c-DFT, and it is one example of an SPM [15, 16]. In all of these models, the electrostatic properties of the solvent are treated as a dielectric continuum with uniform dielectric permittivity, ϵ_r just as in the point-charge model. This section considers the simplest SPM models that include finite size while deferring a discussion of PM electrolytes with second order correlations to the next section.

The pair interactions for a finite sized molecular model of the electrolyte that includes both Coulomb interactions and volume exclusion has

$$u_{ij}(r_{ij}) = u_{ij}^{hs}(r_{ij}) + \frac{z_i z_j e^2}{\epsilon|\mathbf{r}_i - \mathbf{r}_j|} \tag{7.27}$$

where the valence of the solvent is $z_{\text{solvent}} = 0$, and the hard sphere part of the interaction is

$$u_{ij}^{hs}(r_{ij}) = \begin{cases} \infty & r_{ij} < d_{ij} = 0.5(d_i + d_j) \\ 0 & \text{otherwise.} \end{cases} \tag{7.28}$$

The free energy functional for this system can be written

$$\begin{aligned} \Omega[\{\rho\}] &= \mathcal{F}_{id} + \mathcal{F}_{ex} + \sum_i \int \left[V_{n,i}^{\text{ext}}(\mathbf{r}) - \mu_i\right]\rho_i(\mathbf{r})d\mathbf{r} \\ &= \mathcal{F}_{id} + \mathcal{F}_c + \mathcal{F}_{hs} + \sum_i \int \left[V_{n,i}^{\text{ext}}(\mathbf{r}) - \mu_i\right]\rho_i(\mathbf{r})d\mathbf{r} \end{aligned} \tag{7.29}$$

where the Coulomb effects $\mathcal{F}_c$, and the ideal term, $\mathcal{F}_{id}$ were enumerated in the previous section (see equation (7.8)). This approach was first applied to electrolytes in 1990 by Mier-y-Teran and coworkers [11] with further refinement by Tang and coworkers for both RPM electrolyte models [12, 13] and an SPM electrolyte that included a neutral hard sphere solvent the same size as the ions ($d_i = d$) [15, 16]. In these early studies, the hard sphere contribution to the free energy was defined as

$$\mathcal{F}_{hs} = \sum_i \int \bar{\rho}_i^{\nu}(\mathbf{r}) f^{hs}\left(\sum_i \bar{\rho}_i^{\tau}(\mathbf{r})\right) d\mathbf{r} \tag{7.30}$$

where the excess free energy per particle in a bulk fluid is defined by the Carnahan–Starling equation of state to be

$$f^{hs}(\eta) = kT\frac{\eta(4 - 3\eta)}{(1 - \eta)^2} \tag{7.31}$$

where $\eta = (\pi/6)(\sum_i \rho_i d^3)$. Finally, the nonlocal densities in equation (7.30) are

$$\bar{\rho}_i^{\nu,\tau}(\mathbf{r}) = \int \rho_i(\mathbf{r}')w^{\nu,\tau}(|\mathbf{r} - \mathbf{r}'|; \{\bar{\rho}(\mathbf{r})\})d\mathbf{r}'. \tag{7.32}$$

These nonlocal densities essentially act to reduce the energy penalty associated with steep density peaks and oscillatory density profiles. This form for the nonlocal density was first suggested by Tarazona [48], and produces accurate density profiles for hard sphere fluids near hard walls. However, it is difficult to extend to mixtures of arbitrary sizes [49].

At about the same time, Rosenfeld [50] developed a treatment of hard sphere mixtures based on a scaled particle theory (SPT) that is known as fundamental measures theory (FMT). The FMT approach has become the most common basis for c-DFT studies that are defined as perturbations to hard sphere systems because it is a derived (rather than heuristically defined) theory, it is naturally formulated for mixtures with particles of varying sizes, and it is internally consistent from the perspective of sum rules so the thermodynamic route and the microscopic route to force, adsorption, and other properties will be the same. A review paper by Roth [51] details Rosenfeld's original derivation along with subsequent improvements to the theory, and the functionals are briefly summarized here as well.

In FMTs, the excess hard sphere contribution to the free energy is written

$$\beta\mathcal{F}_{hs}[\{\rho_i\}] = \int \Phi(\{n_\alpha, n_{v\alpha}, n_{T\alpha}\})d\mathbf{r} \tag{7.33}$$

where the free energy density, Φ, is a function of a collection of nonlocal densities, $\{n_\alpha, n_{v\alpha}, n_{T\alpha}\}$, that may include vector (subscript v) and tensor (subscript T) in addition to scalar terms The summed nonlocal densities are related to the component density profiles, $\{\rho_i\}$, by linear functionals

$$n_\alpha(\mathbf{r}) = \sum_{i=1}^{N} \int \rho_i(\mathbf{r}')w_i^{(\alpha)}(\mathbf{r} - \mathbf{r}')d\mathbf{r}' \tag{7.34}$$

where the weight functions, w_α^i (which may be scalars, vectors, or tensors), are unique to the specific functional formulation.

The Euler–Lagrange equations for the SPM (or PM) electrolyte are

$$\begin{aligned}\frac{\delta\Omega}{\delta\rho_i} = 0 &= \ln\rho_i(\mathbf{r}) + z_i\tilde{\psi}(\mathbf{r}) + \frac{V^{\text{ext}} - \mu_i}{kT} + \sum_\alpha \int \frac{\partial\Phi}{\partial \mathrm{n}_\alpha}\frac{\delta \mathrm{n}_\alpha}{\delta\rho_i}d\mathbf{r}' \\ &= \ln\rho_i(\mathbf{r}) + z_i\tilde{\psi}(\mathbf{r}) + \frac{V^{\text{ext}} - \mu_i}{kT} + \sum_\alpha \int \frac{\partial\Phi}{\partial n_\alpha}w_i^{(\alpha)}(\mathbf{r} - \mathbf{r}')d\mathbf{r}'.\end{aligned} \tag{7.35}$$

Just as in the previous section, this Euler–Lagrange equation can be coupled with Poisson's equation resulting in a mean field treatment of electrolytes that includes finite size of the fluid particles.

Several different FMTs have been developed since Rosenfeld's original work. Not all FMTs have vector and tensor terms, and these terms will always vanish in uniform fluids. The introduction of this more expansive set of weight functions in the FMT approach is at first glance more complicated than the calculation of the nonlocal densities in equation (7.32). However, the set of FMT weight functions, $\{w_\alpha^i\}$ that define the nonlocal densities, $\{n_\alpha\}$, do not have an embedded dependence on the densities so the FMT formulation is in practice significantly simpler. Three versions of FMT are provided in detail here.

The Rosenfeld fundamental measures theory

The original model of Rosenfeld [50] includes scalar and vector weight functions. It can be written

$$\Phi^{RF}(\{n\}) = -n_0 \ln(1 - n_3) + \frac{n_1 n_2 - \vec{n}_{v1} \cdot \vec{n}_{v2}}{1 - n_3} + \frac{n_2^3 - 3n_2(\vec{n}_{v2} \cdot \vec{n}_{v2})}{24\pi(1 - n_3)^2}, \tag{7.36}$$

where the weight functions, w, that define the nonlocal densities, n, are related to the fundamental geometric characteristics of the particles. Specifically,

$$\begin{aligned} w_i^{(3)}(\mathbf{r} - \mathbf{r}') &= \theta(R_i - |\mathbf{r} - \mathbf{r}'|) \\ w_i^{(2)}(\mathbf{r} - \mathbf{r}') &= 4\pi R_i w_i^{(1)}(\mathbf{r} - \mathbf{r}') = 4\pi R_i^2 w_i^{(0)}(\mathbf{r} - \mathbf{r}') \\ &= \delta(R_i - |\mathbf{r} - \mathbf{r}'|) \\ w_i^{(v2)}(\mathbf{r} - \mathbf{r}') &= 4\pi R_i w_i^{(v1)}(\mathbf{r} - \mathbf{r}') = \frac{\mathbf{r} - \mathbf{r}'}{|\mathbf{r} - \mathbf{r}'|}\delta(R_i - |\mathbf{r} - \mathbf{r}'|). \end{aligned} \tag{7.37}$$

In a uniform bulk fluid, the relationship of the weight functions, n_α, to the fundamental measures of the particles (or the SPT variables, ξ_i) is clear. Specifically, the volume term becomes the packing fraction, $n_3 = \xi_3 = \frac{4\pi}{3}\sum_i \rho_b^i R_i^3$, the $\alpha = 2$ and $\alpha = 1$ terms are related to the total surface area and radius of curvature of the particles respectively with $n_2 = \xi_2 = 4\pi \sum_i \rho_b^i R_i^2$ and $n_1 = \xi_1 = 4\pi \sum_i \rho_b^i R_i$, and finally, the total particle density is $n_0 = \xi_0 = \sum_i \rho_i^b$.

The Rosenfeld FMT was ground breaking; however, the original version of FMT has some shortcomings. Specifically, the equation of state in the bulk reduces to the Percus–Yevick (PY) compressibility equation of state which is known to predict pressures that are too high (in comparison with molecular simulation). The PY pressure can be written in the FMT variables as [52]

$$\beta p_{PY} = \frac{n_0}{1 - n_3} + \frac{n_1 n_2}{(1 - n_3)^2} + \frac{n_2^3}{12\pi(1 - n_3)^3}. \tag{7.38}$$

In addition, the Rosenfeld FMT does not predict freezing of hard spheres. Both of these shortcomings were later addressed by others, and are summarized briefly below.

The White Bear functional

To improve on the underlying equation of state, Roth, Evans, Lang, and Kahl [53] developed functionals they named the White Bear (WB) model (after the local pub in which it was developed). The WB functional is based on the Mansoori–Charnahan–Starling–Leland (MCSL) equation of state for mixtures [54],

$$\beta p_{MCSL} = \frac{n_0}{1 - n_3} + \frac{n_1 n_2}{(1 - n_3)^2} + \frac{n_2^3}{12\pi(1 - n_3)^3} - \frac{n_3 n_2^3}{36\pi(1 - n_3)^3}. \tag{7.39}$$

The WB functional uses the same weight functions, $w_i^{(\alpha)}$ and nonlocal densities, n_α as the Rosenfeld functional defined above. The free energy density for the WB functional used in inhomogeneous c-DFT calculations is

$$\begin{aligned}\Phi^{WB}(\{n\}) = &- n_0 \ln(1 - n_3) + \frac{n_1 n_2 - \vec{n}_{v1} \cdot \vec{n}_{v2}}{1 - n_3} \\ &+ \frac{n_2^3 - 3n_2(\vec{n}_{v2} \cdot \vec{n}_{v2})}{36\pi n_3^2(1 - n_3)^2}(n_3 + (1 - n_3)^2 \ln(1 - n_3)).\end{aligned} \tag{7.40}$$

The White Bear functional—mark II

Hansen-Goos and Roth later demonstrated that the MCSL equation can be improved by generalizing the Carnahan–Starling–Boublik equation of state to mixtures [52]. Their equation of state is

$$\beta p_{CSIII} = \frac{n_0}{1 - n_3} + \frac{n_1 n_2\left(1 + \frac{1}{3}n_3^2\right)}{(1 - n_3)^2} + \frac{n_2^3\left(1 - \frac{1}{3}(2n_3 + n_3^2)\right)}{12\pi(1 - n_3)^3}. \tag{7.41}$$

Hansen-Goos and Roth used this improved equation of state to develop the White Bear mark II functional (WBII) [55]. It again is based on the Rosenfeld weight functions and nonlocal densities. The free energy density of the WBII functional is

$$\begin{aligned}\Phi^{WBII}(\{n\}) = &- n_0 \ln(1 - n_3) \\ &+ \frac{n_1 n_2 - \vec{n}_{v1} \cdot \vec{n}_{v2}}{1 - n_3}\left(1 + \frac{1}{3n_3}(2n_3 - n_3^2 + 2(1 - n_3)\ln(1 - n_3))\right) \\ &+ \frac{n_2^3 - 3n_2(\vec{n}_{v2} \cdot \vec{n}_{v2})}{24\pi(1 - n_3)^2}\left(1 - \frac{1}{3n_3^2}[2n_3 - 3n_3^2 + 2n_3^3 + 2(1 - n_3)\ln(1 - n_3)]\right).\end{aligned} \tag{7.42}$$

Regardless of which model is chosen, the bulk pressure also can be written

$$\beta p = \beta p^{id} + \beta p_{hs}^{\text{ex}} = n_0 - \Phi + \sum_\alpha \frac{\partial \Phi}{\partial n_\alpha} n_\alpha. \tag{7.43}$$

Note that n_0 is the ideal gas contribution to the pressure, and the remaining terms are the excess pressure due to the treatment of the fluid as finite size hard spheres.

Freezing of hard sphere fluids

The Rosenfeld, WB, and WBII functionals all fail to predict freezing because they don't adequately address the crossover to zero dimensions that occurs when the particles are confined to a cavity as is the case in a solid. In the Rosenfeld functional, the third term in equation (7.36) diverges for strongly ordered systems. An empirical modification to the Rosenfeld functional suggested by Rosenfeld, Schmidt, Löwen, and Tarazona [56, 57] to address this crossover effect replaced the third term in equation (7.36) with

$$\Phi^{RF}_{\text{3rd term}} = \frac{n_2^3(1 - |n_{v2}/n_2|)^q}{24\pi(1 - n_3)^2} \tag{7.44}$$

where $q = 3$ worked well to represent crossover without sacrificing accuracy in the three-dimensional fluid.

A more rigorous approach by Tarazona [58] modified the Rosenfeld FMT by introducing a tensor weight function that is consistent for both the ordered solid and the fluid,

$$w^i_{m2}(\mathbf{r} - \mathbf{r}') = \left(\frac{(\mathbf{r} - \mathbf{r}')(\mathbf{r} - \mathbf{r}')}{|\mathbf{r} - \mathbf{r}'|^2} - \frac{1}{3}\mathbf{I}\right) w^i_2(\mathbf{r} - \mathbf{r}'). \tag{7.45}$$

The third term in equation (7.36) is then modified to be

$$\Phi^{RF}_{\text{3rd term}} = \frac{n_2^3 - 3n_2(n_{v2} \cdot n_{v2}) + 9\left(n_{v2}n_{m2}n_{v2} - \frac{1}{2}Tr(n^3_{m2})\right)}{24\pi(1 - n_3)^2}. \tag{7.46}$$

This approach can also be applied to the WB functionals as discussed by Roth [51], and it is an important development that unifies the c-DFT approach for fluids and solids.

Effects of finite size in electrolyte interfaces

Figure 7.5 compares the density profiles for a 0.1M 1:1 electrolyte in a dense solvent with uniform sizes for all species, $d_i = d$. The observed density oscillations arise from the explicit presence of the dense solvent that results in particle packing near the surface. The structure in the solvent is mirrored in the density profiles of the ions. This generic result dates back to the work of Tang, Scriven and Davis [15], and qualitatively improves the monotonic decay of the point–charge (also denoted Poisson–Boltzmann—or PB) model. Note that the longer range electrostatic decay is not significantly affected by the short range structure, as can be seen in the ion profiles at $x/d > 6$.

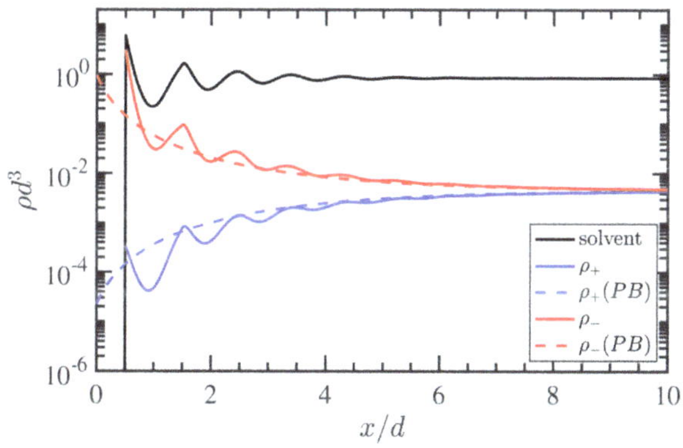

Figure 7.5. Density profiles for a 1:1 electrolyte near a charged planar wall with $q_s d^2/e = 0.3$ for the case where $d_i = d$ for all species, and the bulk densities for the various species are $\rho_{\text{solv}} d^3 = 0.85$, $\rho^{+/-} d^3 = 0.004\,62$ (corresponding to an 0.1 M electrolyte when $d = 0.425$ nm). Solid lines show FMT calculations from the White Bear (I) model while dashed lines show the Poisson–Boltzmann result.

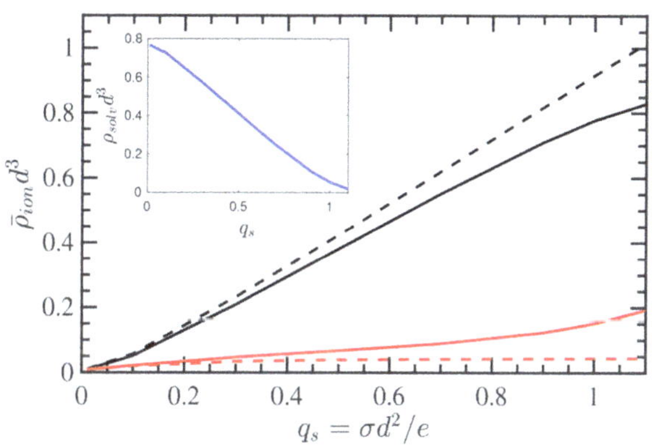

Figure 7.6. Average ion density, $\bar{\rho}_{\text{ions}} d^3 = (\bar{\rho}_+ + \bar{\rho}_-) d^3$, in the first solvation shell (black lines) and second solvation shell (red lines) away from a planar surface as a function of the surface charge, $q_s = \sigma d^2/e$, for the same electrolyte as in figure 7.5. Solid lines show FMT calculations from the White Bear (I) model while dashed lines show the Poisson–Boltzmann result. The inset shows the solvent density in the first solvation shell as a function of q_s. At very high surface charge, the solvent can be displaced from the surface.

Section 7.2 discussed the unphysical densities that can result from the point particle model of the electrolyte. To illustrate how this situation is improved by adding finite size to the model, consider averaging the density in discrete solvation shells away from the surface as

$$\bar{\rho}_j^{\text{shell}} = \sum_i \frac{1}{d} \int_{(j-1)d}^{jd} \rho_i(x) dx. \tag{7.47}$$

This simple averaging gives a picture of the degree of packing predicted by the two models. Figure 7.6 shows how the average density in the first two solvation shells varies as a function of the surface charge for the same 1:1 0.1M electrolyte described in figure 7.5. While the FMT model predicts a very high local densities of solvent and counterions at $x = d/2$, the average density of ions in the first solvation layer $0 < x/d < 1$ is lower than for the point-charge model, and the deviation increases as surface charge increases. Consequently, the second solvation shell of the FMT c-DFT has a higher ion concentration than the point-charge model. Finally, note that freezing in hard spheres occurs at a packing fraction of $\xi = \frac{4\pi}{3}\rho R^3 = 0.494$ (or $\rho d^3 \approx 0.943$). The point-charge model allows $\rho d^3 > 0.943$ while the FMT approach more accurately captures the free energy cost of high densities and pushes ions to the second solvation shell.

The structural features of the finite size model are reflected in the interaction forces between surfaces. Figure 7.7 compares the forces between charged surfaces immersed in a 0.1M 1:1 electrolyte predicted for both the finite size and point particle models. At short range ($H/d < 10$), the forces are dominated by particle packing in the confined space, but at longer range the small remaining structural features are superimposed on the electrostatic decay. This figure also compares forces from thermodynamic route, $f_s = -\partial\Omega^s/\partial H$ and the microscopic structural route based on contact pressures (using equation (7.22)).

As $H \to \infty$, the solvation force, $f_s \to 0$, and the surfaces become independent, exerting no net force on each other. However, the inclusion of finite size introduces many more zeros in f_s. All of the zeros with negative slope correspond to minima in the surface free energy, Ω_s. Section 6.2.3 introduced the potential of mean force (PMF),

$$W(H) = \Omega_s(H) - \Omega_s(H = \infty), \tag{7.48}$$

as a fluid averaged interaction potential between the surfaces that has the property $W(H) \to 0$ as $H \to \infty$. The PMF is important because the global minimum of the

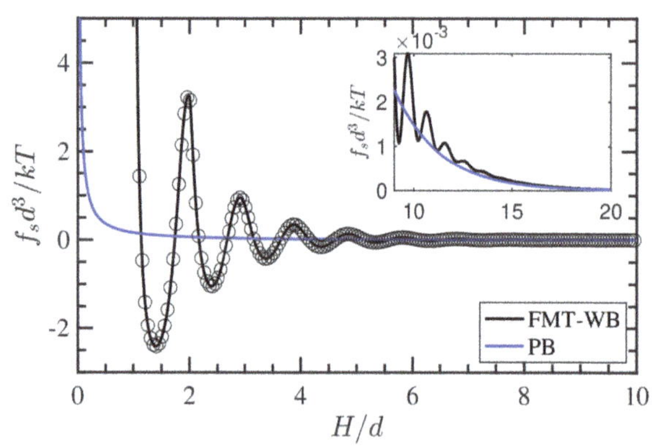

Figure 7.7. The solvation force, $f_s d^3/kT$, as a function of the surface separation, H/d, for a 1:1 electrolyte near a charged wall with $q_s d^2/e = 0.1$ for the case where $d_i = d$ for all species. The bulk densities were $\rho_{\mathrm{solv}} d^3 = 0.75$ and $\rho_{+/-} d^3 = 0.004\,62$ (corresponding to an 0.1 M electrolyte when $d = 0.425$ nm). Black lines and symbols show a finite size model using the FMT-WB functional. Blue lines show the Poisson–Boltzmann (PB) result. The black circles were calculated using the thermodynamic route to the force while the lines were calculated from the contact pressures. The inset emphasizes that electrostatic decay far from the surface.

PMF dictates the equilibrium position of the surfaces (in a situation where they are free to move). While the point-charge model predicts repulsion at all finite surface separations, the finite size model of the electrolyte may allow for other global minima due to the short range oscillations in the solvation force that is linked to many local minima in the PMF. If two states at different surface separation have the same PMF, $W(H_1) = W(H_2)$, and if that value of the PMF is the global minimum, then those two states are in thermodynamic equilibrium. Mathematically, this is an example of a bifurcation where one solution branch splits into two. The topic of these bifurcations, and their importance for constructing phase diagrams will be discussed further in section 7.5 and section 16.2.

Figure 7.8 shows the PMF as a function of surface charge and solvent density for the same electrolyte described in figure 7.7. The top figure (bulk solvent density $\rho_s^b d^3 = 0.75$) shows that at low surface charge, the finite-size effects overwhelm the electrostatic repulsion, and the global minimum of the PMF is found at $H/d = 1.145$. However, as the surface charge increases, the electrostatic effects dominate, and the global minimum of the PMF is found at $H/d = \infty$. The density of the solvent plays a role as well. In the lower pane of figure 7.8, the solvent density is

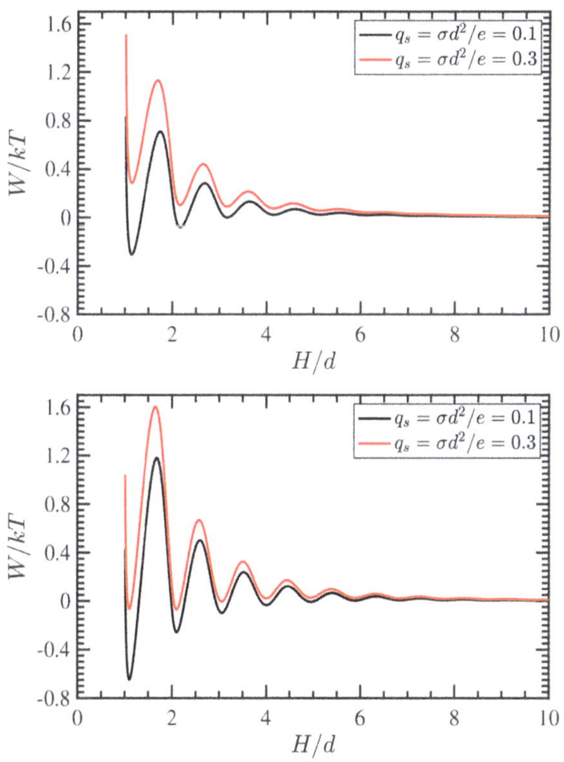

Figure 7.8. Potentials of mean force, W/kT, between charged walls as a function of surface separation, H/d. Both cases are based on the 1:1 electrolyte described in figure 7.7 with solvent densities of $\rho d^3 = 0.75$ (upper figure) and $\rho d^3 = 0.85$ (lower figure). Surfaces charges of $q_s = \sigma d^2/e = 0.1$ (black lines) and $q_s = \sigma d^2/e = 0.3$ (red lines) are included in both figures.

increased to $\rho_s^b = 0.85$. The increase in solvent density results in larger amplitude oscillations in the PMF. While increasing charge still increases the importance of the electrostatics, the case $q_s d^2 = 0.3$ now has two nearly equal short range minima at $H/d \approx 1.1$ and $H/d \approx 2.1$. Both have $W(H) < 0$ so they represent two coexisting stable states.

7.4 Classical-DFT with correlations

Section 6.2.2 discussed how the direct correlation function hierarchy can be leveraged to construct free energy functionals. This section introduces density functionals for electrolytes that include second order correlations. Early studies of the electrical double layer that include second order correlations in a restricted primitive model (RPM) electrolyte date to the 1980s [10] and 1990s [11, 12]. Later investigations on the solvent primitive model (SPM) considered double layer structure and interfacial forces for both systems with particles of identical size, $d_i = d$ [15], and for systems with fluid species of unequal size [17].

Consider an electrolyte with the excess Helmholtz free energy,

$$\mathcal{F}_{ex} = \mathcal{F}_{hs} + \mathcal{F}_c + \mathcal{F}_{corr} \tag{7.49}$$

where $\mathcal{F}_c$ includes the mean field electrostatic interactions and $\mathcal{F}_{hs}$ includes finite size effects as developed in the previous two sections. The formally exact expression for $\mathcal{F}_{corr}$ is found by integrating from a reference system (usually the homogeneous bulk system at the state point of interest) to the inhomogeneous system [45]. Taking a linear path, $\rho_i(\mathbf{r}; \alpha) = \rho_i^{\text{bulk}} + \alpha\Delta\rho_i(\mathbf{r})$,

$$\begin{aligned}\mathcal{F}_{corr}&[\{\rho_i(\mathbf{r})\}]\\ &= -kT\sum_{ij}\int_0^1 d\alpha \int_0^\alpha d\alpha'\left(\int\int \Delta\rho_i(\mathbf{r})\Delta\rho_j(\mathbf{r}')\Delta c_{ij}(\mathbf{r}, \mathbf{r}', \alpha')d\mathbf{r}d\mathbf{r}'\right)\end{aligned} \tag{7.50}$$

where

$$\Delta c_{ij}(\mathbf{r}, \mathbf{r}'; \alpha) = c_{ij}(\mathbf{r}, \mathbf{r}'; \alpha) - c_{ij}^{ES}(\mathbf{r}, \mathbf{r}'; \alpha) - c_{ij}^{hs}(\mathbf{r}, \mathbf{r}'; \alpha). \tag{7.51}$$

The mean field electrostatic (c^{ES}) and hard sphere (c^{hs}) terms are subtracted from the complete direct correlation function to avoid double counting in the construction of $\mathcal{F}_{ex}$ using equation (7.49). In this construct, $\mathcal{F}_{corr}$, is then a correction where Δc is designed to capture in some sense the cross-correlations between hard sphere and electrostatic effects.

7.4.1 Bulk fluid functionals for correlations

In order to proceed, some approximation for Δc is required. The simplest approach is to assume the correlations in the inhomogeneous system are identical to the correlations in a homogeneous bulk fluid,

$$\Delta c_{ij}(\{\rho_i(\mathbf{r})\}, \mathbf{r}, \mathbf{r}', \lambda) = \Delta c_{ij}^{\text{bulk}}(\{\rho_i^b\}, |\mathbf{r} - \mathbf{r}'|). \tag{7.52}$$

The grand free energy for an electrolyte in an external field is then

$$\Omega[\{\rho\}] = \mathcal{F}_{id} + \mathcal{F}_c + \mathcal{F}_{hs} + \mathcal{F}_{corr} - \sum_i \int \left[\mu_i - V_{n,i}^{\text{ext}}(\mathbf{r})\right]\rho_i(\mathbf{r})d\mathbf{r}, \tag{7.53}$$

where

$$\mathcal{F}_{corr} = -\frac{1}{2}kT\sum_{ij}\int\int \rho_i(\mathbf{r})\rho_j(\mathbf{r}')\Delta c_{ij}(\{\rho_i^b\}, |\mathbf{r}-\mathbf{r}'|)d\mathbf{r}d\mathbf{r}'. \tag{7.54}$$

Analytical results for correlation functions (and other thermodynamics properties) of bulk electrolytes were first derived in the early 1970s by Waisman and Lebowitz [59, 60] and Blum [61, 62] who solved the Ornstein–Zernike equations using the mean spherical approximation(MSA). The solution for the RPM is

$$\Delta c_{ij}(\{\rho_i^b\}, |\mathbf{r}-\mathbf{r}'|) = \begin{cases} -\lambda_B z_i z_j\left(2\frac{B}{d} - \left(\frac{B}{d}\right)^2 r - \frac{1}{r}\right) & \text{if } r < d, \\ 0 & \text{otherwise,}\end{cases} \tag{7.55}$$

where $r = |\mathbf{r}-\mathbf{r}'|$, $B(\{\rho_i^b\}) = (\kappa d + 1 - \sqrt{1 + 2\kappa d})/(\kappa d)$, and $\kappa = 1/\lambda_D$. Note that the Debye length and Bjerrum length (λ_D, λ_B) were defined in section 7.2.

Generalizing to ions of arbitrary sizes, d_i, results in a more complex, although still analytically derived, expression [17, 61, 63],

$$\Delta c_{ij}(r) = 2\lambda_B\left(\frac{z_i z_j}{2r} + z_i N_j - X_i(N_i + \Gamma X_i) + \frac{d_i}{3}(N_i + \Gamma X_i)^2\right) \tag{7.56}$$

if $0 \leqslant r_{ij} \leqslant (d_j - d_i)/2$ and $d_i < d_j$, or

$$\Delta c_{ij}(r) = \frac{\lambda_B}{r}\left(\begin{array}{l} z_i z_j + (d_i - d_j)\left\{\begin{array}{r} \frac{X_i + X_j}{4}\left[(N_i + \Gamma X_i) - (N_j + \Gamma X_j)\right] \\ -\frac{d_i - d_j}{16}\left[(N_i + \Gamma X_i + N_j + \Gamma X_j)^2\right] \\ -4N_i N_j \end{array}\right\} \\ -r\left\{\begin{array}{r} (X_i - X_j)(N_i - N_j) + (d_i + d_j)N_i N_j \\ +(X_i^2 + X_j^2)\Gamma - \frac{1}{3}\left[d_i(N_i + \Gamma X_i)^2 + d_j(N_j + \Gamma X_j)^2\right] \end{array}\right\} \\ +r^2\left\{\begin{array}{l} \frac{X_i}{d_i}(N_i + \Gamma X_i) + \frac{X_j}{d_j}(N_j - \Gamma X_j) + N_i N_j \\ -\frac{1}{2}\left[(N_i + \Gamma X_i)^2 + (N_j + \Gamma X_j)^2\right] \end{array}\right\} \\ +r^4\left\{\frac{(N_i + \Gamma X_i)^2}{6d_i^2} + \frac{(N_j + \Gamma X_j)^2}{6d_j^2}\right\} \end{array}\right) \tag{7.57}$$

if $|d_j - d_i|/2 \leqslant r < (d_i + d_j)/2$. The various parameters needed to define the above expressions are

$$
\begin{aligned}
X_i &= \frac{z_i}{1+\Gamma d_i} - \left(\frac{cd_i^2}{1+\Gamma d_i}\right)\left(\frac{\sum_k\{(\rho_k d_k z_k)/(1+\Gamma d_k)\}}{1 + c\sum_k\{(\rho_k d_k^3)/(1+\Gamma d_k)\}}\right),\\
c &= \frac{\pi}{2\left[1 - \frac{\pi}{6}\sum_k \rho_k d_k^3\right]},\\
N_i &= (X_i - z_i)/d_i,\\
\Gamma &= \sqrt{\pi\lambda_B \sum_k \rho_k X_k^2}.
\end{aligned} \tag{7.58}
$$

An iterative calculation is needed to determine the parameters in equation (7.58), but that calculation is preliminary to the iterative solve of Euler–Lagrange and Poisson's equation and causes no additional difficulty.

Figure 7.9 shows density profiles for 1:1 and 2:1 electrolytes comparing point-charge, primitive, and solvent primitive models. The figure compares functionals that include the bulk MSA direct correlation functions from equations (7.57) and (7.58) to a strict mean field (SMF) model that has $\Delta c_{ij} = 0$. All of the hard-sphere interactions in the figure are based on the original White Bear FMT functionals described in section 7.3 [53].

Perhaps the most striking result in figure 7.9 is the observation of charge inversion in the 2:1 electrolyte for some of the models. Consider the 2-component models first; both the PB and the PM cases with the SMF functional result in monotonic ion density and potential profiles for both the 1:1 and 2:1 electrolytes. In contrast, the 2:1 PM electrolyte that includes correlations using the MSA functional, clearly exhibits charge inversion. The anion density profile has a pronounced peak at $x \approx 0.51$ nm, and the electrostatic potential is nonmonotonic. Molecular dynamics simulation results [64] included in the figure are consistent with this result. Charge inversion was identified in early studies [12], and it is an important behavioral domain that depends on the surface charge, the bulk ion density, the relative valences of the ion species, and the sizes of the ion species. When the system of interest falls within this charge inversion domain, it is particularly important to include second order correlations in the c-DFT functionals.

Adding a finite-size dense solvent to the model (the SPM cases in figure 7.9) also qualitatively changes the predictions for the structure of the electrolyte interface. Specifically, the density profiles for all species acquire oscillations due to volume exclusion effects. When the SPM is combined with the bulk MSA treatment of correlations, charge inversion in the 2:1 electrolyte is somewhat amplified from the PM-MSA case. The structure of the solvent is relatively unaffected by the ions (see the insets in the lower panels), and nearly identical solvent density profiles are observed independent of the valence of the electrolyte or whether the MSA bulk correlations are included.

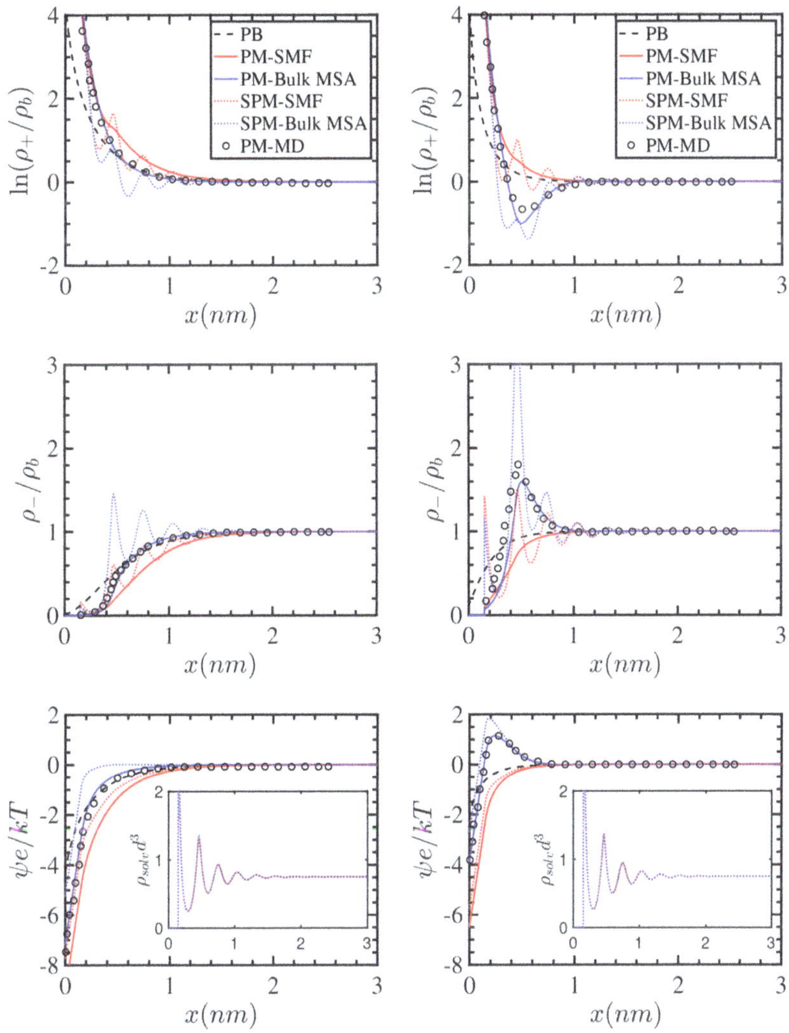

Figure 7.9. Double layer structure at a surface with $q_s(x = 0) = -0.5$ C m^{-2} for 1:1 (left) and a 2:1 (right) electrolytes. Models include: a Poisson–Boltzmann (PB) point charge model, 2 component primitive models (PM), and 3-component solvent primitive models (SPM) with finite size solvent as well as ions. Molecular Dynamics (MD) results were estimated from figure 5 of [64]. The insets show solvent density profiles for SPM cases. Fluid parameters were: $T = 298$ K, $\epsilon_s = 78.4$, $d = 0.3$ nm (for all species), $\rho_+^b = 1M$ and $\rho_s d^3 = 0.75$ (SPMs).

As expected, solvation forces and potentials of mean force (PMF) in systems with two interacting surfaces also depend on the type of electrolyte model and functional applied as shown in figure 7.10.

Charge inversion has a significant effect on the solvation force and the PMF. The 2:1 electrolyte with the PM-MSA functional has an attractive solvation force ($f_s < 0$) for $1.24 < H/d < 2.25$, with a minimum in PMF found at $H/d = 1.25$. In contrast, when correlations are neglected (the PM-SMF case), the solvation force is always repulsive ($f_s > 0$) so the equilibrium position for free surfaces would be $H = \infty$. Thus the charge inversion leads to the possibility of attractions between like charged surfaces and

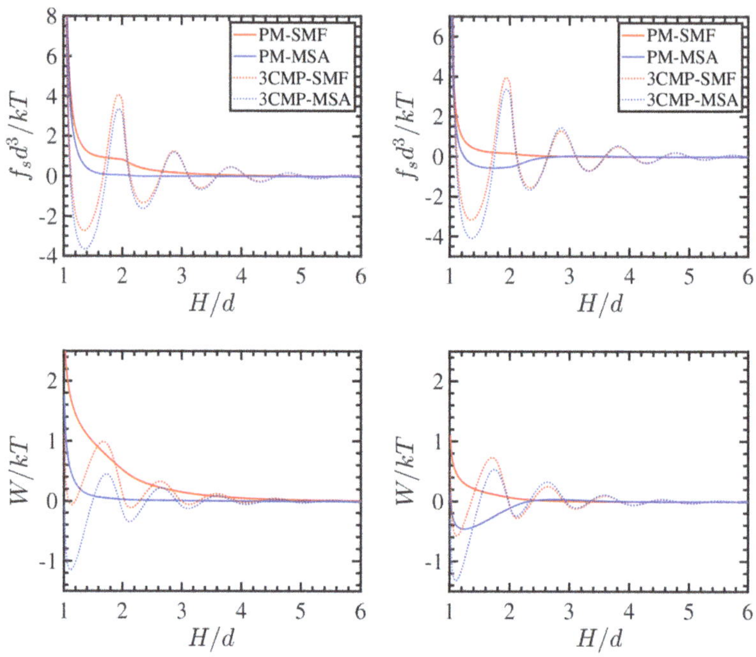

Figure 7.10. Solvation forces, $f_s(H)$, and potentials of mean force, $W(H)$, as a function of the surface separation, H, for a 1:1 (left) and 2:1 (right) electrolyte near a charged wall. All parameters are the same as in figure 7.9.

equilibrium arrangements favoring small separations with only a few solvation layers. The SPM again exhibits oscillatory structure in the force and PMF due to the presence of the finite size solvent. Both SPM-MSA and SPM-SMF cases have multiple short range minima with $W(H) < 0$. When the second order MSA correlations are included in the functional, the minimum at $H/d = 1.11$ becomes the deepest and most favored.

The cases discussed above have relatively short range electrostatic effects (small λ_D) and species of equal size. As the electrostatic interactions become longer range at lower bulk concentrations, the double layer becomes broader. If the various species have different sizes, different packing effects may come into play. To illustrate this complexity, consider a 0.01M 2:1 electrolyte where the cations (counterions) are twice the size of the anions and solvent. Like the previous case, a high surface charge of -0.5 C m^{-2} is considered. Figure 7.11 shows density profiles, forces, and free energies for this case with the same PM and SPM models considered previously. Charge inversion is again predicted for the PM and SPM models with the MSA functional, but unlike figure 7.9, MD simulations do not show charge inversion.

Turning to solvation forces and PMF results (lower panes in figure 7.11), charge inversion again leads to an attractive force and a free energy minimum at $H/d = 3.56$ for the PM-MSA case. When solvent is included, the global minimum shifts to $H/d = 3.08$. There is an additional competitive minima at ($H/d = 2.05$). Both models using strict mean field (SMF) functionals predict the global PMF minimum at $H/d = \infty$.

Finally, density profiles are presented for cases where there is a small gap global minimum in the PMF in figure 7.12. In the 2-component PM-MSA case, $H = 3.56d \approx 1.8d_+$ and the cations are found throughout the gap with a preference

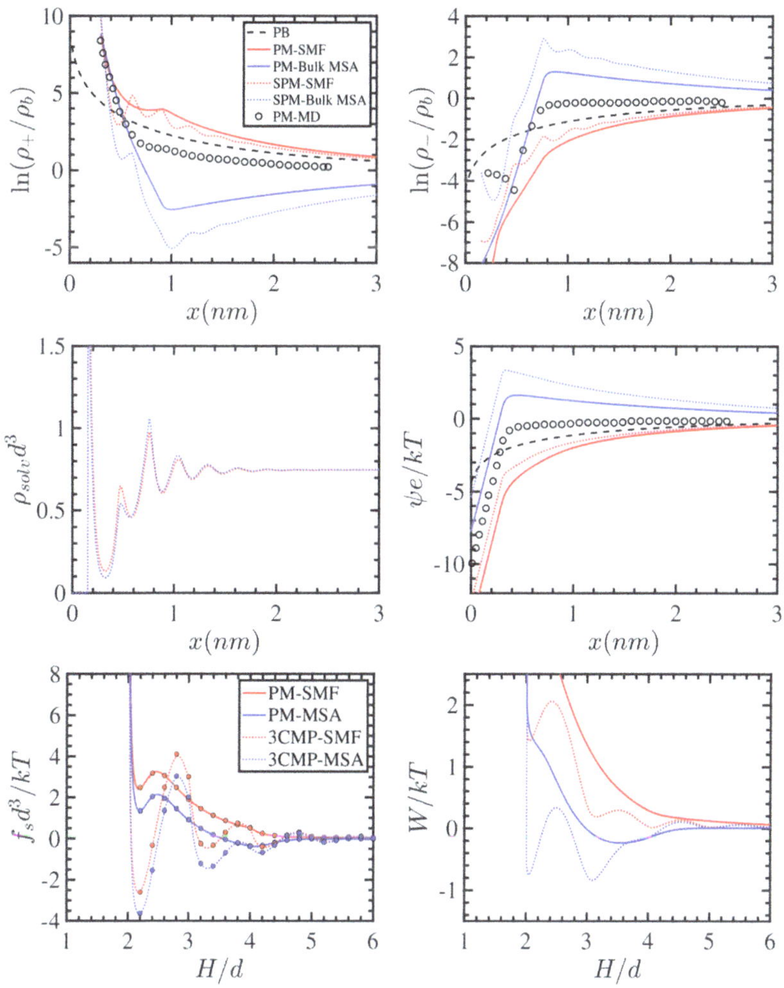

Figure 7.11. Double layer structure at an isolated wall (top 4 panels) along with the solvation forces (f_s), and potential of mean force (W) as function of surface separation (H/d) of two identical walls at a surface with $q_s(x = 0) = -0.5$ C m^{-2} for a variety of models as described in figure 7.9. The parameters are identical to figure 7.9 with the exception that $d_+/d = 2$ where $d_- = d_s = d$, and $\rho_+ = 0.01M$. Molecular dynamics (MD) results were estimated from figure 4 of [64]. The symbols in the lower left panel were calculated from $f_s = -\partial\Omega_s/\partial H$ while the lines were calculated using $f_s = \tilde{p}(H) - \tilde{p}(\infty)$.

for locations near the surfaces. The anions are largely excluded from the gap. The number of any confined species per unit surface area (or adsorption) is easily calculated as $\Gamma_i = \int \rho_i(x)dx$, and for this case $\Gamma_+/\Gamma_- \approx 10^6$. So there will be only one anion for every million cations between the surfaces. Those few anions prefer (have highest density at) positions that also have highest cation density. The centers of the anions can approach the surfaces more closely than the cations due to their smaller size, but those surface locations are less favorable (have lower density).

When finite size solvent is included, in the SPM-MSA case, the optimal spacing decreases to $H/d = 3.08$. The counterions again prefer surface locations with less chance

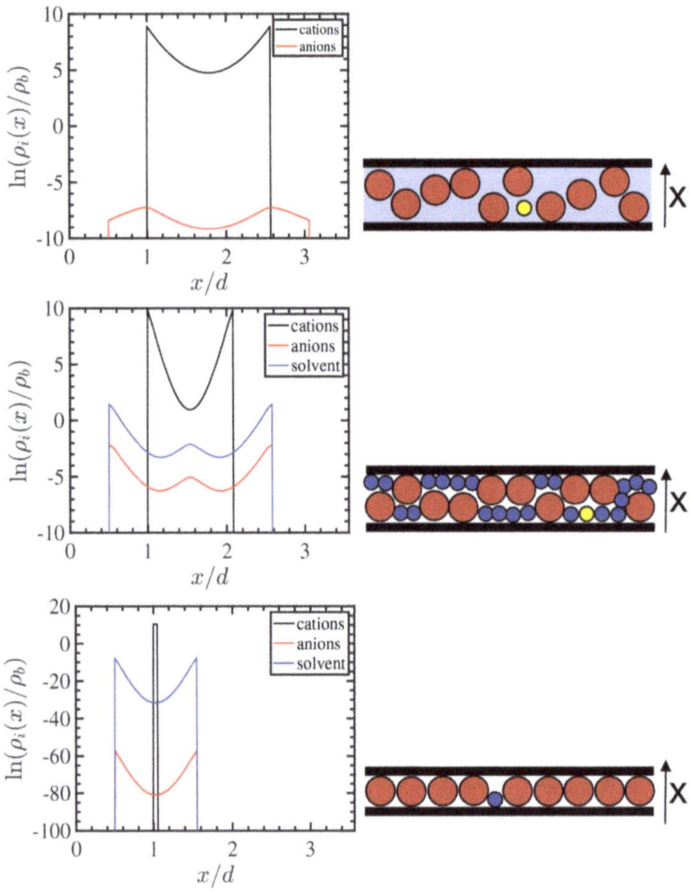

Figure 7.12. Density profiles in the confined systems corresponding to the minima in PMF with $W(H) < 0$ in figure 7.11. Sketches of the various states are shown to the right with red cations, blue solvent, and yellow anions. The uniform dielectric continuum solvent is included as a uniform blue background in the top figure.

they will be found in the center of the gap. This spacing allows for a perfect fit of one cation and one solvent (or anion) particle across the gap, and a signficant amount of solvent will be found in the gap with $\Gamma_s/\Gamma_+ \approx 2.6$. Anions are still scarce with $\Gamma_+/\Gamma_- \approx 2.6 \times 10^4$. Both solvent and anion particles can approach the surface more closely than the cations, and in this case, those small particles prefer the surface locations.

Finally, the other competitive minima is at $H/d = 2.05$ for the SPM-MSA case. This gap effectively excludes anions and allows for only trace solvent adsorption with $\Gamma_+/\Gamma_s \approx 2.8 \times 10^4$. The small numbers of solvent in the gap will most likely be found at the surfaces. The cations are found in a plane with their centers very near the center of the gap. While this case is not the global free energy minimum, it represents a second competing state that could become dominant with changes in the state point.

Finally, note that MD simulations do not predict the charge inversion observed in figure 7.11. In fact the MD results lie between the PM-SMF and PM-MSA results, and are perhaps closer to the PM-SMF case. It has been noted that a SMF approach if often better than expected [65]. In any case, it is clear that using a bulk fluid approximation for the second order correlations does not always yield accurate results. It is also difficult to use this approach in a liquid–vapor interface since an appropriate bulk reference fluid would be different in the two phases with some patchwork required at the interface.

There are at least two approaches that could improve on the bulk-MSA approximation for correlations. One option is to use molecular simulations to determine $c_{ij}(r)$ numerically for bulk systems or for some number of interfacial systems. Then an approximate functional could be constructed using a suitable interpolation scheme [66]. The alternative is to consider more sophisticated theoretical approximations that allow for the direct correlation functional to more properly reflect the inhomogeneities in the interfacial system. Two possibilities are outlined below.

7.4.2 Interfacial fluid functionals for correlations

One natural improvement from the homogeneous bulk approximation in the previous section is to design correlation functionals that are linked to the local conditions in interfacial fluids. Gillespie and coworkers have developed a reference fluid density (RFD) method that defines a reference fluid based on the inhomogeneous density profile in the interface [67, 68].

Consider a nonlocal density that depends on the range of the electrostatic interactions, $R_{ES}(\mathbf{r})$, and a tuning function, $\alpha_i(\mathbf{r})$, as

$$\bar{\rho}_i(\mathbf{r}) = \frac{3}{4\pi R_{ES}^3(\mathbf{r})} \int_{|r-r'|\leqslant R_{ES}(\mathbf{r})} \alpha_i(\mathbf{r}')\rho_i(\mathbf{r}')d\mathbf{r}'. \tag{7.59}$$

The second order correlation function in the RFD approach sets

$$c_{ij}^{(2),ES}(\{\rho_i, (\mathbf{r})\}\mathbf{r}, \mathbf{r}') = c_{\text{bulk}, ij}^{(2),ES}(\{\bar{\rho}_i\}, |\mathbf{r} - \mathbf{r}'|). \tag{7.60}$$

The c-DFT calculation requires finding $\{\alpha_i(\mathbf{r})\}$ and $R_{ES}(\mathbf{r})$ simultaneously with computing $\{\rho_i(\mathbf{r})\}$ and $\psi(\mathbf{r})$ while applying the constraints that (1) $\sum_i \alpha_i(\mathbf{r})\rho_i(\mathbf{r}) = 0$, and (2) a fluid with density profiles $\{\alpha_i(\mathbf{r})\rho_i(\mathbf{r})\}$ has the same ionic strength at all $\mathbf{r}$ as the fluid with density profiles $\{\rho_i(\mathbf{r})\}$. Improved accuracy over the homogeneous bulk-MSA approach has been demonstrated with comparisons to Monte-Carlo simulations [68]. However, the RFD approach is both computationally expensive and somewhat ad-hoc in construction [69].

More recently Roth and Gillespie developed an approach that functionalizes the MSA equations [69, 70]. Their approach interprets the MSA as a description of the interactions between shells of charge as previously suggested by Wei and Blum [71]. The functionalized MSA approach is again constructed using nonlocal densities based on the inhomogeneous fluid. The Roth and Gillespie electrostatic functional is summarized here using their notation.

Consider first a bulk electrolyte within the MSA [61, 71–73]. The internal energy of that bulk electrolyte with number densities, $\{\rho_i\}$, and particle sizes, d_i, is

$$u^{ES}/kT = -\lambda_B \sum_i \rho_i \left(\frac{z_i^2 \Gamma}{1 + \Gamma d_i} + \frac{z_i d_i \eta}{1 + \Gamma d_i} \right) \tag{7.61}$$

where

$$\eta = \frac{1}{H(\Gamma)} \sum_i \frac{z_i \rho_i d_i}{1 + \Gamma d_i}, \tag{7.62}$$

$$H(\Gamma) = \sum_i \frac{\rho_i d_i^3}{1 + \Gamma d_i} + \frac{2}{\pi}\left(1 - \frac{\pi}{6}\sum_i \rho_i d_i^3\right), \text{ and} \tag{7.63}$$

$$\Gamma^2 = \pi\lambda_B \sum_i \rho_i \left(\frac{z_i - \eta d_i^2}{1 + \Gamma d_i}\right)^2. \tag{7.64}$$

The screening parameter, Γ, is derived by minimizing the electrostatic free energy in the MSA. It has the property that as $d_i \to 0$ (point particles), $2\Gamma \to \kappa$. A new length scale, $b_i = R_i + 1/(2\Gamma)$ can be introduced that represents the radius of a spherical shell of smeared charge. As the concentration of ions increases, Γ increases, and this b_i approaches the particle radii, $b_i \to R_i = d_i/2$.

The entropy density in the MSA is

$$s^{ES} = -\frac{k}{3\pi}\Gamma^3. \tag{7.65}$$

Given equations (7.61) and (7.65), it is straightforward to construct the Helmholtz free energy density, $f = u^{ES} - Ts^{ES}$ and the grand free energy density, $\Omega/V = f - \sum_i \mu_i \rho_i$. The MSA pressure is then $p = -\Omega/V$ or

$$p^{MSA}/kT = -\frac{\Gamma^3}{3\pi} - \frac{2\lambda_B \eta^2}{\pi}. \tag{7.66}$$

Roth and Gillespie functionalized the MSA by introducing a nonlocal shell density, n_i^{sh}, for each species, i, and recasting the internal energy in terms of that nonlocal density. Specifically,

$$n_i^{sh}(\mathbf{r}) = \int \rho_i(\mathbf{r}')w_i^{sh}(\mathbf{r}, \mathbf{r}')d\mathbf{r}' \tag{7.67}$$

where the normalized shell weight functions, w_i^{sh}, are

$$w_i^{sh}(\mathbf{r}, \mathbf{r}') = \frac{\delta(|\mathbf{r} - \mathbf{r}'| - b_i)}{4\pi b_i^2}, \tag{7.68}$$

and the shell radius is defined by the bulk as $b_i = R_i + (2\Gamma_{\text{bulk}})^{-1}$. A dimensionless inhomogeneous free energy density functional based on the nonlocal shell densities is defined by equations (7.61) and (7.65) to be

$$\Phi^{ES}(\{n_i^{sh}\}) = -\lambda_B \sum_i n_i^{sh}\left(\frac{z_i^2\Gamma(\mathbf{r})}{1 + \Gamma(\mathbf{r})d_i} + \frac{z_i d_i \eta(\mathbf{r})}{1 + \Gamma(\mathbf{r})d_i}\right) + \frac{\Gamma(\mathbf{r})}{3\pi} \tag{7.69}$$

where $\Gamma(\mathbf{r})$ and $\eta(\mathbf{r})$ are found by replacing ρ_i, Γ, and η in equations (7.62)-(7.64) with n_i^{sh}, $\Gamma(\{n_i^{sh}(\mathbf{r})\})$ and $\eta(\{n_i^{sh}(\mathbf{r})\})$.

The grand free energy functional constructed by Roth and Gillespie combines the $\mathcal{F}_c$ and $\mathcal{F}_{corr}$ terms in equation (7.53) so that

$$\Omega[\{\rho\}] = \mathcal{F}_{id} + \mathcal{F}_{hs} + \mathcal{F}_{ES} - \sum_i \int \left[\mu_i - V_{n,i}^{ext}(\mathbf{r})\right]\rho_i(\mathbf{r})d\mathbf{r}, \tag{7.70}$$

where the excess electrostatic free energy is

$$\begin{aligned}\mathcal{F}_{\mathrm{ES}}[\{\rho_i\}] = &\sum_i z_i e \int \rho_i(\mathbf{r})\tilde{\psi}(\mathbf{r})d\mathbf{r} + kT\int \Phi^{\mathrm{ES}}(\{n_i^{sh}\})d\mathbf{r}\\ &+ \frac{kT}{2}\sum_{ij}\int\int \rho_i(\mathbf{r})\rho_j(\mathbf{r}')\left(u_{ij}^{sh}(|\mathbf{r}-\mathbf{r}'|) - u_{ij}^{c}|\mathbf{r}-\mathbf{r}'|\right)\theta(r - R_{ij})d\mathbf{r}d\mathbf{r}'\end{aligned} \tag{7.71}$$

where the first term is the same mean field electrostatics contribution in equation (7.12). The step function is $\theta(r - R_{ij}) = 1$ for $r \leqslant R_{ij}$ and $\theta = 0$ otherwise. This sets the range of the integration given the interaction radius, $R_{ij} = (R_i + R_j)/2$. The interaction between two shells of charge is given by

$$u_{ij}^{sh} = \frac{z_i z_j e^2}{4\epsilon b_i b_j}\left(2(b_i + b_j) - r - \frac{(b_i - b_j)^2}{r}\right) \tag{7.72}$$

and the pair Coulomb interactions is given by equation (7.1). The difference between the two is

$$u_{ij}^{sh}(r) - u_{ij}^{c}(r) = -\frac{z_i z_j e^2}{4\epsilon b_i b_j}\frac{(r - b_{ij})^2}{r}. \tag{7.73}$$

where $r = |\mathbf{r} - \mathbf{r}'|$ and $b_{ij} = (b_i + b_j)/2$.

As in previous sections, the DFT calculation requires the derivation of a set of Euler–Lagrange equations, $\delta\Omega/\delta\rho_i = 0$. Assuming that an FMT model is again used for the hard cores, the Euler–Lagrange equations are now

$$\begin{aligned}\frac{\delta\Omega[\{\rho(\mathbf{r})\}]}{\delta\rho_i} = 0 = &\ln\rho_i(\mathbf{r}) - \mu_i + z_i\tilde{\psi}(\mathbf{r}) + \sum_\alpha \int \frac{\partial\Phi^{FMT}}{\partial n_\alpha} w_\alpha^i(\mathbf{r} - \mathbf{r}')d\mathbf{r}'\\ &+ \int \frac{\partial\Phi^{\mathrm{ES}}}{\partial n_i^{sh}} w_i^{sh}(|\mathbf{r} - \mathbf{r}'|)d\mathbf{r}'\\ &+ \sum_j \int \rho_j(\mathbf{r}')\left(u_{ij}^{sh}(|\mathbf{r}-\mathbf{r}'|) - u_{ij}^{c}(|\mathbf{r}-\mathbf{r}'|\right)\theta(r - R_{ij})d\mathbf{r}'.\end{aligned} \tag{7.74}$$

Note that the simplicity of the Φ^{ES} term arises from the constraints of the derivation of equation (7.64) and its functional analog. Specifically, that implicit expression for $\Gamma(\mathbf{r})$ arises directly from the requirement that

$$\frac{\partial F_{ex}^{\mathrm{ES}}}{\partial \Gamma} = \frac{\partial \Phi^{\mathrm{ES}}}{\partial \Gamma} = 0. \tag{7.75}$$

Inserting the definitions for $\eta(\mathbf{r})$ and $H(\mathbf{r})$ into equation (7.69) will leave an expression that in an explicit notation is $\Phi^{\mathrm{ES}}(\{n_i^{sh}[\rho_i]\}, \Gamma(\{n_i^{sh}[\rho_i]\}))$. The Euler–Lagrange contribution would be

$$\frac{\delta}{\delta \rho_i}\left(\int \Phi^{\mathrm{ES}} d\mathbf{r}\right) = \int \frac{\partial \Phi^{\mathrm{ES}}}{\partial n_i^{sh}} \frac{\delta n_i^{sh}}{\delta \rho_i} d\mathbf{r}' + \int \frac{\partial \Phi^{\mathrm{ES}}}{\partial \Gamma} \frac{\partial \Gamma}{\partial n_i^{sh}} \frac{\delta n_i^{sh}}{\delta \rho_i} d\mathbf{r}', \tag{7.76}$$

but happily, the second term vanishes due to the constraint in equation (7.75).

From a numerical point of view, the complete system of equations now includes the Euler–Lagrange equations, Poisson's equation, and the functional analogs of equations (7.62)-(7.64).

Along with the development of this functional, Roth and Gillespie presented both DFT and Monte Carlo (MC) calculations on 20 different PM systems across a range of electrolyte concentrations, size ratios, and ion valences [69]. They found reasonably good agreement between the electrostatic shell DFT approach and comparable Monte-Carlo simulations. A further exhaustive investigation identified conditions where the RFD and shell-functional approach are most different [64] and discussed the relative strengths of the two approaches.

7.5 Classical-DFT with cohesive interactions

Since the earliest c-DFT work on fluid systems by Sam and Ebner in the 1970s, there has been interest in predicting the properties of fluids in the vicinity of surfaces [66]. However, hard sphere fluids do not exhibit a liquid–vapor phase transition because there are no cohesive interactions in the molecular model. Therefore, it is quite common to include cohesive interactions in c-DFT investigations of fluids at interfaces. This approach has been used to study surface mediated manifestations of fluid phase transitions including wetting, pre-wetting, drying, and capillary condensation [74–79].

Understanding surface mediated phase transitions in electrolyte systems in particular is of great importance since surface charge can develop from chemical interactions of surfaces with the ubiquitous water vapor present in the environment. Thin liquid films that form on surfaces can speed corrosion and ultimately materials failure. Despite the importance of the applications, there was some delay between the early work on surface mediated fluid phase transitions and the application of the c-DFT approach to liquid–vapor interfaces of electrolytes. This delay was due in part to the challenges of handling the variation in polarization of water across the liquid–vapor interface. At saturation where the liquid and vapor phases are in thermodynamic equilibrium, the relative permittivity of water varies in the liquid phase from $\epsilon_r \approx 87.8$ at $T = 273$ K to $\epsilon_r \approx 7.2$ at $T = 646$ K while in the vapor phase $\epsilon_r \approx 1.000\,07$ at $T = 273$ K and $\epsilon_r \approx 3.75$ at $T = 646$. At room temperature ($T = 298$ K), liquid water has the familiar value of $\epsilon_r \approx 78.5$ while vapor phase water has

$\epsilon_r \approx 1.000\,3$. At the critical point ($T = 647.14$), both liquid ad vapor have $\epsilon_r \approx 5.4$ [80, 81].

c-DFT functionals have been developed to address solvent polarization in a detailed way for electrolytes [17, 18, 24, 43], but a discussion of those functionals is reserved for section 7.7. Here, a simpler approach is taken. Consider a solvent primitive model (SPM) electrolyte with cohesive interactions, but constant relative permittivity, ϵ_r. While the approximation for permittivity is rather extreme, this simple model allows for a straightforward introduction of cohesive forces along with a general description of using c-DFT to investigate surface mediated phase transitions.

Cohesive (attractive) interactions are usually introduced using a short range pair potential. Often the range of the interaction is limited by performing a cut and shift of the potential to ignore long range tails. Common choices for the interaction potentials, u_{ij}, are Lennard–Jones (LJ) interactions, Yukawa interactions, and square well interactions. These pair potentials are defined by some energy scale, A_{ij}, and some length scale(s) which may be a particle diameter, d_{ij}, a van der Waals diameter, σ_{ij}, a decay parameter, λ, or an interaction range parameter, Δ. Specific examples are:

$$u_{ij}(r) = \begin{cases} 4A_{ij}\left[\left(\dfrac{d_{ij}}{r}\right)^{12} - \left(\dfrac{d_{ij}}{r}\right)^{6}\right] & \text{Lennard–Jones} \\ -A_{ij}\dfrac{\exp(-r/\lambda)}{r/d_{ij}} & \text{Yukawa} \\ \begin{cases} \infty & \text{for } r < d_{ij} \\ -A_{ij} & \text{for } d_{ij} \leqslant r \leqslant d_{ij} + \Delta \\ 0 & \text{for } r > d_{ij} + \Delta \end{cases} & \text{Square Well} \end{cases} . \tag{7.77}$$

These neutral pair interactions may be included in the c-DFT functionals at a mean field level as described in section 6.2.2. They are treated as a perturbation of a hard sphere system in order to make use of the accurate FMT functionals that describe the hard sphere systems (see section 7.3). The interaction is modified to reflect this split in physical properties. Consider a Lennard–Jones interaction. Using the Weeks, Chandler Anderson approach described in section 5.3.2, the attractive mean field pair potential used in c-DFT calculations will be

$$u_{ij}^{att}(r) = \begin{cases} u_{ij}^{LJ}(r_{\min}) & r \leqslant r_{\min} \\ u_{ij}^{LJ}(r) - u_{ij}^{LJ}(r_c) & r_{\min} < r < r_c \\ 0 & r \geqslant r_c \end{cases} \tag{7.78}$$

where $r_{\min}$ is the distance where the LJ potential is a minimum, and r_c is the cutoff distance to be applied.

The total excess free energy is written

$$\mathcal{F}_{ex} = \mathcal{F}_{hs} + \mathcal{F}_{\text{ES}} + \mathcal{F}_{att} \tag{7.79}$$

where the electrostatic functional may include both the Coulombic mean field term and if desired second order correlations ($\mathcal{F}_{\text{ES}} = \mathcal{F}_c + \mathcal{F}_{corr}$). The neutral cohesive mean-field interactions are included in $\mathcal{F}_{att}$ where

$$\mathcal{F}_{att}[\{\rho_i(\mathbf{r})\}] = \frac{1}{2}\sum_{ij}\int\int u_{ij}^{att}(|\mathbf{r} - \mathbf{r}'|)\rho(\mathbf{r})\rho(\mathbf{r}')d\mathbf{r}d\mathbf{r}' \tag{7.80}$$

and the neutral mean-field contribution to the Euler–Lagrange equation is

$$\frac{\delta \mathcal{F}_{att}}{\delta \rho_i(\mathbf{r})} = \sum_j \int \rho_j(\mathbf{r}')u_{ij}(|\mathbf{r} - \mathbf{r}'|)dr'. \tag{7.81}$$

Bulk phase coexistence

To study surfaces that may have adsorbed thin films, it is first necessary to have an understanding of the phase coexistence of the bulk fluid. It is simplest to begin with an understanding of the pure solvent. So, consider first a fluid of Lennard–Jones particles with parameters $A/kT = 1.25$ and where the cutoff distance is set to $r_c = 3d$. As was discussed in sections 5.4 and 6.2.3, fluids at temperatures below their critical points will exhibit a phase separation where the order parameter (density) becomes discontinuous as the chemical potential is varied. Figure 7.13 shows how the grand potential (or pressure) varies with fluid chemical potential for a pure solvent. The results were generated using an arc-length continuation (ALC) algorithm [82] to allow mapping of the turning points and unstable branches in addition to the desired intersecting stable branches of the curve. The ALC algorithm is discussed further in section 16.2.1. At the intersection of the stable branches, the two fluid phases (liquid and vapor) have equal chemical potentials and so are in coexistence. At that point, the vapor phase has a density of $\rho_v d^3 \approx 0.000\,67$ while the liquid phase has a density of $\rho_l d^3 \approx 0.767$. The chemical potential and pressure at coexistence are $\beta\mu_{\text{coex}} \approx -5.099$ and $\beta p_{\text{coex}} d^3 \approx 0.006\,39$.

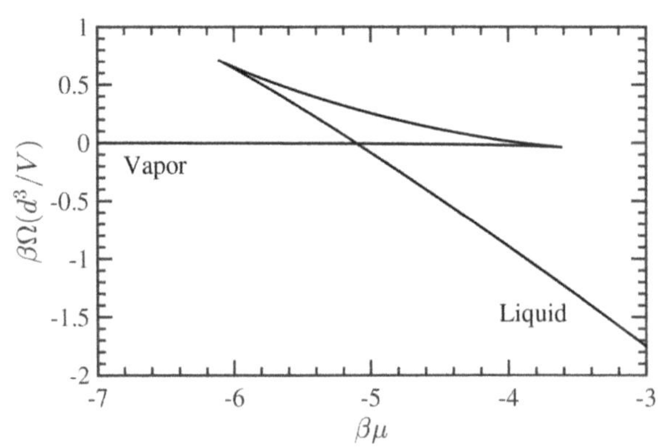

Figure 7.13. Arc length continuation calculations of a bulk solvent. The grand potential, $\Omega = -pV$ is reported as a function of chemical potential, μ. The two intersecting thermodynamically stable branches are liquid and vapor phases as indicated.

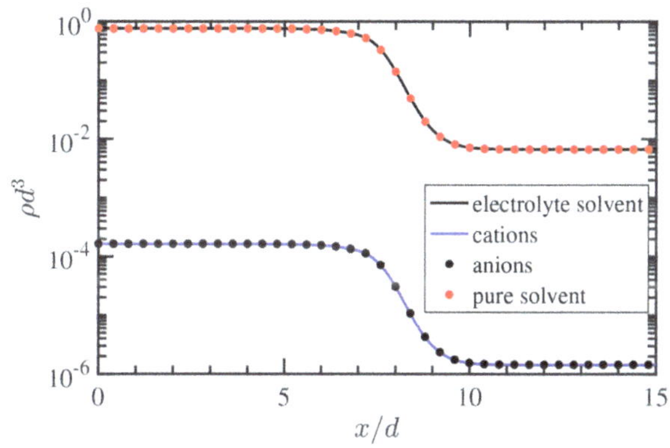

Figure 7.14. A comparison of liquid–vapor interfaces for a pure solvent and an electrolyte where the liquid electrolyte has a concentration of $0.1M$. Note that cation and anion profiles are identical, and the solvent interface between the pure system and the electrolyte are also nearly indistinguishable.

Of course to study an electrolyte, it is necessary to introduce the ions to the system. The various parameters are set to be $\epsilon_r = 78.5$, $T = 298$ K, all $d_{ij} = d = 0.3$ nm, and $A_{ij}/kT = 1$ for all interactions except the solvent–solvent interactions (which remains at $A_{ss}/kT = 1.25$). The liquid phase ion concentration at coexistence is set to be 0.1M (or $\rho_+^{liq} d^3 = \rho_-^{liq} d^3 = 0.001\,65$). The vapor phase ion densities are then determined by the requirement of coexistence ($\mu_i^{vap} = \mu_i^{liq}$), and are found to be $\rho_+^{vap} d^3 = \rho_-^{vap} d^3 = 1.445 \times 10^{-6}$. Figure 7.14 shows how the fluid density varies across a liquid–vapor interface for both the electrolyte and the pure solvent described above. In this simple model, the solvent profile of the electrolyte is nearly identical to the pure fluid. Furthermore, the cation and anion profiles are identical at all locations through the interface so there is no net local charge. However, this is not generally the case, and a nonzero charge distribution is more often expected [18].

Adsorption at a charged surface

When a surface is exposed to a vapor phase, molecules from the vapor can adsorb on the surface forming a film. Those films can range from a sparse monolayer to a macroscopically thick condensed film in the limit of saturation (the pressure, $p \to p_{sat}$) where the free energy of the liquid and vapor phases are identical, and complete wetting occurs [83]. A rich phase space involving the addition of liquid-like layers may also exist below the saturation pressure depending on the strength of fluid-wall interactions. These transitions are called layering transitions [45].

Consider a surface that interacts with the same vapor phase electrolyte described above in the situation where $p < p_{sat}$. Assuming the surface–fluid interactions are also LJ type, where the 12-6 LJ potential is integrated over the surface to yield an external field of the form

$$V_i^{\text{ext}}(x) = \frac{2\pi}{3}\epsilon_{wf}^{solv}\left[\frac{2}{15}\left(\frac{d}{x}\right)^9 - \left(\frac{d}{x}\right)^3\right]. \tag{7.82}$$

As the energy parameter ϵ_{wf} increases, the surface becomes more attractive to the fluid, and the adsorbed fluid layer(s) become more dense. Multiple fluid layers are possible although the extent of the adsorbed layer is limited below the saturation point because the vapor phase is of lower free energy than the bulk liquid. Once the film is thick enough that it has bulk-like properties, it will be unstable relative to the vapor state.

The situation is a bit more complicated if the surface carries a charge and is immersed in an electrolyte vapor. In that case counterions will be required in the adsorbed layers (or in the vapor phase double layer) to neutralize that charge. In many real situations, surfaces in contact with humid environments have surface groups that dissociate to create both the surface charge and at least some of the needed counterions near the surface. These surface mediated chemical processes are discussed in section 7.6.1. Here counterions required for charge neutrality must be adsorbed from the vapor phase.

The results presented here used a cut and shifted version of equation (7.82) where the solvent–surface interactions have $x_c = 3d$, while ion–wall interactions have $x_c = d$. Furthermore, $\epsilon_{wf,s}$ varies for solvent–wall interactions, but is set to $\epsilon_{wf,+}/kT = \epsilon_{wf,-}/kT = 1$ for ion–wall interactions. Hence, the solvent is attracted to the surfaces based on LJ interactions while counterions adsorb as a result of surface charge. Coions can be expected to be displaced from the surface relative to the bulk as they generally are in double layers. The bulk solvent density is set to $\rho_{\text{solv}}^{b} d^3 = 0.005\,99$ which corresponds to $p/p_{\text{sat}} \approx 0.9$.

Figure 7.15 shows how the surface free energy, Ω_s, varies with changes in the wall–solvent interactions for several values of surface charge. In all cases shown, two discontinuous phase transitions are observed at the relatively small surface charges. Note that in this model, the Debye length is very long with $\lambda_D/d = 107.5$. The

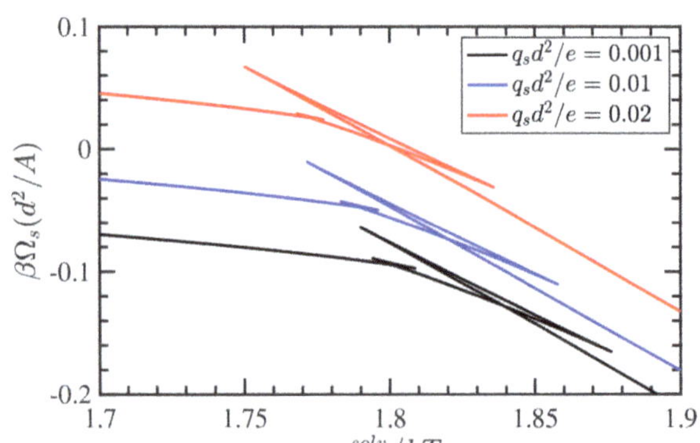

Figure 7.15. Surface free energy, Ω_s, as a function of the solvent wall interaction parameter, ϵ_{wf}^{solv}, for three different surface charges. Discontinuous transitions are observed at slope discontinuities along the low free energy path.

calculations presented here were performed in a computational domain of $L_x/d = 1000$ in order to allow for sufficient electrostatic decay away from the surfaces. In more sophisticated models (see section 7.7), the electrostatic permittivity should be small in the vapor phase, reducing λ_D by nearly two orders of magnitude.

The order parameter used to describe surface phases and surface mediated phase transitions is the excess adsorption,

$$\Gamma_i^{ex} = \int \left[\rho_i(\mathbf{r}) - \rho_i^b\right] d\mathbf{r}. \tag{7.83}$$

As in the bulk case, the order parameters will exhibit a discontinuous change at any first order phase transition. Figure 7.16 shows the variation of Γ_{solv}^{ex} with solvent–wall interaction parameter ϵ_{wf}^{solv} and surface charges ranging from $q_s d^2/e = 0.001$ to 0.3 Phase transitions associated with a discontinuous change in the order parameters are indicated with vertical lines for several of the cases. Unstable branches were again traced out using an arc-length continuation algorithm.

Examining the density profiles of various cases elucidates the structural changes in the adsorbed layers that accompany the observed discontinuous transitions in Ω_s and Γ^{ex}. Density profiles for a low charge case, $q_s d^2/e = 0.02$, and a high charge case, $q_s d^2/e = 0.3$, are shown in figure 7.17. The red curves in all plots indicate solutions in the vicinity of $\epsilon_{\rm wf}/kT \approx 1.78$ where at low charge there are two distinct phase transitions. At higher charge, the first transition disappears (the state point is supercritical with respect to that transition), and the second shifts to lower $\epsilon_{\rm wf}$ indicating that higher surface charge facilitates easier formation of the adsorbed liquid layers. The very long (and unphysical!) electrostatic decays are quite clear in the ion profiles in figure 7.17. However, note that in all cases where liquid-like layers have formed, ion densities in the adsorbed layers are 10–1000 times greater than the ion densities just outside the adsorbed layer. Thus, even with the very long electrostatic decay, there is strong counterion adsorption close to the surface whenever the adsorbed layer is liquid-like. A proper treatment of the vapor phase permittivity can be expected to enhance this effect since charge neutrality requirements must be almost entirely satisfied within the adsorbed layers in that case.

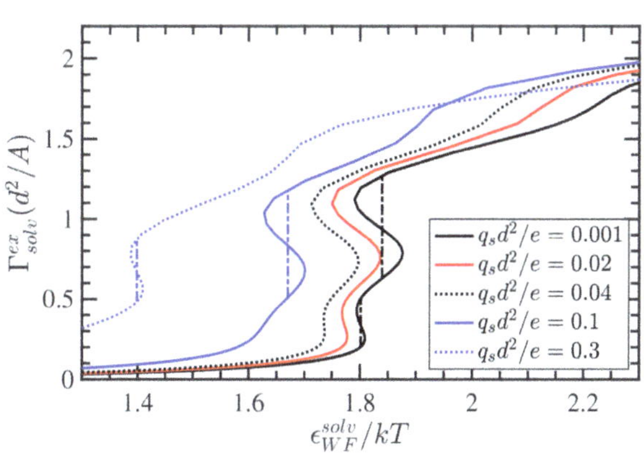

Figure 7.16. Excess adsorption of solvent, Γ_{solv}^{ex} as a function of ϵ_{wf}. Discontinuous changes in order the parameter are indicated by the dashed-dotted vertical lines for three cases.

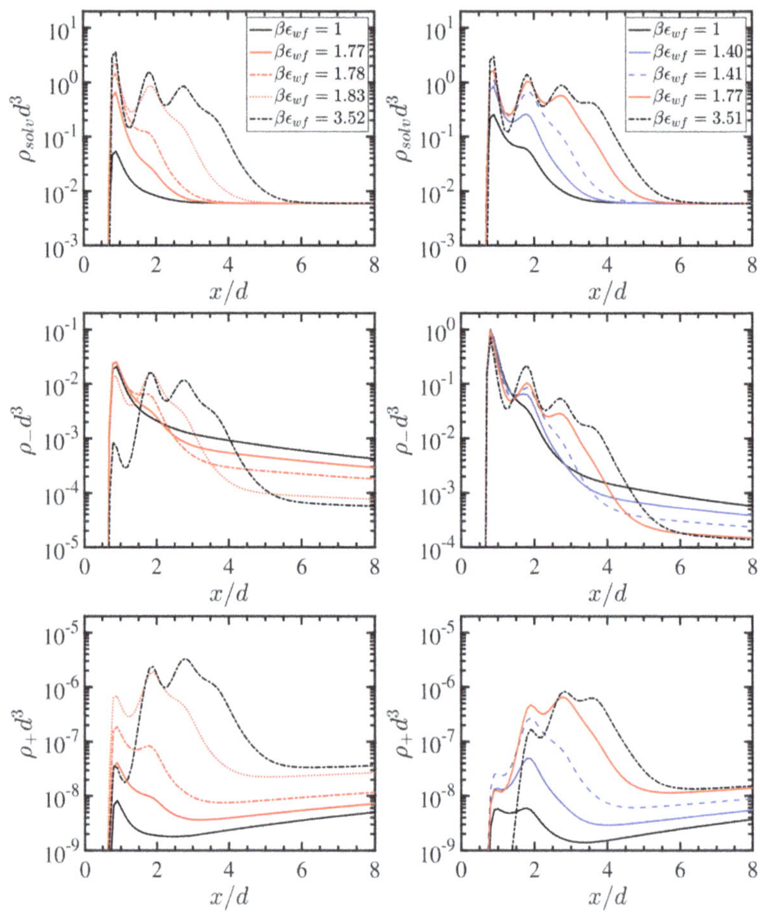

Figure 7.17. Density profiles for solvent (top), counterions (middle), and coions(bottom) at a low surface charge (left: $q_s d^2/e = 0.02$) and a high surface charge (right: $q_s d^2/e = 0.3$). Red curves are all near $\epsilon_{wf} \approx 1.8$. The red curves on the left show representative profiles for the adsorbed layer(s) on either side of two phase transitions. Blue curves (right) show profiles on either side of the one remaining discontinuous transition at higher charge at $\epsilon_{wf} \approx 1.4$. Black curves compare the cases $\epsilon_{wf} = 1$ (solid) and $\epsilon_{wf} \approx 3.5$ (dash–dot).

Studies of this kind can be used to develop phase diagrams that provide a full picture of the possible stable fluid phases as a function of the operational state parameters. For the simple case above, the relevant set of state parameters is $\{\epsilon_{wf}^{solv}, q_s, \{\mu_i\}, \text{and } T\}$. By fixing $\{\mu_i\}$ and T, one slice of the this state space is being explored. The phase diagram for this slice is shown in figure 7.18. It shows where the layering transitions occur and the regions where different interfacial fluid morphologies exist. Developing a complete understanding of any system requires a more complete investigation of the state parameters, and the phase diagrams are often more complex than figure 7.18. Algorithms that can track transition points in state space (see section 16.2.2) help to improve the efficiency of the process.

There are many situations where the state space will be even more complex than the simple case considered here. For example consider a surface that is created with a

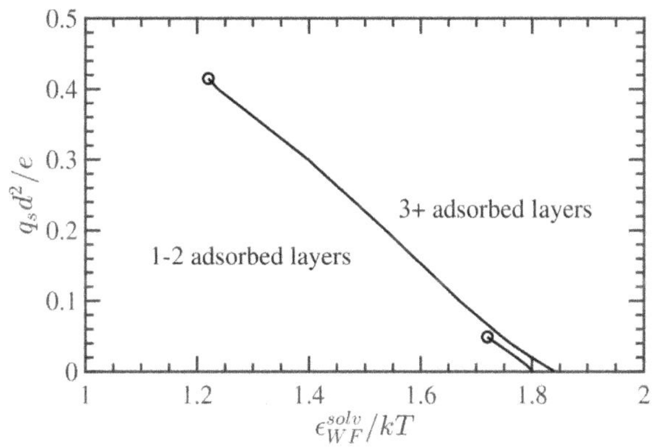

Figure 7.18. Phase diagram for solvation layers of electrolyte at a charged wall with cohesive interactions between the wall and vapor phase. The solid lines show discontinuous (first order) transitions that terminate in critical points (continuous or second order phase transitions). At very low surface charge, there is a discontinuous transition from 1 to 2 adsorbed layers. At higher surface charge, the only discontinuous transition occurs when the surface layer exceeds 2 adsorbed layers. At high enough charge, no discontinuous transitions are observed.

pattern on a nanoscopic scale with regions that exhibit different fluid–surface interactions. In that case, state parameters are also needed to capture the length scales of the heterogeneity and the chemical differences among the surface regions [84–86].

Cohesive forces are important because they make it possible to tune the strength of different types of interactions (surface, solvent, ions). Including these intricacies in the molecular model allows access to a wide variety of surface induced phase transitions that occur in fluids adsorbed at surfaces. Part II of this book provides further illustrations from detailed investigations of double layers that include not only cohesive interactions, but also chemically active surfaces. Including those surface reactions in c-DFT approaches is discussed further in the next section.

7.6 Classical-DFT for systems with active surfaces

The surface side of the interfacial system has, until this point, been quite passive (simple geometry, constant surface charge, uniform permittivity matching the fluid). Surfaces are usually not so straightforward. Surface chemistry determines surface charge in many important (charge-regulating) cases, a dielectric mismatch at the fluid–surface boundary should be expected, and there is a dynamic interplay between double layer structure, (induced) polarization effects, and surface charge [87]. This section discusses how surface chemistry and electronic effects can be included in c-DFT calculations.

7.6.1 Surface chemistry

Charge regulation is important in many systems including colloids [88, 89], oxide layers [90], and proteins [91]. In these kinds of systems, the constant charge (CC) and constant potential (CP) boundary conditions are not the most natural. Rather a charge regulating (CR) boundary condition is required. An understanding of surfaces with CR boundaries began to develop in the 1970s with the work of Ninham and Parsegian [92] and Chan and coworkers [93]. Reiner and Radke later developed a variational approach for charge surfaces [35]. More recently, a detailed understanding of short-range forces in charge regulating systems has been developed [42] and CR boundaries have been combined with detailed c-DFT calculations to investigate the interplay of finite size effects and cohesive interactions near CR boundaries [94–96].

Introducing surface chemistry results in an apriori unknown external field for the c-DFT calculation. A self-consistent field calculation is required where the surface charge (which is controlled by the chemistry) must be found along with the electrolyte density profiles. This can be achieved with a careful treatment of the boundary conditions. The first step is to define the surface chemistry of interest. In general, surfaces may composed of a collection of site types, $\{A_j\}$, with valence, z_j, that may be involved in reversible chemical reactions with free species (defined as the set $\{F_i\}$) found in an adjoining fluid phase. That fluid phase may be a liquid, a vapor, or a condensed layer of microscopic thickness depending on the application of interest. The distribution of reaction sites may be nonuniform (with a surface density, $\rho^s_{A_j}(\mathbf{r}_s)$) although much of the literature has focused on the homogenous case $\rho^s_{A_j}(\mathbf{r}_s) = \rho^s_{A_j}$. Reaction sets may be single site association/dissociation reactions, multiple site competitive reactions, or coupled sets of surface and bulk reactions to name a few options.

For the purposes of illustration, this section describes a simple pair of simultaneous surface reactions in detail. The two surface reactions do not compete for the available free species, but they do compete in controlling the surface charge. More specifically, consider a surface that has two surface sites, A_1 and A_2. Each of the two sites can be charged after association with free ions, F_i. If the two reactions are independent, they can be written

$$\begin{aligned} A_1 + F_1^+ &\rightleftharpoons A_3^+ \\ A_2 + F_2^- &\rightleftharpoons A_4^- . \end{aligned} \tag{7.84}$$

Each reaction, will have a unique equilibrium constant related to the equilibrium concentrations (or densities) of the various species. For the reactions in equation (7.84), the pair of equilibrium constants is

$$\begin{aligned} K_+ &= \frac{[A_1][F_1]_s}{[A_3]} = \frac{\rho^s_{A_1}\rho_{F_1}(\mathbf{r}_s)}{\rho^s_{A_3}}, \text{ and} \\ K_- &= \frac{[A_2][F_1]_s}{[A_4]} = \frac{\rho^s_{A_2}\rho_{F_2}(\mathbf{r}_s)}{\rho^s_{A_4}}. \end{aligned} \tag{7.85}$$

In the literature, nomenclature using molar concentrations, $[A]$, is common. For consistency with this text, we use number densities, ρ, where ρ_i^b indicates a bulk free species, ρ_i^s indicates a surface bound group, $\rho_t^s = \sum_i \rho_i^s$ is the total density of surface sites, and $\rho_i(\mathbf{r}_s)$ indicates the density of a fluid phase species at a surface.

Following the development of Chan [93] for the independent equations described above, the surface charge will be $q_s = q_s^+ - q_s^-$ where the two contributions to the surface force are

$$q_s^+(\mathbf{r}_s) = e\theta\rho_t^s\left(\frac{\rho_{A_3^+}}{\rho_{A_3^+} + \rho_{A_1}}\right) = e\theta\rho_t^s\left(\frac{\rho_{F_1^+}(\mathbf{r}_s)/K_+}{1 + \rho_{F_1^+}(\mathbf{r}_s)/K_+}\right)$$
$$q_s^-(\mathbf{r}_s) = e(1-\theta)\rho_t^s\left(\frac{\rho_{A_4^-}}{\rho_{A_4^-} + \rho_{A_2}}\right) = e(1-\theta)\rho_t^s\left(\frac{\rho_{F_2^-}(\mathbf{r}_s)/K_-}{1 + \rho_{F_2^-}(\mathbf{r}_s)/K_-}\right), \tag{7.86}$$

where θ in this context is the fraction of surface sites that can participate in the cation reaction while the remainder are available to the anion reaction. Note that equation (7.85) was used to recast the surface densities to explicitly include equilibrium constants and the free ionic species. The surface ion densities, $\rho_{F_i}(\mathbf{r}_s)$ are related to their bulk densities, $\rho_{F_i}^b$ in a point charge electrolyte by

$$\rho_{F_i}(\mathbf{r}_s) = \rho_{F_i}^b \exp(-z_i\psi^*(\mathbf{r}_s))) \tag{7.87}$$

where $\psi^* = \psi e/kT$. Defining parameters $\alpha_+ = \ln K_+$ and $\alpha_- = \ln K_-$, the net surface charge can be written

$$q_s(\mathbf{r}_s)/e = \rho_t^s\left(\frac{\theta\rho_{F_1^+}^b}{\rho_{F_1^+}^b + e^{\alpha_+ + \psi^*(\mathbf{r}_s)}} - \frac{(1-\theta)\rho_{F_2^-}^b}{\rho_{F_2^-}^b + e^{\alpha_- - \psi^*(\mathbf{r}_s)}}\right). \tag{7.88}$$

An alternate development discussed in detail by Reiner and Radke [35] begins with the statement of chemical equilibrium which requires that

$$\mu_{A_1} + \mu_{F_1^+} = \mu_{A_3^+} \tag{7.89}$$

$$\mu_{A_2} + \mu_{F_2^-} = \mu_{A_4^-}. \tag{7.90}$$

It is straightforward to show that their approach also leads directly to equation (7.88) for the surface charge.

Table 7.1 presents a few different reaction types from the literature and the resulting surface charge descriptions. The presentation in the table is consistent with the text; however it is also natural to introduce hyperbolic trigonometric functions so the notation in the literature varies.

Charge regulation in point charge electrolytes

Charge regulating (CR) surfaces add significant complexity to the state space for double layers. In the case of two simple independent reactions detailed above, the state parameters include $\{\rho_i^b\}$, T, θ, α_+, and α_-. Figure 7.19 shows how the surface charge and surface potential vary at an isolated planar surface as the fraction of sites available to the cation reaction, θ, and the surface chemistry (set by the ratio α_-/α_+) vary.

When $\alpha_+ = -6$ (the case in the figure), the charge saturates at $\alpha_-/\alpha_+ < 0$ where the free anions bind only weakly to the surface. In this case, the (positive) surface charge

Table 7.1. Examples of surface reactions, and the resulting boundary condition. To clarify the notation, the electrostatic potential at the surface is written $\psi^s = e\psi(\mathbf{r}_s)/kT$, and the bulk hydrogen density is written $\rho_H^b = \rho_{H^+}^b$. In the chemical equations, surface species are indicated with a superscript *s*; otherwise, chemical group notation is taken from the literature. Finally, note that the last case neglects Stern layer considerations discussed in [90].

Reaction(s)	$(q_s d^2/e)/\rho_t^s$	Description
$AB^s \rightleftharpoons A^{+,s} + B^-$	$1/\left(1 + \rho_{B^-}^b e^{(\psi^s - \alpha)}\right)$	Dissociation of ionizable group $\alpha = \ln K$ where K is the dissociation constant [42].
$A^s + B^+ \rightleftharpoons AB^{+,s}$	$\rho_{B^+}^b / \left(\rho_{B^+}^b + e^{(\psi^s + \alpha)}\right)$	Charging of a neutral surface group with $\alpha = \ln K$ [42].
$A_1^s + B_1^+ \rightleftharpoons A_1B_1^{+,s}$ $A_2^s + B_2^- \rightleftharpoons A_2B_2^{-,s}$	$\theta\rho_{B_1^+}^b / \left(\rho_{B_1^+}^b + e^{(\psi^s + \alpha_+)}\right)$ $+ (1 - \theta)\rho_{B_2^-}^b / \left(\rho_{B_2^-}^b + e^{(\psi^s + \alpha_-)}\right)$	Charging two independent neutral surface groups with $\alpha_i = \ln K_i$ [42].
$AH_2^{+,s} \rightleftharpoons AH^s + H^+$ $AH^s \rightleftharpoons A^{-,s} + H^+$	$\rho_H^b(t_1 - t_2) / \left(1 + \rho_H^b(t_1 + t_2)\right)$ with terms: $t_1 = \exp(-\psi^s - \alpha_+)$, and $t_2 = \exp(-\psi^s + \alpha_-)$	Competing +/− ionization of a neutral surface by dissociation of H^+ with $\alpha_{+/-} = \ln K_{+/-}$ [93].
$AH_2^{+,s} \rightleftharpoons AH^s + H^+$ $AH^s \rightleftharpoons A^{-,s} + H^+$ $AH_2^{+,s}X^- \rightleftharpoons AH_2^{+,s} + X^-$ $A^{-,s}M^+ \rightleftharpoons A^{-,s} + M^+$	$(t_1t_2 - t_3t_4)/(1 + t_1t_2 + t_3t_4)$ with: $t_1 = \rho_H^b e^{(-\psi^s - \alpha_+)}$ $t_2 = 1 + \rho_{X^-}^b e^{(\psi^s - \alpha_X)}$ $t_3 = e^{(\psi^s + \alpha_-)}/\rho_H^b$ $t_4 = 1 + \rho_{M^+}^b e^{(-\psi^s - \alpha_M)}$	Complexation of surface groups with M^+X^- electrolyte along with ionization by dissociation of H^+ with $\alpha_{+/-} = \ln K_{+/-}$ and $\alpha_{M/X} = \ln K_{M/X}$ [90].

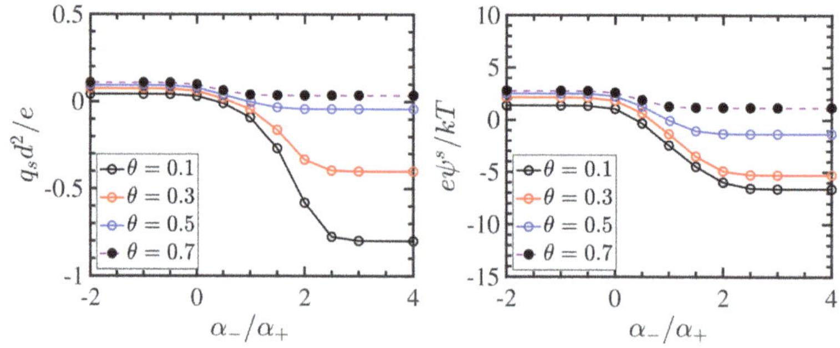

Figure 7.19. Surface charge, q_s (left), and surface potential, ψ^s (right), at a planar surface with CR boundaries. The bulk ion concentrations are $[B_1^+] = [B_2^-] = 0.01M$ (or $\rho_{B_i}d^3 = 0.007\,527\,5$ with $d = 0.5$ nm. $\alpha_+ = -6$ ($pK_+ \approx 2.48$). The chemistry of the anion reaction (α_-) and the fraction of surface sites participating in the cation reaction (θ) vary.

arises from the binding of free cations. The surface charge also saturates at $\alpha_{-}/\alpha_{+} \geqslant 3$ where the anions bind strongly to the surface filling up all available sites. For most of the cases shown, the saturation of anions leads to a negatively charged surface. The one exception is where there are too few of the anion binding sites available ($\theta = 0.7$). In that case, the surface charge is positive for all values of α_{-}/α_{+}. The variations in the surface potential mirror the observations for surface charge, and both surface properties are presented for completeness. The sensitivity of CR surfaces to chemical equilibria shows that systems with ionizable groups (like proteins!) operating near the isoelectric point (where $q_s = 0$) can be expected to be highly sensitive to changes in state point.

Interacting CR surfaces are particularly interesting. Figure 7.20 shows how the surface charge and solvation force vary with the distance between two planar surfaces. All types of boundary conditions (CR, CC, and CP) are included for comparison. The state points shown correspond to two cases from figure 7.19. In the first case ($\theta = 0.7$ and $\alpha_{-}/\alpha_{+} = 4$), the anions bind strongly, but there are few sites available. In the second ($\theta = 0.1$ and $\alpha_{-}/\alpha_{+} = 1.5$), the anions bind less strongly, but there are more sites available to them. The q_s (CC) and ψ^s (CP) boundary values were set to the isolated wall CR results in figure 7.19. Both cases demonstrate that as two CR surfaces approach each other, the surface charge decreases with $q_s \to 0$ as $H \to 0$. In this way CR surfaces are like CP surfaces. However, the forces tell a

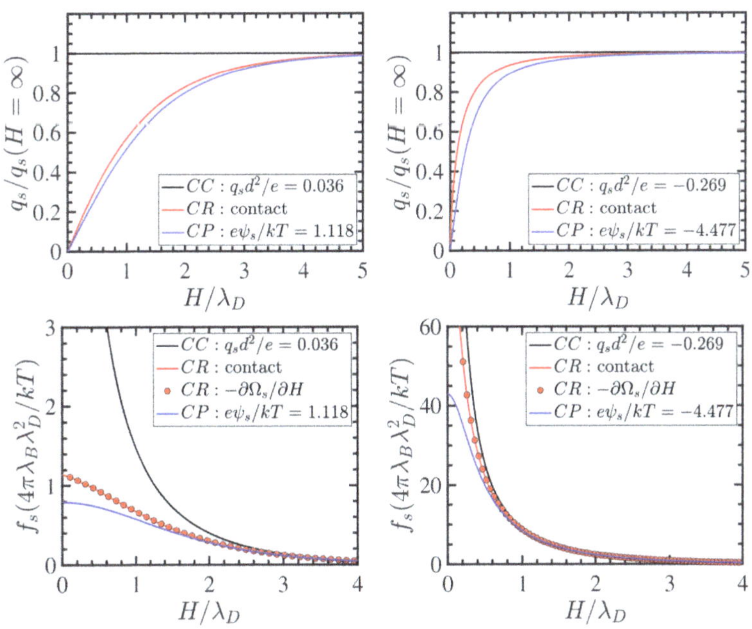

Figure 7.20. Surface charge, q_s and solvation forces, f_s, between two planar surfaces separated by H. CC, CP, and CR boundaries are compared for two systems. CR conditions were $\alpha_{+} = -6$ ($pK_{+} \approx 2.48$), $\alpha_{-}/\alpha_{+} = 4$ and $\theta = 0.7$ (left) and $\alpha_{+} = -6$, $\alpha_{-}/\alpha_{+} = 1.5$ and $\theta = 0.1$ (right). The CC and CP boundary values are given in the figures. The solvation force is scaled by $4\pi\lambda_B\lambda_D^2$ to facilitate comparison with figure 3 in [42]. The bulk ion concentrations are $[B_1^+] = [B_2^-] = 0.01$ M (or $\rho_{B_i}d^3 = 0.007\,527\,5$ with $d = 0.5$ nm).

different story and depending on chemistry, they can saturate at small H (like the CP case) or they can increase exponentially (like the CC case).

Free energies in charge regulating systems

Note that figure 7.20 includes both the contact route and the thermodynamic route to the force for the CR case. Calculating the surface free energy for the latter requires the treatment of CR boundaries developed by Reiner and Radke [34, 35]. They showed that there is a contribution to the grand potential due to surfaces that can be written generally as

$$\int_S Q_s^{RR}(\psi^s, \mathbf{r}_s)d\mathbf{r}_s \tag{7.91}$$

where

$$Q_s^{RR}(\psi_s, \mathbf{r}_s) = \int_0^{\psi^s} q_s(\psi^s, \mathbf{r}_s)d\psi. \tag{7.92}$$

This term accounts for the energy associated with charging up the interface. When CC or CP boundaries are considered at uniform planar surfaces, this term is straightforward with

$$Q_s^{RR}(\mathbf{r}_s) = \begin{cases} q_s\psi^s & \text{CC boundary} \\ 0 & \text{CP boundary} \end{cases}. \tag{7.93}$$

However, when the surface is charge regulating. The specific chemistry defines the surface term. For the case considered above with two independent reaction sites, equation (7.88) defines $q(\mathbf{r}_s)$. The surface term, Q_s^{RR} requires integrating over ψ as in equation (7.92). This definite integral can be performed analytically with the result

$$\begin{aligned} Q_s^{RR}(\mathbf{r}_s) = \theta\left(\psi^s - \ln\left[\frac{\rho_{F_1^+}^b + e^{\alpha_1+\psi^s}}{\rho_{F_1^+}^b + e^{\alpha_1}}\right]\right) \\ - (1-\theta)\left(\psi^s - \ln\left[\frac{\rho_{F_2^-}^b + e^{\alpha_2-\psi^s}}{\rho_{F_2^-}^b + e^{\alpha_2}}\right]\right). \end{aligned} \tag{7.94}$$

A change in variables

The charge regulating boundary conditions described above are written with surface charge as a function of the electrostatic potential,

$$q_s(\psi^s, \mathbf{r}_s) = q_s(\psi(\mathbf{r})\delta(|\mathbf{r} - \mathbf{r}_s|)). \tag{7.95}$$

However, it can be anticipated that a density formulation,

$$q_s(\{\rho_i^s\}, \mathbf{r}_s) = q_s(\{\rho_i(\mathbf{r})\}\delta(|\mathbf{r} - \mathbf{r}_s|)), \tag{7.96}$$

might be preferred when using c-DFT for more detailed fluid models.

This change of variables is straightforward for point charge systems. Considering the same 2-site reaction model, the surfaces charge can be written as

$$q_s(\{\rho_i^s\}, \mathbf{r}_s)/e = \rho_t^s\left(\frac{\theta\rho_{F_1^+}(\mathbf{r}_s)}{K_+ + \rho_{F_1^+}(\mathbf{r}_s)} - \frac{(1-\theta)\rho_{F_2^-}(\mathbf{r}_s)}{K_- + \rho_{F_2^-}(\mathbf{r}_s)}\right). \tag{7.97}$$

The free energy surface term, Q_s^{RR} requires modifying equation (7.92)

$$Q_s^{RR}(\{\rho_i^s\}, \mathbf{r}_s) = \sum_i \int_{\{\rho_i^b\}}^{\{\rho_i^s\}} q_s(\{\rho_i^s(\mathbf{r})\}, \mathbf{r}_s)\left(-\frac{1}{z_i}\frac{1}{\rho_i}\right)d\rho_i \tag{7.98}$$

where the change in integration parameters consistent with the Boltzmann distribution requires

$$d\psi = -\sum_i \frac{1}{z_i}\frac{1}{\rho_i}d\rho_i. \tag{7.99}$$

After integration, the free energy contribution to the 2-site reaction model becomes

$$\begin{aligned} Q_s^{RR}(\mathbf{r}_s)/e &= \int_{\rho_{F_1^+}^b}^{\rho_{F_1^+}^s} \frac{q_s^+(\rho_{F_1^+}, \mathbf{r}_s)}{-\rho_{F_1^+}} d\rho_{F_1^+} - \int_{\rho_{F_2^-}^b}^{\rho_{F_2^-}^s} \frac{q_s^-(\rho_{F_2^-}, \mathbf{r}_s)}{\rho_{F_2^-}} d\rho_{F_2^-} \\ &= -\theta\left\{\ln\left[\frac{\rho_{F_1^+}^s(\mathbf{r}_s) + K_+}{\rho_{F_1^+}^b + K_+}\right]\right\} - (1-\theta)\left\{\ln\left[\frac{\rho_{F_2^-}^s(\mathbf{r}_s) + K_-}{\rho_{F_2^-}^b + K_-}\right]\right\}. \end{aligned} \tag{7.100}$$

Charge regulation with finite-size electrolytes

In a c-DFT based approach with more sophisticated model fluids and functionals, the treatment of CR boundaries becomes more challenging. The density formulation of the surface charge (as in equation (7.97)) is the most practical approach when $\mathcal{F}_{ex} \neq 0$. In addition, finite-size fluid models shift the plane of fluid centers away from the plane of surface charge. If the ions have size (but no cohesive forces), the first peak(s) in the density profile will occur a distance $d_i/2$ away from the surface, and the source term for the CR boundaries is

$$q_s(\mathbf{r}_s) = q_s\big(\{\rho_i(\mathbf{r})\}\delta(|\mathbf{r} - (\mathbf{r}_s + \{(d_i/2)\}\mathbf{n})|)\big), \tag{7.101}$$

where $\mathbf{n}$ is the unit normal to the surface. Figure 7.21 compares density profiles of cations for three models near an isolated charged surface including the point charge model, a restricted primitive model (RPM), and a solvent primitive model (SPM). The reaction set used to define the surface is the same 2-site model as in equation (7.97). The parameters are set so that K_- is small and the anion reaction essentially contributes a constant $(1 - \theta)$ to the surface charge. While the point charge and RPM models have similar surface densities (and hence similar surface charges), the SPM model has an enhanced cation peak due to solvent structure.

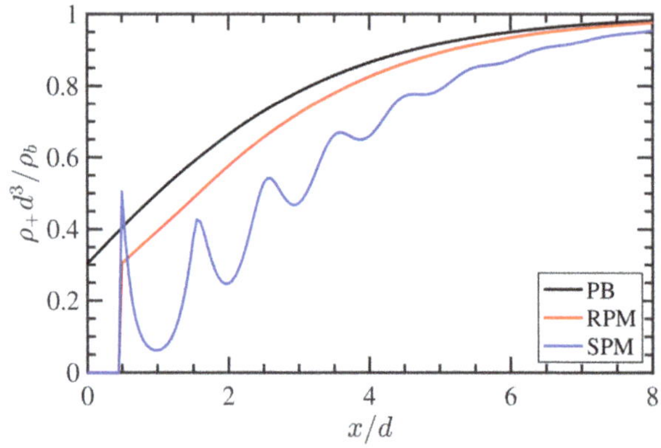

Figure 7.21. Cation density profiles for three model electrolytes at a 2-site CR surface described by equation (7.97) with $\alpha_+ = -6$, $\alpha_- = -24$, and $\theta = 0.7$. The surface densities are $\rho_b(\mathbf{r}_s)/\rho_+^b \approx 0.304\,8$, $0.305\,4$, and 0.505 for PB, RPM, and SPM respectively. The resulting surface charges are $q_s d^2/e \approx 0.036\,5$, $0.036\,8$, and 0.124 for PB, RPM, and SPM respectively. In all cases, $\rho_+^b d^3 = 0.007\,53$ ($0.01\,M$ with $d = 0.5$ nm). The total charge site density is $\rho_t^s = 1$, and in the SPM model, the bulk solvent density is $\rho_{solv}^b d^3 = 0.75$.

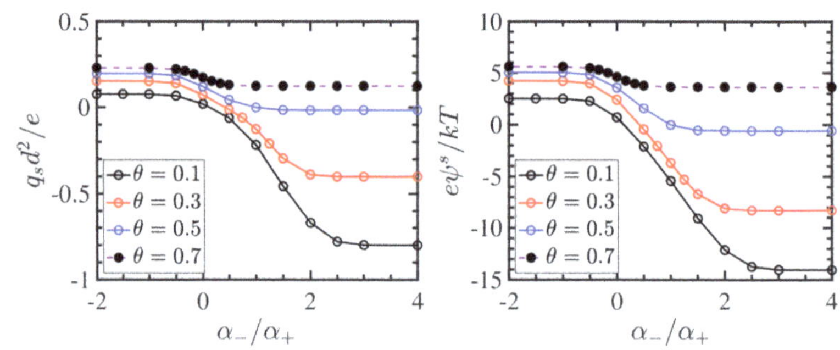

Figure 7.22. Surface charge, q_s (left), and surface potential, ψ^s (right), at a planar surface with CR)boundaries. The bulk ion concentrations are $[B_1^+] = [B_2^-] = 0.01\,M$, or $\rho_{B_i} d^3 = 0.007\,527\,5$ with $d = 0.5$ nm. Also note that $\alpha_+ = -6$ ($pK_+ \approx 2.48$). The chemistry of the anion reaction (α_-), and the fraction of surface sites participating in the cation reaction (θ) vary.

To further highlight the effects of including the dense solvent in the system, figure 7.22 shows how the surface charge and surface potential vary with surface chemistry in the SPM model. These results can be compared to the point-charge model results shown in figure 7.19. The same general behavior is observed because the reaction limits of strong and weak anion bonding at the extremes are unchanged. However, the range of observed surface charges (and surface potential) is larger with solvent in the system. In addition, the transition region between the limits is a bit broader.

Figure 7.23 shows the behavior of two interacting CR surfaces. It includes both the charge and solvation force as a function of surface separation for various models. These results can be compared to the CR results in figure 7.20. The state points are identical, but the force is now strongly oscillatory (and of a larger

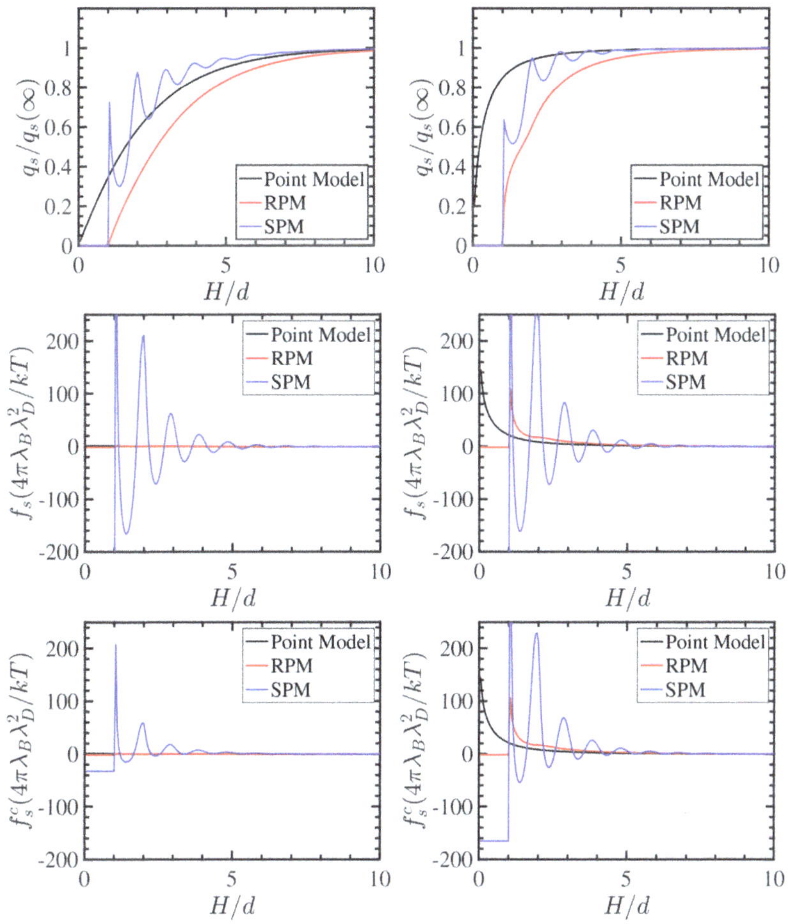

Figure 7.23. Surface charge, q_s, solvation forces, f_s, and the ion contribution to the solvation forces, f_s^c, between planar surfaces separated by H. CR conditions for figures on the left were $\alpha_+ = -6$ ($pK_+ \approx 1.48$), $\alpha_-/\alpha_+ = 4$ and $\theta = 0.7$. Figures on the right have $\alpha_+ = -6$, $\alpha_-/\alpha_+ = 1.5$ and $\theta = 0.1$. Forces are scaled by $4\pi\lambda_B\lambda_D^2$ for comparison with figure 7.20. Bulk ion concentrations were the same as figure 7.22. The solvent density for the SPM cases is $\rho_{solv}^b d^3 = 0.75$.

magnitude) due to the dense finite-size solvent. In order to assess the contribution of the ions alone, the solvation force (equations (7.22) and (7.24)) is easily decomposed into contributions from the neutral solvent species and charged ions. For a system with hard walls, the contact theorem can be applied separately as

$$\begin{aligned} f_s &= f_s^n + f_s^c \\ f_s^n &= kT \sum_{i,\text{neutral}} \rho_i(x_s) \\ f_s^c &= kT \sum_{j,\text{ions}} \rho_j(x_s) - \frac{\epsilon}{8\pi}\left[-\frac{d\psi(x)}{dx}\Big|_{x_s}\right]^2. \end{aligned} \tag{7.102}$$

The two lower panels in figure 7.23 show that the force due to ions alone also exhibit strong oscillatory features, and is significantly larger in magnitude than the force due to ions in the primitive or point-charge models.

Considering free energies, it is not possible to avoid the complexity introduced by the molecular models and functionals when performing a change of variables as in equations (7.98). In that case, the change of variables was relatively straightforward with application of the Boltzmann distribution in (7.99). More detailed functionals have $\mathcal{F}_{ex} \neq 0$. The change of variables needed to find the surface contribution to the free energy in density units (as in equations (7.98)–(7.100)) would require application of the entire Euler–Lagrange equation. Specifically,

$$\delta\psi = -\sum_i \frac{1}{z_i}\left(\frac{1}{\rho_i} + \frac{\delta}{\delta\rho_i}\frac{\delta\mathcal{F}_{ex}}{\delta\rho_i}\right)\delta\rho_i \tag{7.103}$$

reflecting all of the possible contributions (finite size, correlations, polarization, cohesive forces, etc) to the surface charge.

Part II of this book will present extensive results for charge regulating systems that also include cohesive interactions. In that case, there are no density discontinuities, and the surface charge density may be better approximated by taking a nonlocal approach,

$$q_s(\mathbf{r}_s) = q_s(\{\bar{\rho}_i^s, \mathbf{r}_s\}). \tag{7.104}$$

with

$$\bar{\rho}_i^s(\mathbf{r}_s) = \frac{1}{V_f}\int_{\text{fluid}} \rho_i(\mathbf{r})\theta(|\mathbf{r} - (\mathbf{r}_s + \Delta)|)d\mathbf{r}. \tag{7.105}$$

The integral is taken on the fluid side of the surface, and θ is now the usual step function. The range of the integral extends a small distance Δ away from the surface, and the integral is normalized by the volume over which it is taken.

7.6.2 Surface electrostatics

Surfaces can also be active from an electrostatic point of view. Real systems are certainly more nuanced than the simple geometries and constant permittivity approximation applied in much of this chapter. This section briefly introduces cases with surface embedded sources of charge, spatially varying permittivity, and induced polarization effects.

In the case of a spatially varying relative permittivity, Poisson's equation becomes

$$-\epsilon_o\nabla[\epsilon_r(\mathbf{r})\nabla\psi(\mathbf{r})] = q(\mathbf{r}). \tag{7.106}$$

The charge density, $q(\mathbf{r})$, now includes both mobile charges due to the ions and fixed charge sources. Those charge sources may be confined to surfaces as discussed previously or may exist at discrete locations within the surface(s).

For example in studies of solvated (macro)molecules, fixed charges may be set at specific known atomic locations within a surface region defined as the volume occupied by the macromolecule. This surface region will have a net relative permittivity that differs from the fluid. In these cases, a topologically complex 3D surface must be defined or determined [97–100], and electrostatic continuity must be enforced at the boundaries.

Charge source terms can also be sensitive to the state of the system. The charge regulating boundaries are one example, the semiconductor interface problem first introduced in section 3.3 is another. In the planar n-doped semiconductor/oxide interface discussed by Fleharty and coworkers [101], the dielectric properties are constant within each part of the domain with relative permittivities of ϵ_{SC}, ϵ_{ox}, and ϵ_{el} for the semiconductor, oxide, and electrolyte respectively. The charge sources also differ among the different regions of the domain with

$$q(\mathbf{r})/e = \begin{cases} N_d[1 - \chi F_{1/2}(\beta\mu - \tilde{\psi}(\mathbf{r}))] & \text{n-doped semiconductor} \\ 0 & \text{oxide} \\ \sum_i z_i\left(\rho_i^b exp(z_i\tilde{\psi}(\mathbf{r})\right) & \text{electrolyte} \end{cases} \tag{7.107}$$

where N_d is the electron donor concentration, μ is the Fermi level, and $F_{1/2}$ is the Fermi integral of order 1/2 (defined previously in section 3.3).

In order to solve the electrostatics problem, equation (7.106) is simplified by setting $\epsilon(\mathbf{r}) = \epsilon_{\text{region}}$ within each material region. Electrostatic continuity requires that the electrostatic potential is continuous throughout the domain and that interface charges match. Specifically, the boundary condition between the oxide and electrolyte is

$$q_s = \mathbf{n} \cdot [\epsilon_{ox}\nabla\psi_{ox}(\mathbf{r}_{s,oe}) - \epsilon_{el}\nabla\psi_{el}(\mathbf{r}_{s,oe})], \tag{7.108}$$

while the boundary condition between the oxide and semiconductor is

$$0 = \mathbf{n} \cdot [\epsilon_{sc}\nabla\psi_{sc}(\mathbf{r}_{s,so}) - \epsilon_{ox}\nabla\psi_{ox}(\mathbf{r}_{s,so})]. \tag{7.109}$$

Far from the interface, at the domain boundaries both in the semiconductor and in the electrolyte, the electrostatic potential will become spatially invariant so the boundary condition applied is

$$\nabla\psi(|\mathbf{r} - \mathbf{r}_s| \to \infty) = 0. \tag{7.110}$$

Of course, the electrostatic situation is even more complicated than described above because dielectric materials are responsive to the local electric field. Therefore, when charges are brought near a dielectric material, there is an induced polarization effect. If the dipole density in a dielectric material is, ρ_{dp}, then the dipole moment density defines the polarization vector,

$$\mathbf{P}(\mathbf{r}) = \rho_{dp} q_{dp} \vec{\boldsymbol{\delta}} \tag{7.111}$$

where q_{dp} and $\vec{\delta}$ are the local (molecular) charges and the vector between the induced dipoles in the dielectric material. Assuming a linear response to the local electric field, $\mathbf{E}_{loc}(\mathbf{r})$,

$$\mathbf{P}(\mathbf{r}) = \epsilon_0 \chi_e \mathbf{E}_{loc}(\mathbf{r}) = -\epsilon_0 \chi_e \nabla \psi, \tag{7.112}$$

where χ_e is the electric susceptibility of the medium. χ_e is directly related to the electric permittivity, $\epsilon = \epsilon_0(1 + \chi_e)$. The relative electric permittivity is then, $\epsilon_r = \epsilon/\epsilon_0 = (1 + \chi_e)$. A polarization charge can be defined as

$$q^{pol}(\mathbf{r}) = -\nabla \cdot \mathbf{P}(\mathbf{r}). \tag{7.113}$$

Poisson's equation including polarization is then

$$\nabla^2 \tilde{\psi}(\mathbf{r}) = -4\pi\lambda_B\left(q(\mathbf{r}) + Q_s(\mathbf{r}) + q^{pol}(\mathbf{r})\right). \tag{7.114}$$

A variational method for applying electrostatics with induced polarization effects to purely classical systems with complex boundaries was developed by Allen, Hansen, and Melchionna [102] and later applied to studies of the calcium ion channel [103, 104]. Smaller solvated molecules have long been treated using quantum mechanics for the inner region coupled to a classical treatment of the surrounding electrolyte [105–108]. These studies also include induced polarization of the molecule in the inner region of the domain. Recent work along these lines has considered solvation of ions in electrolytes [107] and solid surfaces with electrochemical double layers [108].

7.7 Classical-DFT for water

In all of the sections described above, the solvent (water) is treated as either as a dielectric continuum or at most as a collection of finite sized particles with volume exclusion and some short range mean field interactions. Of course all solvents are significantly more complex. Water in particular is a multi-site molecule with a permanent dipole that exhibits significant hydrogen bonding. The dielectric permittivity of water is a property that follows from its molecular dipolar nature, and can be computed from statistical mechanical models of the fluid [109]. This section outlines two approaches for improving the water model developed by Oleksy and Hansen [17, 18, 24, 43] and focused on the polar nature of water. Other c-DFT based approaches for water include associating fluid functionals that emphasize the hydrogen bonding nature of water [28–31] and multi-site molecular models [25–27]. Those topics are beyond the scope of the current text, but they are also important models to consider when a detailed model of water (or other solvent) is required.

7.7.1 A semi-primitive model for water

The simplest treatment of solvent polarization allows permittivity to vary spatially so that $\epsilon(\mathbf{r})$. This is the solvent semi-primitive model (SSPM) in figure 7.1. Oleksy and Hansen developed SSPMs to study electrolyte interfaces [18]. They considered two models where permittivity was a function of a nonlocal solvent density, $\epsilon(n_s(\mathbf{r}))$ where

$$n_s(\mathbf{r}) = \frac{6}{\pi d_s^3} \int \rho_s(\mathbf{r}')\theta(d_s/2 - |\mathbf{r} - \mathbf{r}'|)d\mathbf{r}'. \tag{7.115}$$

One model was based on a Clausius–Mossotti expression (usually suitable to small dipole moments, m) where the dielectric permittivity was

$$\epsilon(\mathbf{r}) = \frac{(8\pi/9kT)m^2 n_s(\mathbf{r}) + 1}{(4\pi/9kT)m^2 n_s(\mathbf{r}) - 1}. \tag{7.116}$$

Unfortunately, water has a dipole moment that is too high for this approach, and will lead to the 'dielectric catastrophe' at high density [109]. A second a phenomenological approach was also developed for the particular case of the liquid–vapor interface where

$$\epsilon(\mathbf{r}) = 1 + \frac{f(T)}{1 + exp[-a(n_s(\mathbf{r})d^3)] - \rho_s^{mid}(T)d^3)]}. \tag{7.117}$$

In equation (7.117), $\rho_s^{mid}(T)$ is the midpoint of the liquid and vapor densities at a given temperature T, and a is an adjustable parameter that controls the width (and hence the steepness) of the permittivity profile across the liquid–vapor interface. The function, $f(T)$ was a fit of the bulk coexistence of the hard sphere Yukawa model to water between $T = 0°C$ and $T = 50°C$ [18]. Introducing this spatially varying permittivity requires a modifications of Poisson's equation which is now given by equation (7.106).

While the phenomenological approach in equation (7.117) is useful for studying the details of the wetting and drying transitions for electrolytes at charged surfaces, the phenomenological nature of the model is not completely satisfying. To improve on this approach, the polar nature of water must be included in the functionals.

7.7.2 A 'civilized' model for water

To more properly account for polarization in an electrolyte, the solvent model can be constructed with an explicit permanent dipole. This *civilized* model includes a molecular treatment of solvent polarization that leads to screening of ionic charges in electrolytes. Unfortunately, there is no free lunch, and an explicit treatment of dipoles in the c-DFT functionals significantly increases the complexity of the problem.

While some earlier work on dipolar fluids can be found in the literature [19, 20, 22], this discussion follows the more recent presentation by Oleksy and Hansen [43] with a few notation changes for consistency with the current text. The brief introduction here is included to illustrate the complexity that accompanies introduction of the polarization vector. The reader is referred to the literature for further details and discussion of results.

Consider a system that includes both hard-core Yukawa and electrostatic interactions for a system with spherical ions and spherical solvent with an embedded point dipole. The dipole is a vector that will depend on position, $\mathbf{m}(\mathbf{r})$. The magnitude of the dipole is $m = |\mathbf{m}|$, and the unit vector in the direction of $\mathbf{m}$ is $\hat{\mathbf{m}}(\mathbf{r})$. While the orientation of $\mathbf{m}$ varies with position in the fluid, the magnitude of the dipole is assumed to be constant so no induced polarization effects (beyond orientation of the permanent dipoles) are included in this model. The hard core diameters of the particles are d_i. The molecular interactions for the system are

$$u_{ij}(\mathbf{r}_{12}, \mathbf{m}_i, \mathbf{m}_j) = \begin{cases} \infty, & r_{12} < (d_i + d_j)/2 = d_{ij} \\ u_{ij}^Y(\mathbf{r}_{12}) + u_{ij}^{ES}(\mathbf{r}_{12}, \mathbf{m}_i, \mathbf{m}_j) & r_{12} \geqslant d_{ij} \end{cases} \tag{7.118}$$

where the Yukawa and electrostatic parts of the interaction are

$$u_{ij}^Y(\mathbf{r}_{12}) = -\frac{A}{r_{12}}\exp[-\lambda(r_{12}/d_{ij} - 1)] \tag{7.119}$$

$$\begin{aligned} u_{ij}^{ES}(\mathbf{r}_{12}, \mathbf{m}_i, \mathbf{m}_j) = & \frac{z_i z_j e^2}{r_{12}} + \frac{z_i e}{r_{12}^3}\hat{\mathbf{m}}_j \cdot \mathbf{r}_{12} \\ & - \frac{m_i m_j}{r_{12}^3}\left[3(\hat{\mathbf{m}}_i \cdot \hat{\mathbf{r}}_{12})(\hat{\mathbf{m}}_j \cdot \hat{\mathbf{r}}_{12}) - \hat{\mathbf{m}}_i \cdot \hat{\mathbf{m}}_j\right], \end{aligned} \tag{7.120}$$

and where $r_{12} = |\mathbf{r}_{12}| = |\mathbf{r}_1 - \mathbf{r}_2|$ is the distance between the centers of two particles.

The c-DFT is constructed at the mean field level with

$$\mathcal{F}[\{\rho_i\}] = \mathcal{F}_{id}[\{\rho_i\}] + \mathcal{F}_{HS}[\{\rho_i\}] + \mathcal{F}_Y[\{\rho_i\}] + \mathcal{F}_{\mathrm{ES}}[\{\rho_i\}] \tag{7.121}$$

where the $\mathcal{F}_{HS}$ can be treated as discussed in section 7.3, but the other terms now include integrals over solvent dipole orientation since the solvent density is formally $\rho_0 = \rho_0(\mathbf{r}, \hat{\omega})$, where $\hat{\omega} = (\theta, \phi)$ is a unit vector that represents the orientation of the dipoles relative to a fixed frame of reference. The various terms are as follows:

$$\begin{aligned} \mathcal{F}_{id}/kT[\{\rho_i\}] = & \sum_{i,\text{ions}} \int \rho_i(\mathbf{r})[\ln\left(\Lambda_i^3\rho_i(\mathbf{r})\right) - 1]d\mathbf{r} \\ & + \int\int \rho_s(\mathbf{r}, \hat{\omega})[\ln\left(\Lambda_s^3\rho_s(\mathbf{r}, \hat{\omega})\right) - 1]d\hat{\omega}d\mathbf{r} \end{aligned} \tag{7.122}$$

$$\mathcal{F}_Y/kT[\{\rho_i\}] = \frac{1}{2}\sum_{ij}\int\int u_{ij}^Y(|\mathbf{r} - \mathbf{r}'|)\rho_i(\mathbf{r})\rho_j(\mathbf{r}')d\mathbf{r}d\mathbf{r}' \tag{7.123}$$

$$\mathcal{F}_{\mathrm{ES}}/kT[\{\rho_i\}] = \frac{1}{2}\sum_{ij}\int d\omega \int\int \rho_i(\mathbf{r}, \hat{\omega})u_{ij}^{\mathrm{ES}}(\vec{\mathbf{r}}_{12}, \mathbf{m}_i, \mathbf{m}_j)\rho_j(\mathbf{r}', \hat{\omega})d\mathbf{r}d\mathbf{r}' \tag{7.124}$$

where the ion–ion contributions to $\beta\mathcal{F}_{ES}$ do not have any orientational dependence, but ion–dipole and dipole–dipole interactions do require the ω integral. Finally, note that the solvent density in the Yukawa term is

$$\rho_s(\mathbf{r}) = \int \rho_s(\mathbf{r}, \hat{\omega})d\hat{\omega}. \tag{7.125}$$

The grand potential is written

$$\Omega[\{\rho_i\}] = \mathcal{F}[\{\rho_i\}] + \sum_i \int d\hat{\omega} \int \left[V_i^{\text{ext}}(\mathbf{r}, \hat{\omega}) - \mu_i\right]\rho_i(\mathbf{r}, \hat{\omega})d\mathbf{r}, \tag{7.126}$$

where the solvent dipoles also interact with the external field.

As usual, the c-DFT proceeds by minimizing the free energy,

$$\frac{\delta\Omega[\{\rho_i\}]}{\delta\rho_i(\mathbf{r}, \hat{\omega})} = 0, \tag{7.127}$$

and then solving the resulting Euler–Lagrange equations along with Poisson's equation. In this case, the coupled Poisson's equation include a source term due the solvent dipole. It is

$$\nabla^2\psi(\mathbf{r}) = -4\pi(q(\mathbf{r}) - \nabla \cdot \mathbf{P}(\mathbf{r})). \tag{7.128}$$

where $\mathbf{P}(\mathbf{r})$ is the local polarization vector.

The details of the Euler–Lagrange equation and its solution are left to [43], but we note that to solve the system of equations described above numerically, a discretization of both space, $\mathbf{r}$, and the orientations of the dipoles, $\hat{\omega}(\mathbf{r})$, would be required. In a cartesian three-dimensional system for arbitrary geometry, the number of discretization points would be $N_x N_y N_z N_\theta N_\phi$. If 100 points were required in the spatial dimensions and 10 points were used for each orientation dimension, the problem size would contain 10^8 discretization points. Clearly, this approach will become quickly intractable. By restricting the system to flat planar surfaces with the usual translational symmetry in the plane of the surface and also azimuthal symmetry for the dipoles, the problem is reduced in size.

Oleksy and Hansen found generally good agreement between the detailed polarization model and the phenomenological SSPM approach that had a varying dielectric permittivity, $\epsilon(\mathbf{r})$ [24, 43]. However, both approaches have limitations on dipole moment, m, due to the mean field approximation and cannot reach values appropriate for water [109].

Finally, as noted earlier, molecular simulations can be used to generate correlation functions in bulk fluids that are internally consistent since they do not rely on any closure relation. Those correlation functions can be combined with a c-DFT derived from the direct correlation function hierarchy to predict the properties of inhomogeous systems. Ramirez and coworkers used a hybrid simulation-DFT approach to model a Stockmayer polar solvent (a Lennard–Jones fluid with embedded permanent dipoles) [110]. They did not have ions in the system, but found good agreement with simulations for density and polarization profiles at a charged planar surface. Their fluid had a relative dielectric permittivity of $\epsilon_r \approx 69$ while the mean field fluids are limited to $\epsilon_r \approx 10$.

7.8 Classical-DFT for electrokinetic systems

There are many interfacial fluid systems where diffusion, hydrodynamic forces (driven flow) or time-dependent stochastic motion. For example, electrolytes play a crucial role in charge transport in both biological and engineered systems including ion separation membranes, batteries, nano-transistors, micro and nanofluidic devices, and ion channel proteins (see a comprehensive survey by Wei and coworkers [100]). On a historical note, Nernst–Planck kinetic theory was developed around 1890 [111, 112] and was applied to transport through membranes with a specific consideration of biological systems around 1935 [113, 114].

Modeling systems with electrokinetic (or other dynamic) effects has been tackled from various perspectives. The Poisson–Nernst–Planck (PNP) theory plays a central role in many device level (and ion channel protein) studies of electrokinetic phenomena where the electrolyte is treated with the point charge model [115, 116]. Augmenting PNP with a variety of physical effects has led to the Poisson–Nernst–Planck–Bikerman (PNPB) model detailed in a recent review by Liu and Eisenberg [117] and summarized briefly in appendix A. Along different lines, a variational multiscale approach has been developed by Wei and coworkers for systems with charge transport [100]. Their models treat electrolytes using the free energy formulation of Sharp and Honig [33] extended to include mean field interactions as well. They have applied these variational models to both geometry optimization for macromolecules and to a variety of multiscale applications.

Since the 1980s there has also been a significant effort to develop dynamic-DFTs (DDFTs) for various situations where diffusion, hydrodynamic flow, and stochastic effects are important. The earliest phenomenological DDFT for c-DFTs was suggested by Evans in the 1970s [118] which, preceded the development of time-dependent quantum-DFT by Runge and Gross [119]. A recent comprehensive review of DDFTs has been contributed by Vrugt, Löwen, and Wittkowski that includes a broad discussion of theory, computational methods, and applications of DDFTs [9]. As with the variational methods described above, DDFTs can be extended to multiscale systems. One example is the hybrid model of Cheung and coworkers that combined classical-DFT with a device level PNP model [115].

7.8.1 Steady state transport

The simplest illustration of dynamic systems involves steady state diffusive transport. Figure 7.24 shows sketches of a few physical systems that will exhibit diffusion due to chemical potential gradients between two control regions. Two systems (top) have a planar permeable membrane (one with unidirectional diffusion and one with counter diffusion). The lower panels show nanoscale pores embedded in impermeable membranes. One case has a regular pore geometry as in an engineered nanotube, the other has rough boundaries and exhibits single file diffusion. In all cases, the porous material may be chemically functionalized in order to control the transport properties of the system via preferential adsorption of a particular species due to steric, electrostatic, or polarization effects.

In diffusive systems, the electrochemical potential varies spatially, $\tilde{\mu}_i(\mathbf{r})$, and a transport equation is needed. In a continuum mean-field description (ignoring fluctuations and center of mass motion), the time dependent density profiles are

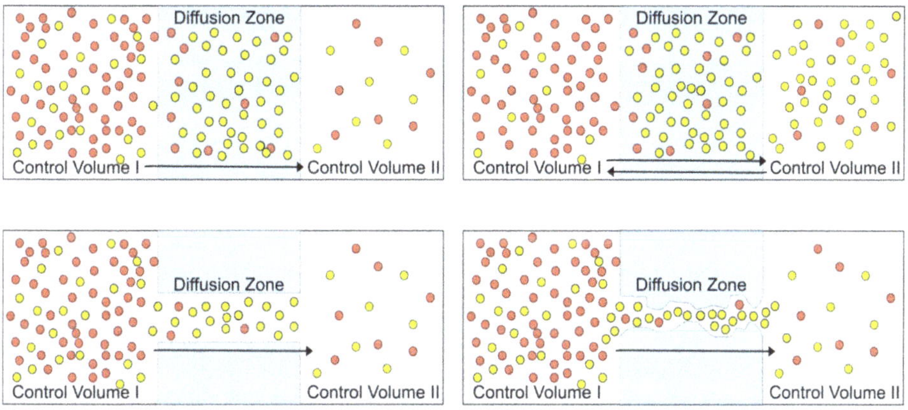

Figure 7.24. Sketches of systems with diffusive transport where all figures are cross sections of three-dimensional systems. Control volumes I and II have $\{\tilde{\mu}_i^{I}\}$ and $\{\tilde{\mu}_i^{II}\}$ respectively. The upper cases have a permeable slab membrane where the left panel shows unidirectional diffusion and the right panel shows counter diffusion. The lower panels show selective nano-scale pores held in place by an impermeable surface. The left panel shows a smooth cylindrical pore, and the right panel shows a rough pore with single file diffusion. Two fluid species are shown and distinguished by color as a generic representation of physical characteristics (size, charge, etc). The arrows indicate direction of diffusion.

$$\frac{\partial \rho_i(\mathbf{r},\,t)}{\partial t} = \nabla \cdot \vec{J}_i \tag{7.129}$$

where the flux of species i is

$$\begin{aligned}\vec{J}_i(\mathbf{r},\,t) &= -\,D_i\rho_i(\mathbf{r},\,t)\nabla\tilde{\mu}_i(\mathbf{r},\,t)\\ &= -\,D_i\rho_i(\mathbf{r},\,t)\nabla\big(\beta\mu_i(\mathbf{r},\,t) + z_i\tilde{\psi}(\mathbf{r},\,t)\big).\end{aligned} \tag{7.130}$$

The diffusion (or mobility) coefficients, D, are written as constants, but could be spatially varying.[1] To consider steady state situations, set $\partial\rho(\mathbf{r},\,t)/\partial t = 0$ in equation (7.129). Then the transport equation is

$$\nabla \cdot \vec{J}_i = 0 = \nabla\big(\rho_i(\mathbf{r})\nabla\big(\mu_i(\mathbf{r}) + z_i\tilde{\psi}(\mathbf{r})\big)\big). \tag{7.131}$$

In an equilibrium system of (ideal) point ions free of external fields, the chemical potential for a given species, i, is simply $\mu_i/kT = \ln\rho_i$, and consequently,

$$\nabla\mu_i = kT\nabla\rho_i/\rho_i. \tag{7.132}$$

In a bulk or neutral system of ideal particles, $\psi = 0$, and a simple Fick's law, $\vec{J}_i(\mathbf{r}) = -D_i\nabla\rho_i(\mathbf{r})$, is recovered.

[1] This is the generalized Fick's Law approach first suggested by Evans [118], and captured by the "standard" DDFT equation (equation (71) in [9]), $\frac{\partial\rho(\mathbf{r},\,t)}{\partial t} = D\nabla\cdot\left(\rho(\mathbf{r},\,t)\nabla\frac{\partial\mathcal{F}}{\partial\rho(\mathbf{r},\,t)}\right)$.

Substituting equation (7.132) into equation (7.131) yields

$$\nabla \cdot \vec{J_i} = 0 = \nabla \cdot \left[\left(\nabla \rho_i(\mathbf{r}) + z_i \rho_i(\mathbf{r}) \nabla \tilde{\psi}(\mathbf{r}) \right) \right]. \tag{7.133}$$

Poisson–Nernst–Planck (PNP) theory is then the combination of the transport equation in equation (7.133) and Poisson's equation (see equation (7.16)) for the special case of point ions.

It is straightforward to extend this approach to include more detailed fluid models with c-DFT since the Euler–Lagrange equation can be systematically improved for various models. A general construct for these steady state electrokinetic systems requires the simultaneous solution of three types of equations; Poisson's equation (for $\psi(\mathbf{r})$), a set of Euler–Lagrange equations (for $\rho_i(\mathbf{r})$), and a set of transport equations (for $\mu_i(\mathbf{r})$). From this perspective, the PNP method is a special case where it is easy to eliminate the Euler–Lagrange equation and recast the transport equations in terms of densities. Finally, we note that in situations where some species experience transport while others are essentially at equilibrium, the transport equation need not be solved for all species since μ_i will be constant for the equilibrium species [120].

In order to calculate fluxes, J_i, the diffusion coefficient is needed (even if it is constant). Equilibrium diffusion coefficients may be easy to calculate for the ions in bulk control regions (see figure 7.24), but values in membranes or nanopores are likely unknown apriori. Diffusion coefficients may be treated as adjustable parameters or estimated from molecular simulations or experiments [121, 122].

Classical-DFT with steady state transport has been compared with molecular simulations using a dual control volume grand canonical molecular dynamics (DCV-GCMD) method [121]. In DCV-GCMD, grand canonical Monte-Carlo (GCMC) is performed in two control domains at either ends of a simulation box to set the chemical potentials of each species in those domains. A diffusion domain is located between the control volumes, and the simulation runs with pure molecular dynamics (MD) in the diffusion domain [123–125]. The goal of the studies in [121] was to compare density and chemical potential profiles from molecular simulation and transport-c-DFT calculations. Figure 7.25 presents one result from [121]. Specifically, $\rho(x)$ and $\mu(x)$, are shown for both DCV-GCMD and transport c-DFT calculations for the case of color diffusion through a membrane where one species is more attracted to the membrane than the other. The fluid particles were neutral Lennard–Jones particles and the c-DFT included finite-size and cohesive effects. The excellent agreement between the simulation and theory demonstrates that accurate flux predictions can be expected from transport c-DFT approach given properly calibrated diffusion coefficients. In order to obtain this agreement, it was critical to also account for center of mass motion (a hydrodynamic effect) in the system as is discussed further in [121].

7.8.2 A transport application—ion channel proteins

Ion channel proteins are one example of a fascinating and important class of problems where c-DFT has been successfully applied. In turn, these systems have

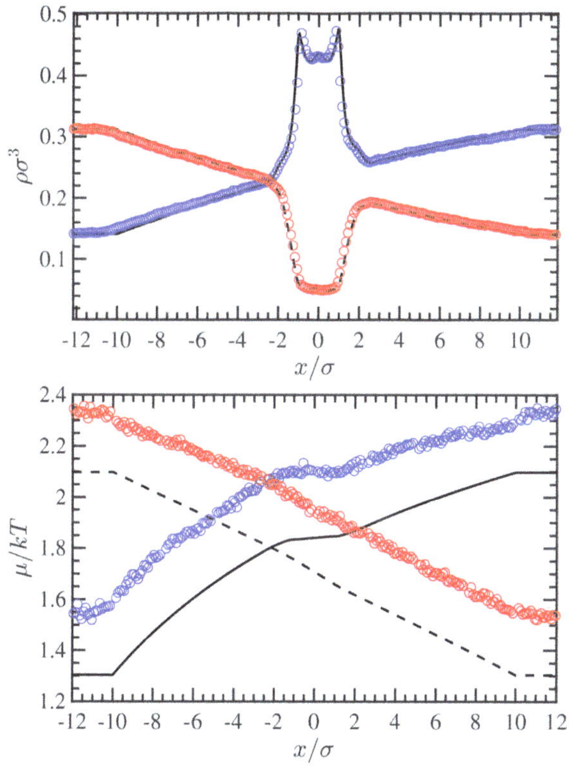

Figure 7.25. Density profiles (top) and chemical potential profiles (bottom) for binary color diffusion through a membrane where one species is more attractive to the membrane than the other. The black lines (solid and dashed) are the transport c-DFT results while the symbols (blue and red) are the GCMD results.

inspired new developments in c-DFT both in numerical methods and in the development of new functionals (section 7.4). A comprehensive review of this large and complex field is beyond the scope of this text, but a brief overview is provided to illustrate the contributions of c-DFT in this arena.

Ion channels are transmembrane proteins that regulate the level of ions in cells. Since the intracellular environment is different from the extracellular environment, the situation is similar to the lower right sketch in figure 7.24 where an ion channel pore has complex geometry and charge distribution resulting from the three-dimensional (3D) protein structure. Experiments (specifically patch–clamp experiments) can be done to measure the current through single isolated channels as a function of the applied voltage (or more generally electrochemical potential gradient) across the membrane [126]. These experiments demonstrate that the open channel current measured through a given channel is constant. It is therefore reasonable to begin with a steady state diffusion model. 3D protein structures and charge distributions for some ion channels are available in the Protein Data Bank (PDB) [127], and the availability of those structures makes it desirable to perform 3D calculations where possible.

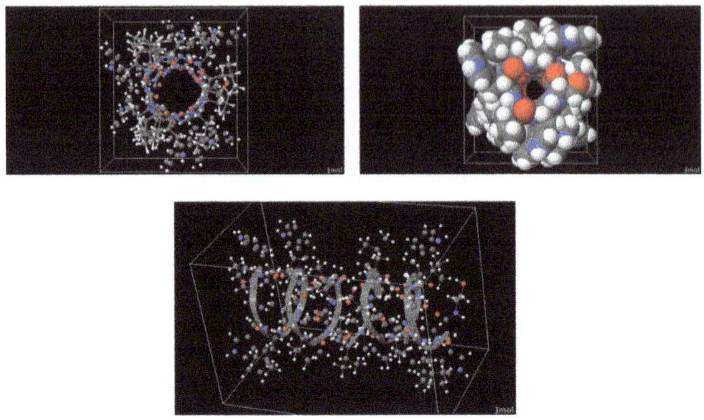

Figure 7.26. Three images of the Gramicidin A channel (1mag PDB structure). The upper figures show views down the Gramicidin pore. The upper left shows a typical ball and stick representation while the upper right includes van der Waals shells for the atoms. The lower figure includes a ribbon representation of the secondary structure in addition to the ball and stick representation in a transverse view spanning the membrane.

Gramicidin (a small antibiotic pore forming peptide [128]) is a model channel with only 272 atoms (see the structure 1mag in the PDB [127, 129]). Three representations of the channel are shown in figure 7.26 including views looking both down the axis of the channel and along the channel across the membrane. The ball and stick representation is compared with a representation that includes the Van der Waals radii of the atoms. Clearly the accessible channel dimensions are smaller than might be inferred from the ball and stick model, and this is a case with a one-dimensional transport region embedded in a three-dimensional system.

It has been understood for over a century that the molecular details of ion channels would be important to their function. However, the necessary computational tools only started to become available in the 1990s. At the time, 3D structures were available for some, but not many channels. As a result, some modeling studies were grounded in a known structure, while others were developed more heuristically. Historically interesting summaries of modeling approaches from the perspective of the late 1990s were presented by Jakobsson [130] and Levitt [131].

Work from the 1990s and early 2000s included both theory and molecular simulation. Molecular dynamics (MD) simulations were performed to study ion transport in gramicidin [132, 133] and larger systems like the potassium channel and OmpF porin [134]. Brownian dynamics simulations (without explicit water) were used to reach longer timescales than was possible with all atom MD [135, 136]. Many investigations at the level of a 1D-PNP model were performed for systems like the acetylcholine receptor channel [116, 137] and the L-type calcium channel [138]. The first (lattice based) 3D-PNP calculation was performed for the gramicidin channel by Kurnikova in 1999 [139]. Liquid state theory was introduced to ion channel models (using an Orstein–Zernike/MSA [61, 62] approach) in 2000 while c-DFT for ion channels was introduced in 1D systems by Gillespie, Nonner and Eisenberg in 2002 [140]. That same year, the first (rather coarse) 3D c-DFT calculations were performed for the gramicidin channel [141]. Two helpful reviews from that period discussed general theoretical approaches [142] and provide computational results for several ion channels [143].

In subsequent years, computing power and numerical methods have continued to improve, just as new ion channel structures have been cataloged in the PDB. 1D PNP/DFT studies have been applied to very large channels such as the ryanodine receptor calcium channel [144, 145] which contains more than 100000 atoms. The PNP method has also been modified to includes steric (and other) effects [146, 147] (see also appendix A).

The early 3D c-DFT calculations in [141] mentioned above were facilitated by the use of (real space) parallel computing algorithms for solving large scale matrix problems on massively parallel computers [148] (as detailed in chapter 16). Further developments for c-DFT to make 3D calculations more practical on smaller computing platforms have been in part motivated by the ion channel application. Those numerical methods have included applying fast Fourier transforms for convolutions [149–152] and using GPUs for further acceleration [153]. Gußmann and Roth [154] have recently performed 3D c-DFT calculations to explore bubble gating in the potassium KcsA channel.

Simulation of ion channels also continue. A dual control volume GCMC similar to the method described in section 7.8 has recently been coupled with BD simulations to study transport through membrane pores [122, 155]. Finally, we note that multiscale modeling approaches have also been applied to ion channels and other nanofluidic systems [100, 115, 122, 156].

References

[1] Gouy G 1910 Sur la constitution de la charge électrique à la surface d'un électrolyte *J. Physique* **9** 457–68

[2] Gouy G 1917 Sur la fonction électrocapillaire *Ann. Phys.* **7** 129–84

[3] Chapman D L 1913 A contribution to the theory of electrocapillarity *Phil. Mag.* **25** 475–81

[4] Hansen J-P and Löwen H 2000 Effective interactions between electric double layers *Annu. Rev. Phys. Chem.* **51** 209–42

[5] Forsman J, Woodward C E and Szparaga R 2015 Classical density functional theory of ionic solutions *Computational Electrostatics for Biological Applications* ed W Rocchia and M Spagnuolo (Berlin: Springer) ch 2 pp 17–38

[6] Wu J and Li Z 2007 Density-functional theory for complex fluids *Ann. Rev. Phys. Chem.* **58** 85–112

[7] Messina R 2009 Electrostatics in soft matter *J. Phys. Cond. Matter* **21** 113102

[8] Evans R, Oettel M, Roth R and Kahl G 2016 New developments in classical density functional theory *J. Phys. Cond. Matter* **28** 240401

[9] te Vrugt M, Löwen H and Wittkowski R 2020 Classical dynamical density functional theory: from fundamentals to applications *Adv. in Phys.* **69** 121–247

[10] Grimson M J and Rickayzen G 1982 Forces between surfaces in electrolyte solutions *Chem. Phys. Lett.* **86** 71–5

[11] Mier L 1990 y Teran, S.H. Suh, H.S. White, and H. T. Davis. A nonlocal free-energy density-functional approximation for the electrical double layer *J. Chem. Phys.* **92** 5087–98

[12] Tang Z, Mier-Y-Teran L, Davis H T, Scriven L E and White H S 1990 Non-local free-energy density-functional theory applied to the electrical double layer *Mol. Phys.* **71** 369–92
[13] Tang Z, Scriven L E and Davis H T 1992 Interactions between primitive electrical double layers *J. Chem. Phys.* **97** 9258–66
[14] Pizio O and Sokołowski S 2014 Solvent primitive model of an electric double layer in slit-like pores: microscopic structure, adsorption and capacitance from a density functional approach *Cond. Matter Phys.* **17** 23603
[15] Tang Z, Scriven L E and Davis H T 1992 A three-component model of the electrical double layer *J. Chem. Phys.* **97** 494–503
[16] Tang Z, Scriven L E and Davis H T 1994 Effects of solvent exclusion on the force between charged surfaces in electrolyte solution *J. Chem. Phys.* **100** 4527–30
[17] Oleksy A and Hansen J-P 2006 Towards a microscopic theory of wetting by ionic solutions. i. surface properties of the semi-primitive model *Mol. Phys.* **104** 2871–83
[18] Oleksy A and Hansen J-P 2009 Microscopic density functional theory of wetting and drying of a solid substrate by an explicit solvent model of ionic solutions *Mol. Phys.* **107** 2609–24
[19] Carnie S L and Chan D Y C 1980 The structure of electrolytes at charged surfaces: Ion-dipole mixtures *J. Chem. Phys.* **73** 2949–57
[20] Augousti A T and Rickayzen G 1984 Solvation forces in a model fluid mixture of ions and dipoles *J. Chem. Soc., Faraday Trans.* **2** 141–56
[21] Moradi M and Rickayzen G 1989 An approximate density functional for an inhomogeneous dipolar fluid *Mol. Phys.* **68** 903–15
[22] Biben T, Hansen J P and Rosenfeld Y 1998 Generic density functional for electric double layers in a molecular solvent *Phys. Rev.* E **57** R3727–30
[23] Warshavsky V and Marucho M 2016 Polar-solvation classical density-fucntional theory for electrolyte aqueous solutions near a wall *Phys. Rev.* E **93** 042607
[24] Oleksy A and Hansen J-P 2011 Wetting and drying scenarios of ionic solutions *Mol. Phys.* **109** 1275–88
[25] Zhao S, Ramirez R, Vuilleumier R and Borgis D 2011 Molecular density functional theory of solvation: from polar solvents to water *J. Chem. Phys.* **134** 194102
[26] Levesque M, Marry V, Rotenberg B, Jeanmairet G, Vuilleumier R and Borgis D 2012 Solvation of complex surfaces via molecular density functional theory *J. Chem. Phys.* **137** 224107
[27] Jeanmairet G, Levy N, Levesque M and Borgis D 2016 Molecular density functional theory of water including density-polarization coupling *J. Phys.: Condens. Matter* **28** 244005
[28] Chapman W G, Gubbins K E, Jackson G and Radosz M 1989 Saft: equation-of-state solution model for associating fluids *Fluid Phase Equilibria* **52** 31–8
[29] Hughes J, Krebs E and Roundy D 2013 A classical density-functional theory for describing water interfaces *J. Chem. Phys.* **138** 024509
[30] Malheiro C, Mendiboure B, Miguez J-M, Piñeiro M M and Miqueu C 2014 Nonlocal density functional theory and grand canonical monte carlo molecular simulations of water adsorption in confined media *J. Phys. Chem.* C **118** 24914–7905
[31] Christelle M and Grégoire D 2020 Estimation of pore pressure and phase transitions of water confined in nanopores with non-local density functional theory *Mol. Phys.* **118** e1742935
[32] van Roij R 2010 Electrostatics in liquids: From electrolytes and suspensions towards emulsions and patchy surfaces *Physica* A **389** 4317–31

[33] Sharp K A and Honig B 1990 Calculating total electrostatic energies with the nonlinear Poisson-Boltzmann equation *J. Phys. Chem.* **94** 7684–92
[34] Reiner E S and Radke C J 1990 Variational approach to the electrostatic free energy in charged colloidal suspensions : General theory for open systems *J. Chem. Soc. Faraday Trans.* **86** 3901–12
[35] Reiner E S and Radke C J 1993 Double layer interactions between charge-regulated colloidal surfaces: Pair potentials for spherical particles bearing ionogenic surface groups *Adv. Colloid Interface. Sci.* **47** 59–147
[36] Forsman J 2004 A simple correlation-corrected Poisson-Boltzmann theory *J. Phys. Chem.* B **108** 9236–45
[37] Stern O 1924 Zur theorie der elektrolytischen doppelschicht *Z. Electrochem* **30** 508–16
[38] Derjaguin B and Landau L 1941 Theory of the stability of strongly charged lyophobic sols and of the adhesion of strongly charged particles in solution of electrolytes *Acta. Physicochim. URSS* **14** 633–62 Reprinted in English in 1993 *Prog. Surface Sci.* **43** 30–59
[39] Verwey E J W and Overbeek J T G 1948 *Theory of the Stability of Lyophobic Colloids. The Interaction of Particles Having an Electric Double Layer* (Amsterdam: Elsevier)
[40] Henderson D, Blum L and J L Lebowitz 1979 An exact formula for the contact value of the density profile of a system of charged hard spheres near a charged wall *J. Electroanaly. Chem.* **102** 315–9
[41] Martin P A 1988 Sum rules in charged fluids *Rev. Mod. Phys.* **60** 1075–127
[42] Markovich T, Andelman D and Podgornik R 2016 Charge regulation: a generalized boundary condition? *Europhys. Lett.* **113** 26004
[43] Oleksy A and Hansen J-P 2010 Wetting of a solid substrate by a 'civilized' model of ionic solutions *J. Chem. Phys.* **132** 204702
[44] Derjaguin B V, Churaev N V and Muller V M 1987 *Surface Forces* (New York: Springer)
[45] Evans R 1992 Density functionals in the theory of nonuniform fluids *Fundamentals of Inhomogeneous Fluids* ed D Henderson (New York: Dekker) ch 3 pp 85–175
[46] Henderson D, Bryk P, Sokolowski S and Wasan D T 2000 Density-functional theory for an electrolyte confined by thin charged walls *Phys. Rev.* E **61** 3896–903
[47] Henderson D 2001 Simulation and theory of the electrochemical double layer for strong ionic interactions *J. Molec. Liquids* **92** 29–39
[48] Tarazona P 1985 Free-energy density functional for hard spheres *Phys. Rev.* A **31** 2672–9
[49] Tan Z, Marini Bettolo Marconi U, van Swol F and Gubbins K E 1989 Hard-sphere mixtures near a hard wall *J. Chem. Phys.* **90** 3704–12
[50] Rosenfeld Y 1989 Free-energy model for the inhomogeneous hard-sphere fluid mixture and density functional theory of freezing *Phys. Rev. Lett.* **63** 980–3
[51] Roth R 2010 Fundamental measure theory for hard-sphere mixtures: a review *J. Phys.: Cond. Matter* **22** 063102
[52] Hansen-Goos H and Roth R 2006 A new generalization of the carnahan-starling equation of state to additive mixtures of hard spheres *J. Chem. Phys.* **124** 154506
[53] Roth R, Evans R, Lang A and Kahl G 2002 Fundamental measure theory for hard-sphere mixtures revisited: the white bear version *J. Phys.: Cond. Matter* **14** 12063–78
[54] Mansoori G A, Carnahan N F, Starling K E Jr. and Leland T W 1971 Equilibrium thermodynamic properties of the mixture of hard spheres *J. Chem. Phys.* **54** 1523–5
[55] Hansen-Goos H and Roth R 2006 Density functional theory for hard-sphere mixtures: the white bear version mark ii *J. Phys.: Cond. Matter* **18** 8413–25

[56] Rosenfeld Y, Schmidt M, Löwen H and Tarazona P 1996 Dimensional crossover and the freezing transition in density functional theory *J. Phys.: Cond. Matter* **8** L577–81

[57] Rosenfeld Y, Schmidt M, Löwen H and Tarazona P 1997 Fundamental-measure free-energy density functional for hard spheres: Dimensional crossover and freezing *Phys. Rev.* E **55** 4245–63

[58] Tarazona P 2000 Density functional for hard sphere crystals: A fundamental measure approach *Phys. Rev. Lett.* **84** 694–7

[59] Waisman E and Lebowitz J L 1972 Mean spherical model integral equation for charged hard spheres i. method of solution *J. Chem. Phys.* **56** 3086–93

[60] Waisman E and Lebowitz J L 1972 Mean spherical model integral equation for charged hard spheres ii. results *J. Chem. Phys.* **56** 3093–9

[61] Blum L 1975 Mean spherical model for asymmetric electrolytes i. method of solution *Mol. Phys.* **30** 1529–35

[62] Blum L and Hoeye J S 1977 Mean spherical model for asymmetric electrolytes. 2. thermodynamic properties and the pair correlation function *J. Phys. Chem.* **81** 1311–6

[63] Hiroike K 1977 Supplement to blum's theory for asymmetric electrolytes *Mol. Phys.* **33** 1195–8

[64] Voukadinova A, Valisko M and Gillespie D 2018 Assessing the accuracy of three classical density functional theories of the electrical double layer *Phys. Rev.* E **98** 012116

[65] Archer A J, Chacko B and Evans R 2017 The standard mean-field treatment of interparticle attraction in classical dft is better than one might expect *J. Chem. Phys.* **147** 034501

[66] Saam W F and Ebner C 1977 Density-functional theory of classical systems *Phys. Rev.* A **15** 2566–8

[67] Gillespie D, Nonner W and Eisenberg R S 2003 Density functional theory of charged, hard-sphere fluids *Phys. Rev.* E **68** 031503

[68] Gillespie D, Valiskó M and Boda D 2005 Density functional theory of the electrical double layer: the rfd functional *J. Phys. Cond. Matter* **17** 6609–26

[69] Roth R and Gillespie D 2016 Shells of charge: a density functional theory for charged hard spheres *J. Phys.: Cond. Matter* **28** 244006

[70] Roth R and Gillespie D 2017 Corrigendum: Shells of charge: a density functional theory for charged hard spheres *J. Phys. Cond. Matter* **29** 449501

[71] Wei D and Blum L 1987 Internal energy in the mean spherical approximation as compared to Debye-Hückel theory *J. Phys. Chem.* **91** 4342–3

[72] Blum L 1980 Solution of the ornstein-zernike equation for a mixture of hard ions and yukawa closure *J. Stat. Phys.* **22** 661–72

[73] Nonner W, Catacuzzeno L and Eisenberg B 2000 Binding and selectivity in l-type calcium channels: A mean spherical approximation *Biophysical J.* **79** 1976–92

[74] Tarazona P and Evans R 1984 A simple density functional theory for inhomogeneous liquids: Wetting by gas at a solid-liquid interface *Mol. Phys.* **52** 847–57

[75] Evans R and Tarazona P 1984 Theory of condensation in narrow capillaries *Phys. Rev. Lett.* **52** 557–60

[76] Evans R, Marini Bettolo Marconi U and Tarazona P 1986 Capillary condensation and adsorption in cylindrical and slit-like pores *J. Chem. Soc., Faraday Trans.* **2** 1763–87

[77] Tarazona P, Marini Bettolo Marconi U and Evans R 1987 Phase equilibria of fluid interfaces and confined fluids: non-local versus local density functionals *Mol. Phys.* **60** 573–95

[78] Peterson B K, Walton J P R B and Gubbins K E 1986 Fluid behaviour in narrow pores *J. Chem. Soc., Faraday Trans.* **2** 1789–800
[79] Evans R 1990 Fluids adsorbed in narrow pores: phase equilibria and structure *J. Phys.: Condens. Matter* **2** 8989–9007
[80] Chistyakov A D 2007 The permittivity of water and water vapor in saturation states *Russian J. Phys. Chem.* **81** 5–8
[81] Malmberg C G and Maryott A A 1956 Dielectric constant of water from 0° to 100°C *J. Res. Natl. Bur. of Stand. (U.S.)* **56** 1–8
[82] Salinger A G and Frink L J D 2003 Rapid analysis of phase behavior with density functional theory. i. novel numerical methods *J. Chem. Phys.* **118** 7457–65
[83] van Swol F and Henderson J R 1989 Wetting and drying transitions at a fluid-wall interface: Density functionl theory versus computer simulation *Phys. Rev.* A **40** 2567–78
[84] Frink L J D and Salinger A G 1999 Wetting of a chemically heterogeneous surface *J. Chem. Phys.* **110** 5969–77
[85] Yatsyshin P, Parry A O, Rascón C and Kalliadasis S 2017 Classical density functional study of wetting transitions on nanopatterned surfaces *J. Phys.: Condens Matter* **29** 094001
[86] Scheemilch M, Quirke N and Henderson J R 2003 Wetting of nanopatterned surfaces: The striped surface *J. Chem. Phys.* **118** 816–29
[87] Griffiths D J 2013 Introduction to Electrodynamics *Introduction to Electrodynamics; 4th edition* (Boston, MA: Pearson) Re-published by Cambridge University Press in 2017
[88] Boz˘ič A L and Podgornik R 2018 Anomalous multipole expansion: Charge regulation of patchy inhomogeneously charged spherical particles *J. Chem. Phys.* **149** 163307
[89] Majee A, Bier M and Podgornik R 2018 Spontaneous symmetry breaking of charge-regulated surfaces *Soft Matter* **14** 985–91
[90] Chan D Y C, Healy T W, Supasiti T and Usui S 2006 Electrical double layer interactions between dissimilar oxide surfaces with charge regulation and stern-grahame layers *J. Colloid and Interface Sci.* **296** 150–8
[91] Lund M and Jönsson B 2005 On the charge regulation of proteins *Biochemistry* **44** 5722–7
[92] Ninham B W and Parsegian V A 1971 Electrostatic potential between surface bearing ionizable groups in ionic equilibrium with physiologic saline solution *J. Theor. Biol.* **31** 405–28
[93] Chan D, Perram J W, White L R and Healy T W 1975 Regulation of surface potential at amphoteric surfaces during particle-particle interaction *J. Chem. Soc., Faraday Trans.* **1** 1046–57
[94] Fleharty M E, van Swol F and Petsev D N 2014 The effect of surface charge regulation on conductivity in fluidic nanochannels *J. Colloid Interface Sci.* **416** 105–11
[95] Fleharty M E, van Swol F and Petsev D N 2016 Solvent role in the formation of electric double layers with surface charge regulation: A bystander or a key participant? *Phys. Rev. Lett.* **116** 048301
[96] van Swol F and Petsev D N 2018 Solution structure effects on the properties electric double layers with surface charge regulation assessed by density functional theory *Langmuir* **34** 13808–20
[97] Bardhan J P 2015 Boundary-integral and boundary-element methods for biomolecular electrostatics: Progress, challenges, and important lessons from CEBA 2013 *Computational Electrostatics for Biological Applications* ed W Rocchia and M Spagnuolo (Berlin: Springer) ch 6 pp 121–42

[98] Patané G and Spagnuolo M 2015 State-of-the-art and perspectives of geometric and implicit modeling for molecular surfaces *Computational Electrostatics for Biological Applications* ed W Rocchia and M Spagnuolo (Berlin: Springer) ch 8 pp 157–76
[99] Dias S E D and Gomes A J P 2015 Triangulating gaussian-like surfaces of molecules with millions of atoms *Computational Electrostatics for Biological Applications* ed W Rocchia and M Spagnuolo (Berlin: Springer) ch 9 pp 177–98
[100] Wei G-W, Zheng Q, Chen Z and Xia K 2012 Variational multiscale models for charge transport *SIAM Review* **54** 699–754
[101] Fleharty M E, van Swol F and Petsev D N 2015 Charge regulation at semiconductor-electrolyte interfaces *J. Colloid and Interface Sci.* **449** 409–15
[102] Allen R, Hansen J-P and Melchionna S 2001 Electrostatic potential inside ionic solutions confined by dielectrics: a variational approach *Phys. Chem. Chem. Phys.* **3** 4177–86
[103] Boda D, Gillespie D, Nonner W, Henderson D and Eisenberg B 2004 Computing induced charges in inhomogeneous dielectric media: Application in a monte carlo simulation of complex ionic systems *Phys. Rev.* E **69** 046702
[104] Boda D, Valisko M, Eisenberg B, Nonner W, Henderson D and Gillespie D 2006 The effect of protein dielectric coefficient on the ionic selectivity of a calcium channel *J. Chem. Phys.* **125** 034901
[105] Tomasi J, Mannucci B and Cammi R 2005 Quantum mechanical continuum solvation models *Chem. Rev.* **105** 2999–3094
[106] Andreussi O, Dabo I and Marzari N 2012 Revised self-consistent continuum solvation in electronic-structure calculations *J. Chem. Phys.* **136** 064102
[107] Dupont C, Andreussi O and Marzari N 2013 Self-consistent continuum solvation (sccs): The case of charged systems *J. Chem. Phys.* **139** 214110
[108] Nattino F, Truscott M, Marzari N and Andreussi O 2019 Continuum models of the electrochemical diffuse layer in electronic structure calculations *J. Chem. Phys.* **150** 014722
[109] Høye J S and Stell G 1974 Statistical mechanics of polar systems *J. Chem. Phys.* **61** 562–72
[110] Ramirez R, Gebauer R, Mareschal M and Borgis D 2002 Density functional theory of solvation in a polar solvent: Extracting the functional from homogeneous solvent simulations *Phys. Rev.* E **66** 031206
[111] Nernst W 1889 Die elektromotorische wirksamkeit der jonen *Z. Phys. Chem.* **4** 129–81
[112] Planck M 1890 die erregung von electricităt und warme in electrlyten *Ann. Der Phys.* **275** 161–86
[113] Teorell T 1935 Studies on the 'diffusion effect upon' ionic distribution i. some theoretical considerations *Proc. Natl. Acad. Sci. USA* **21** 152–61
[114] Meyer K H and Sievers J F 1936 La perméabilité des membranes i. théorie de la perméabilité ionique *Helv. Chim. Acta* **19** 649–64
[115] Cheung J, Frischknecht A L, Perego M and Bochev P 2017 A hybrid, coupled approach for modeling charged fluids from the nano to the mesoscale *J. Comp. Phys.* **348** 364–84
[116] Levitt D G 1991 General continuum theory for multiion channel. i. theory *Biophys. J.* **59** 271–7
[117] Liu J-L and Eisenberg B 2020 Molecular mean-field theory of ionic solutions: A Poisson-Nernst-Planck-Bikerman model *Entropy* **22** 550
[118] Evans R 1979 The nature of the liquid-vapour interface and other topics in the statistical mechanics of non-uniform, classical fluids *Adv. Phys.* **28** 143–200

[119] Runge E and Gross E K U 1984 Density-functional theory for time-dependent systems *Phys. Rev. Lett.* **52** 997–1000
[120] Zheng Q and Wei G W 2011 Poisson-Boltzmann-Nernst-Planck model *J. Chem. Phys.* **134** 194101
[121] Frink L J D, Thompson A and Salinger A G 2000 Applying molecular theory to steady-state diffusing systems *J. Chem. Phys.* **112** 7564–71
[122] Valiskó M, Bartlomiej Matejczyk, Ható Z, Kristóf T, Mádai E, Fertig D, Gillespie D and Boda D 2019 Multiscale analysis of the effect of surface charge pattern on a nanopore's rectification and selectivity properties: from all-atom model to Poisson–Nernst–Planck *J. Chem. Phys.* **150** 144703
[123] Heffelfinger G S and van Swol F 1994 Diffusion in lennard-jones fluids using dual control volume grand canonical molecular dynamics simulation (dcv-gcmd) *J. Chem. Phys.* **100** 7548
[124] Heffelfinger G S and Ford D M 1998 Massively parallel dual control volume grand canonical molecular dynamics with ladera i. gradient driven diffusion in lennard-jones fluids *Mol. Phys.* **94** 659–71
[125] Thompson A P, Ford D M and Heffelfinger G S 1998 Direct molecular simulation of gradient-driven diffusion *J. Chem. Phys.* **109** 6406–14
[126] Sakmann B and Neher E (ed) 1995 *Single Channel Recording* (New York: Springer)
[127] Berman H M, Westbrook J, Feng Z, Gilliland G, Bhat T N, Weissig H, Shindyalov I N and Bourne P E 2000 The protein data bank *Nucleic Acids Research* **28** 235–42
[128] Chadwich D J and Cardew G (ed) 1999 *Gramicidin and Related Ion Channel-forming Peptides* (New York: Wiley)
[129] Ketchem R R, Lee K C, Huo S and Cross T A 1996 Macromolecular structural elucidation with sold-state nmr-derived orientational constraints *J. Biomolec. NMR* **8** 1–14
[130] Jakobsson E 1998 Using theory and simulation to understand permeation and selectivity in ion channels *Methods* **14** 342–51
[131] Levitt D G 1999 Modeling of ion channels *J. Gen. Physiol.* **113** 789–94
[132] Roux B, Prodhom B and Karplus M 1995 Molecular dynamics study of single and double occupancy *Biophysics* J **68** 876–92
[133] Woolf T B and Roux B 1997 The binding site of sodium in the gramicidin a channel: comparison of molecular dynamics with solid-state nmr data *Biophysics* J **72** 1930–45
[134] Sansom M S P, Shrivastava I H, Ranatunga K M and Smith G R 2000 Simulations of ion channels-watching ions and water move *Trends in Biochem. Sci.* **25** 368–74
[135] Chung S H, Hoyles M, Allen T and Kuyucak S 1998 Study of ionic currents across a model membrane channel using brownian dynamics *Biophysics* J **75** 793–809
[136] Bek S and Jakobsson E 1994 Brownian dynamics study of a multiply-occupied cation channel: application to understanding permeation in potassium channels *Biophys. J.* **66** 1028–38
[137] Levitt D G 1991 General continuum theory for multiion channel. ii. application to aceylcholine channel *Biophys. J.* **59** 278–88
[138] Nonner W and Eisenberg B 1998 Ion permeation and glutamate residues linked by Poisson-Nernst-Planck theory in l-type calcium channels *Biophys. J.* **75** 1287–305
[139] Kurnikova M G, Coalson R D, Graf P and Nitzan A 1999 A lattice relaxation algorithm for three-dimensional Poisson-Nernst-Planck theory with application to ion transport through the gramicidin a channel *Biophys. J.* **76** 642–56

[140] Gillespie D, Nonner W and Eisenberg R S 2002 Coupling Poisson–Nernst–Planck and density functional theory to calculate ion flux *J. Phys.: Cond. Matter* **14** 12129–45

[141] Frink L J D, Salinger A G, Sears M P, Weinhold J D and Frishknecht A L 2002 Numerical challenges in the application of density functional theory to biology and nanotechnology *J. Phys. Cond. Matter* **14** 12167–87

[142] Henderson D, Busath D D and Rowley R 2001 Fluids near surfaces and in pores and membrane channels *Progress Surf. Sci.* **68** 279–95

[143] Roux B, Allen T, Berneche S and Im W 2004 Theoretical and computational models of biological ion channels *Quarterly Reviews of Biophysics* **37** 15–03

[144] Gillespie D, Xu L, Wang Y and Meissner G 2005 de)constructing the ryanodine receptor: Modeling ion permeation and selectivity of the calcium release channel *J. Phys. Chem.* B **109** 15598–610

[145] Gillespie D, Xu L and Meissner G 2014 Selecting ions by size in a calcium channel: They ryanodine receptor case study *Biophys J.* **107** 2263–73

[146] Liu J L and Eisenberg B 2015 Numerical methods for a Poisson-Nernst Planck-Fermi model of biological ion channels *Phys. Rev.* E **92** 012711

[147] Chen J-H, Chen R-C and J-L Liu 2018 A GPU Poisson-Fermi solver for ion channel simulations *J. Chem. Phys.* **229** 99–105

[148] Frink L J D and Salinger A G 2000 Two- and three-dimensional nonlocal density functional theory for inhomogeneous fluids: I. Algorithms and parallelization *J. Comp. Phys.* **159** 407–24

[149] Sears M P and Frink L J D 2003 A new efficient method for density functional theory calculations of inhomogeneous fluids *J. Comp. Phys.* **190** 184–200

[150] Knepley M G, Karpeev D A, Davidovits S, Eisenberg R S and Gillespie D 2010 An efficient algorithm for classical density functional theory in three dimensions: Ionic solutions *J. Chem. Phys.* **132** 124101

[151] Bernet T, Piñeiro M M, Plantier F and C Miqueu 2020 A 3d non-local density functional theory for any pore geometry *Mol. Phys.* **118** e1767308

[152] Edelmann M and Roth R 2016 A numerical efficient way to minimize classical density functional theory *J. Chem. Phys.* **144** 074105

[153] Stopper D and Roth R 2017 Massively parallel GPU-accelerated minimization of classical density functional theory *J. Chem. Phys.* **147** 064508

[154] Gußmann F and Roth R 2017 Bubble gating in biological ion channels: A density functional theory study *Phys. Rev.* E **95** 062407

[155] Berti C, Furini S, Gillespie D, Boda D, Eisenberg R S, Sangiorgi E and Fiegna C 2014 Three-dimensional brownian dynamics simulator for the study of ion permeation through membrane pores *J. Chem. Theory and Comput.* **10** 2911–26

[156] Ható Z, Valiskó M, Kristóf T, Gillespie D and Boda D 2017 Multiscale modeling of a rectifying bipolar nanopore: explicit water versus implicit-water simulations *Phys. Chem. Chem. Phys.* **19** 17816–26

Part II

Structure of a single electric double layer: effects due to surface charge regulation and non-Coulombic interactions

IOP Publishing

Molecular Theory of Electric Double Layers

Dimiter N Petsev, Frank van Swol and Laura J D Frink

Chapter 8

Molecular properties of a single electric double layer

8.1 Classical density functional theory model of a single flat electric double layer

The focus of this chapter is on application of classical density functional theory (c-DFT) to determine the structure and properties of a single charged interface that forms between a substrate and electrolyte solution (i.e., an EDL). The surface charge is due to a charge regulation surface chemical equilibrium and the solution molecular structure that accounts for the contributions from all species such as ions and solvent. The analysis, presented below, is limited to a mean field approximation model. Also, any dipole effects that might stem from the solvent polarity are not included. We use an approach colloquially known as 'semi-primitive', in which the solvent is present but its polarity is not explicitly accounted by the interactions. Instead, the Coulombic interactions are scaled with the average bulk dielectric permittivity of the solution. The semi-primitive models have been proven to be very successful in demonstrating the effect of the liquid structure on the properties of charged electrolyte interfaces [1–15]. These approximations lead to considerable computational efficiency, while allowing one to obtain important insights regarding the solution structure and surface chemistry effects on the properties of EDLs.

The system under consideration is a single, flat, and charged interface in contact with an infinite electrolyte solution. Hence, the representative characteristic functional is the grand thermodynamic potential, which has the form (see chapter 7)

$$\begin{aligned}\Omega\left[\{\rho_i(z)\}\right] &= \mathcal{F}^{id}\left[\{\rho_i(z)\}\right] + \mathcal{F}^{\text{ex}}_{HS}\left[\{\rho_i(z)\}\right] + \mathcal{F}^{\text{ex}}_{long}\left[\{\rho_i(z)\}\right] \\ &+ 2\pi \sum_{i=1}^{M} \int RdR \int dz\, \rho_i(R, z)\left[V_i^{ext}(R, z) - \mu_i\right]\end{aligned} \tag{8.1}$$

where M is the number of solution species and R is the radial coordinate that runs parallel to the charged interface. The free energy terms account for the ideal

$$\begin{aligned}\mathcal{F}^{id}\left[\{\rho_i(z)\}\right] = {} & 2\pi\, k_B T \sum_{i=1}^{M} \int R dR \\ & \times \int dz\, \rho_i(R, z)\left\{\ln\left[\Lambda_i^3 \rho_i(R, z)\right] - 1\right\},\end{aligned} \tag{8.2}$$

hard sphere

$$\mathcal{F}_{HS}^{ex}\left[\{\rho_i(z)\}\right] = 2\pi\, k_B T \int R dR \int dz\, \phi_{HS}\{n_\alpha(R, z)\}, \tag{8.3}$$

and mean field long-range

$$\begin{aligned}\mathcal{F}_{long}^{ex}\left[\{\rho_i(z)\}\right] = {} & \frac{\pi}{2} \sum_{i=1}^{M} \sum_{j=1}^{M} \int R dR \\ & \times \int dz \int dz' \rho_i(R, z) \rho_j(R, z')\, \phi_{LR}(R, |z - z'|)\end{aligned} \tag{8.4}$$

contributions. The function $\phi_{HS}\{n_\alpha(R, z)\}$ is derived from the fundamental measure theory (see chapters 6 and 7, and references [16, 17]). It has the remarkable property of being independent on the number of components M. The long-range interaction energy $\phi_{LR}(R, |z - z'|) = \phi_{LR}(r_{ij})$ depends on the distance $r_{ij} = r_{ij}(R, |z - z'|)$ between species i and j. It has a non-Coulombic (Lennard–Jones (LJ))

$$\phi_{LJ}(r_{ij}) = 4\epsilon_{ij}\left[\left(\frac{d_{ij}}{r_{ij}}\right)^{12} - \left(\frac{d_{ij}}{r_{ij}}\right)^{6}\right],\ r_{ij} > d_{ij} \tag{8.5}$$

and a Coulombic

$$\phi_{el}(r_{ij}) = \frac{q_i q_j}{4\pi\varepsilon\varepsilon_0 r_{ij}},\ r_{ij} > d_{ij}, \tag{8.6}$$

contribution [5, 7–10] where $d_{ij} = (d_i + d_j)/2$, d_i and d_j are the diameters of species i and j. The energy parameter ϵ_{ij} determines the depth of the attractive energy well. Hence, the long-range pair interactions energy reads

$$\phi_{lr}(r_{ij}) = \phi_{LJ}(r_{ij}) + \phi_{el}(r_{ij}). \tag{8.7}$$

The long-range interactions with the substrate wall also have LJ

$$\phi_{LJw}(z) = \epsilon_{iw}\left[\frac{2}{15}\left(\frac{d_s}{z}\right)^{9} - \left(\frac{d_s}{z}\right)^{3}\right],\quad z > d_i/2 \tag{8.8}$$

and electrostatic

$$\phi_{elw}(z) = \frac{q_i \sigma z}{2\varepsilon\varepsilon_0},\quad z > d_i/2 \tag{8.9}$$

components. The total long-range interaction with the charged wall is then

$$\phi_{lrw}(r_{ij}) = \phi_{LJw}(z) + \phi_{elw}(r_{ij}). \tag{8.10}$$

The structure of the EDL is found by minimizing the grand thermodynamic potential (8.1), or solving the equation (see equation (8.1))

$$\frac{\delta\Omega[T, V, \{\mu_i\}]}{\delta\rho_j(z)} = 0 \tag{8.11}$$

to obtain the density distribution of each solution component $\rho_j(z)$, including all ions and the solvent. Equation (8.11) is solved together with a charge regulation boundary condition. In our subsequent analysis we use the charge regulation condition defined by equations (2.4)–(2.6).

Knowing the density distributions of all species, including the positive and negative ions, allows one to find the total charge in the fluid $\rho_{el}(z)$. This charge is exactly balanced by the surface charge σ, or

$$\sigma = -\int_0^\infty dz\ \rho_{el}(z), \tag{8.12}$$

where the charge density equals the sum over all charged species, multiplied by their charge, or $\rho_{el}(z) = \sum_i q_i \rho_i(z)$.

8.2 Solution structure in an electric double layer with surface charge regulation

The solvent molecules are typically neutral, and therefore, are not involved in Coulombic interactions. However, they are often the most numerous components in electrolyte solutions, and thus, have a significant impact on the properties of EDLs. The solvent molecules contribute to the liquid stricture of the electrolyte solution, which is predominantly determined by excluded volume and other short range interactions [18]. They fill most of the volume in the solution and the dissolved ions may fit only in void, unoccupied (by the solvent molecules) spaces. This leads to density distributions (for all components) that exhibit multiple peaks near the wall, which reflect the local solution structure.

A few examples of the density profiles and local structuring are presented in figures 8.1, 8.2, and 8.3. The analysis is performed using the semi-primitive mean field model described above. All ions in these examples are monovalent. The solvent molecules interact among themselves, as well as with the dissolved ions, which corresponds to ionic solvation. The inter-ionic interaction is a combination of Coulombic and non-Coulombic (LJ) contributions. The LJ energy parameter of fluid interaction is $\epsilon_{ij}/k_BT = 1$ for all species. The charged wall is considered to be hard, or $\epsilon_{iw}/k_BT = 0$. The solution contains solvent, positive PDIs, positive non-PDIs, and a common negative counterion for both. The concentration of the potential determining ions is given by a 'pH' value. In accordance with the semi-primitive model, we use the bulk dielectric permittivity to scale the Coulombic interactions, or $\varepsilon = 78.5$ at

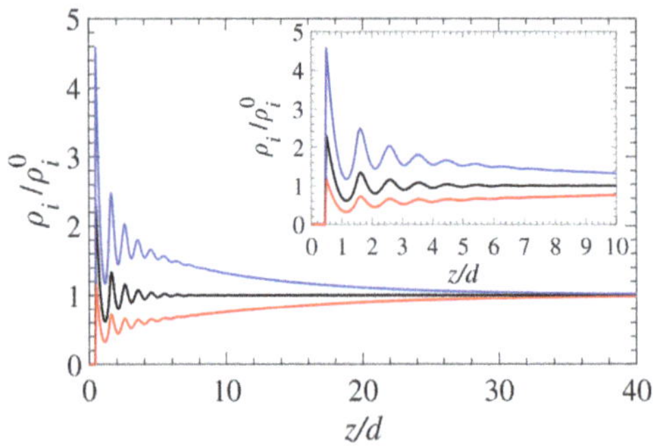

Figure 8.1. Normalized solutions components density profiles $\rho_i(z)$ near a charge regulating interface. ρ_i^0 is the bulk density of component *i*, far from the surface. All species have the same diameter *d*. The surface is negatively charged. The solution parameters are pH = 4 and $I = 0.01$ M. The black curves corresponds to the neutral solvent, the blue curve depicts the positive ions, and the red curve—the negative ions. (Reprinted with permission from [11], Copyright (2018) American Chemical Society.)

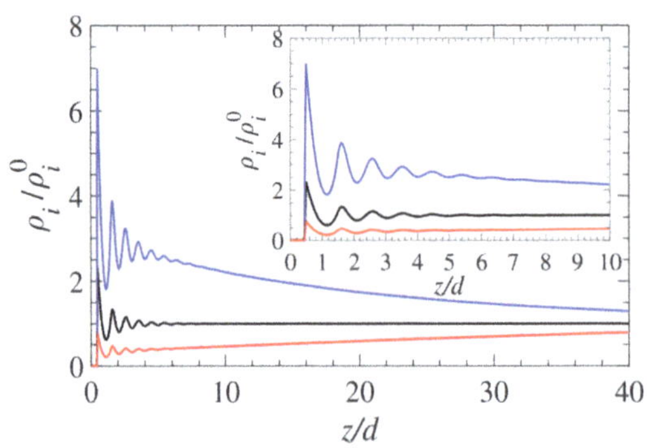

Figure 8.2. Normalized solutions components density profiles $\rho_i(z)$ near a charge regulating interface. ρ_i^0 is the bulk density of component *i*, far from the surface. All species have the same diameter *d*. The surface is negatively charged. The solution parameters are pH = 4 and $I = 0.001$ M. The black curves correspond to the neutral solvent. The black curves corresponds to the neutral solvent. The black curves corresponds to the neutral solvent, the blue curve depicts the positive ions, and the red curve—the negative ions. (Reprinted with permission from [11], Copyright (2018) American Chemical Society.)

temperature $T = 298$ K. The surface charge is determined by the charge regulation conditions (2.4)–(2.6) with parameters $\Gamma d^2/e = 0.66$ (Γ being the number of surface reactive groups per unit area), $pK_- = 6$, and $pK_+ = -2$. It is also assumed that σ is represented by a uniform surface charge density (per unit area). The concentration of PDIs near the surface is determined by equation (2.15) with $w(z) = 1$. The overall fluid density is $\rho d^3 = 0.8$. For further simplicity, all species are assumed to have the same diameter *d*.

Figure 8.1 shows the density profiles of all components for an electrolyte solution having a pH = 4, and total ionic strength $I = 0.01$ M. The latter corresponds to a Debye screening length $(\kappa d)^{-1} = 10.56$. The densities are scaled with their respective values, far away from the charged interface. Since the positive PDIs and non-PDIs have exactly the same size, they perfectly overlap and are represented by a single (blue) curve.

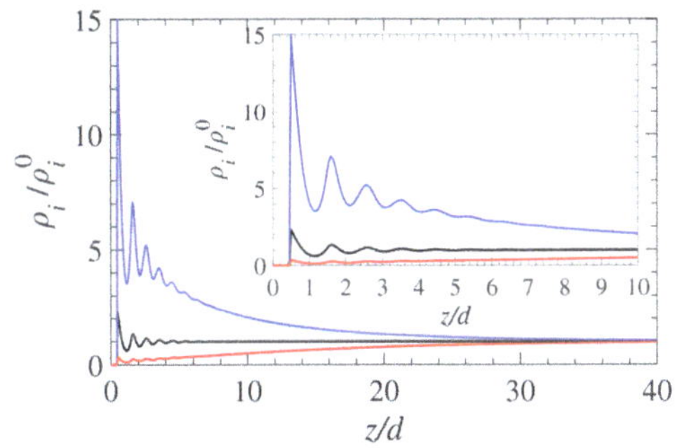

Figure 8.3. Normalized solutions components density profiles $\rho_i(z)$ near a charge regulating interface. ρ_i^0 is the bulk density of component i, far from the surface. All species have the same diameter d. The surface is negatively charged. The solution parameters are pH = 5 and $I = 0.01$ M. The black curves corresponds to the neutral solvent. The black curves corresponds to the neutral solvent. The black curves corresponds to the neutral solvent, the blue curve depicts the positive ions, and the red curve—the negative ions. (Reprinted with permission from [11], Copyright (2018) American Chemical Society.)

Their concentration increases near the charged wall surface due to electrostatic attraction (the surface charge is negative). The peaks near the wall indicate a liquid-like structuring, facilitated by the presence of the large number of solvent molecules.

The negative ions (red curve) show similar structuring near the wall, but their density increases away from the surface because of the electrostatic repulsion exerted by the negative surface.

The solvent density profile (black curve) is unaffected by the surface charge and only exhibits the local structuring effect induced by the presence of the hard wall. The inset offers a more detailed picture of the solution structure near the charged surface. Interestingly, if the long-range electrostatic interaction with the charged wall is subtracted, the ion density curves collapse onto that for the neutral solvent [7, 8]. The surface-induced liquid structure decay with distance can be examined following the approach suggested by Martynov for the analysis of radial distribution functions [19]. According to this approach, the structural maxima (or minima) fall on an exponential $\sim [1 + A\exp(-\beta z)]$, where A is the peak amplitude and β^{-1} is the characteristic length of structural decay. The characteristic length for the case depicted in figure 8.1 is estimated to be $(\beta d)^{-1} \simeq 1.3$ [7, 8].

Figure 8.2 presents the density profiles for an electrolyte solution that is ten times more diluted than the one discussed above. The overall ionic strength is $I = 0.001$ M while the remaining parameters are the same. The Debye screening length in this case is $(\kappa d)^{-1} = 33.39$. This indicates that relative density variations for the positive and negative ions would propagate to about three times greater distance from the charged surface. This is clearly evident in figure 8.2, which shows deviations for both the positive and negative ions, from their bulk values even at a distance of 40 ionic/molecular diameters.

The density of the positive ions (blue curve) at the surface, represented by the first peak, appears higher than in the higher ionic strength case shown in figure 8.1. The reason for this observation is that all plotted densities are normalized by their respective bulk values away from the surface. In reality, the absolute density next to the surface is substantially lower for I = 0.001 M (figure 8.2) than it is for I = 0.01 M (figure 8.1).

The density of the negative ions (red curve) is also much lower near the surface, which is not only because of the lower overall density, but also due to the electrostatic repulsion from the negatively charged surface.

The solvent density profile (black curve) in figure 8.2 is identical to the one in figure 8.1. In addition, after subtracting the long-range wall interaction contributions, both the positive and negative ionic density profiles curves collapse onto the one for the solvent. This uniform curve is indistinguishable from the one for I = 0.01 M, which means that liquid structure is unaffected by the background salt concentration and its impact on the Coulombic interactions with the charged wall.

Figure 8.3 illustrates a case where the ionic strength is I = 0.01 M (same as in figure 8.1) but the surface is more negatively charged. The latter is achieved by increasing the solution pH to 5. As a result, the positive ions (blue curve) experience stronger attraction to the charged surface, which is seen from the higher peaks near the wall. The negative ions (red curve) are expelled more efficiently because of the increased repulsion, while the neutral solvent structure (black curve) is unaffected.

The precise density profiles in the EDL couple to the surface charge by affecting the charge regulation chemical equilibrium (see equations (2.4)–(2.6)). However, the charge itself determines the densities of the ions, and most importantly that of the PDIs. Based on that reasoning, the description of the EDL requires solving equation (8.11) together with the surface charge regulation condition (2.5), or equivalently (2.6) [7]. It turns out that the precise solution structure strongly couples with the surface chemistry.

The effect of the solvent on the surface charged is demonstrated in figure 8.4. It demonstrates the result of a virtual experiment where the solvent contribution is gradually 'switched on' as its molecular diameter d_s (relative to that of the ions, d) changes from zero to one. For $d_s/d \to 0$ all curve coalesce at the same value for the surface charge $\sigma d^2/e = 0.002\,94$.

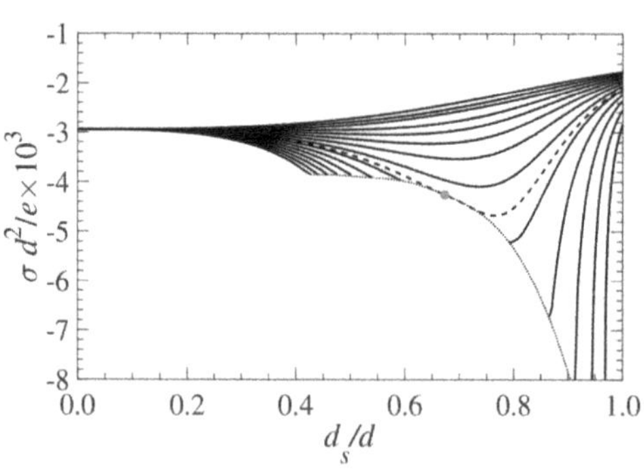

Figure 8.4. Surface charge of an EDL as a function of the solvent molecular diameter. The solid curves correspond different values of the LJ parameter ϵ_{ij}/k_BT, which starts at zero (top curve) and increases by increments of 0.1. The dashed curve is for $\epsilon_{ij}/k_BT = 0.76$. As ϵ_{ij}/k_BT increases, the curves exhibit a transition from monotonic to non-monotonic behavior, and finally break at the spinodal (dotted curve). The blue dot corresponds to the critical point. All remaining parameters are the same as in figure 8.1. (Reprinted from [8], Copyright (2017), with permission from Elsevier.)

In absence of any explicit solvent, and at moderate salt concentrations, the electrolyte solution behaves as an ideal gas of ions, placed in an external electric field. This is also reinforced by the fact that the solvent LJ interactions are substantially reduced or vanish with the molecular diameter (see equation (8.5)).

As the solvent molecular diameter increases, the fluid non-Coulombic interactions become more important and lead to a distinction between the curves based on the values of the LJ interaction parameter ϵ_{ij}. Note that increasing the solvent diameter is equivalent to an increase of the fluid volume fraction which, in the presence of attractive forces, leads to a phase separation and the formation of spinodal phase boundary indicated by the dotted curve. The dot in the figure corresponds to the critical point. This phase transition is not physical in the sense that an experiment, based on a gradual variation in solvent molecular size, is not feasible. However, it clearly indicates how unrealistic the models are that ignore the solvent structural contribution. These include the PB equation, as well as more advanced statistical-mechanical models that account for the ionic dimensions, but still ignore the solvent structure [20–24]. The physically relevant domain is on the right of the spinodal envelope where the solvent and ionic diameters are comparable.

8.3 Conclusions

The equilibrium properties of charged interfaces are governed by the thermodynamic equilibrium between the electrolyte solution and the reactive charge regulating surface. While the surface chemical reaction involves only the PDIs, their local density and proximity to the substrate depends on all species in the solution. The focus in this chapter is on the role of the solvent molecules, which is extremely important. Typically, the solvent density is much greater than the densities of the dissolved ions. Hence, it is the solvent that is primarily important for the overall structure of the electrolyte solution. The solvent molecules are usually neutral and therefore involved in non-Coulombic interactions (LJ, excluded volume). The solvent exhibits these types of interactions with itself, the dissolved ions, and the charged regulating surface. The solution structure couples to the charge regulating chemical reaction, which determines the surface charge and potential. Hence, an accurate account of the particular surface chemistry and the precise solution structure near the wall are both essential for the analysis of the EDL.

References

[1] Tang Z, Scriven L E and Davis H T 1991 Density-functional perturbation theory of inhomogeneous simple fluids *J. Chem. Phys.* **95** 2659–68

[2] Tang Z, Scriven L E and Davis H T 1992 A three-component model of the electrical double layer *J. Chem. Phys.* **97** 494–503

[3] Tang Z, Scriven L E and Davis H T 1994 Effects of solvent exclusion on the force between charged surfaces in electrolyte solution *J. Chem. Phys.* **100** 527–4530

[4] Ted Davis H, Zhang L and White H S 1992 Simulations of solvent effects on confined electrolytes *J. Chem. Phys.* **98** 5793–9

[5] Frink L J D and van Swol F 1994 Solvation forces and colloidal stability–a combined Monte-Carlo and density–functional theory approach *J. Chem. Phys.* **100** 9106–16

[6] Forsman J, Woodward C E and Szparaga R 2015 ed W Rocchia and M Spagnuolo *Computational Electrostatics for Biological Applications* (Cham: Springer) ch 2 pp 17–38

[7] Fleharty M E, van Swol F and Petsev D N 2016 Solvent role in the formation of electric double layers with surface charge regulation: A bystander or a key participant? *Phys. Rev. Lett.* **116** 048301

[8] Vangara R, Brown D C R, van Swol F and Petsev D N 2017 Electrolyte solution structure and its effect on the properties of electric double layers with surface charge regulation *J. Colloid Interface Sci.* **488** 180–9

[9] Vangara R, van Swol F and Petsev D N 2017 Solvation effects on the potential and charge distributions in electric double layers *J. Chem. Phys.* **147** 214704

[10] Vangara R, van Swol F and Petsev D N 2018 Solvophilic and solvophobic surfaces and non-coulombic surface interactions in charge regulating electric double layers *J. Chem. Phys.* **148** 044702

[11] van Swol F and Petsev D N 2018 Solution structure effects on the properties electric double layers with surface charge regulation assessed by density functional theory *Langmuir* **34** 13808–20

[12] Vangara R, Stoltzfus K, van Swol K, York M and Petsev D N 2019 Coulombic and non-coulombic effects in charge-regulating electric double layers *Mater. Res. Express* **6** 086331

[13] Oleksy A and Hansen J-P 2006 Towards a microscopic theory of wetting by ionic solutions. i. surface properties of the semi-primitive model *Mol. Phys.* **104** 2871–83

[14] Oleksy A and Hansen J-P 2009 Microscopic density functional theory of wetting and drying of solid substrate by an explicit solvent model of ionic solutions *Mol. Phys.* **107** 2609–24

[15] Gillespie D, Petsev D N and van Swol F 2020 Electric double layers with surface charge regulation using density functional theory *Entropy* **22** 132

[16] Rosenfeld Y 1989 Free–energy model for the inhomogeneous hard–sphere fluid mixture and density functional theory of freezing *Phys. Rev. Lett.* **63** 980–3

[17] Roth R 2010 Fundamental measure theory for hard-sphere mixtures: a review *J. Phys.: Condens. Matter* **22** 063102

[18] Hansen J P and McDonald I R 2006 *Theory of Simple Liquids* (New York: Academic)

[19] Martynov G A 2008 Power and exponential asymptotic forms of correlation functions *Theor. Math. Phys.* **156** 1356–64

[20] Plischke M 1988 Pair correlation functions and density profiles in the primitive model of the electric double layer *J. Chem. Phys.* **88** 2712–8

[21] Yu Y X, Wu J Z and Gao G H 2004 Density–functional theory of spherical electric double layers and zeta potentials of colloidal particles in restricted–primitive model electrolyte solutions *J. Chem. Phys.* **120** 7223–33

[22] Tang Z, Striven L E and Davis H T 1992 Interactions between primitive electrical double layers *J. Chem. Phys.* **97** 9258–66

[23] Pizio O and Sokolowski S 2006 On the effects of ion–wall chemical association on the electric double layer: A density functional approach for the restricted primitive model at a charged wall *J. Chem. Phys.* **125** 024512

[24] Blum L and Henderson D 1992 Statistical mechanics of electrolytes at interfaces Henderson D *Fundamentals of Inhomogeneous Fluids* (New York: Dekker) ch 6 pp 239–76

IOP Publishing

Molecular Theory of Electric Double Layers

Dimiter N Petsev, Frank van Swol and Laura J D Frink

Chapter 9

Ionic solvation effects and solvent–solvent interactions

Ionic solvation is a term that describes the interaction of ions with the solvent molecules. In aqueous solutions, that would be an ion–dipole type of interaction. Our model is based on the 'semi-primitive' approach [1–3], which mimics the solvent–ion interactions using the LJ energy (8.5), and a bulk dielectric permittivity $\varepsilon = 78.5$. While approximate, this approach considerably simplifies the computations and provides many important insights about the solution structure effects on the properties of EDLs [4–6]. Recent experimental studies of the ionic solvation effects on the structure of EDLs are available in [7, 8].

9.1 Solvation of the potential determining ions

Since the surface charge $\tilde{\sigma}$ is determined by a chemical equilibrium such as (2.4)–(2.6), it is clear that the solvation of the PDIs should have a major impact on the EDL properties. Strong solvation interactions reduce the PDI concentration in the subsurface layer, shift the equilibrium towards greater surface dissociation, and therefore, increase the negative charge magnitude. The opposite case of poor solvation leads to an accumulation of PDIs at the surface and neutralizing of the negative surface charge as described by equations (2.4)–(2.6).

Examples for the PDI density profiles near a charged wall are shown in figure 9.1. The solvation interaction energy varies between $\epsilon_{s-pdi}/k_BT = 0.25$ and $\epsilon_{s-pdi}/k_BT = 5$. All other fluid components interact with LJ energy $\epsilon_{ij}/k_BT = 1$. The PDI density profiles for $\epsilon_{s-pdi}/k_BT < 1$ exhibit very pronounced peaks near the charged wall. Hence, if the solvation energy ϵ_{s-pdi}/k_BT is less than the typical fluid interactions energies between the remaining solution components, the PDIs tends to accumulate in the first layer next to the EDL charged interface with the substrate. At $\epsilon_{s-pdi}/k_BT = 1$, there is still a small peak near the wall. However, as the LJ parameter solvation energy parameter ϵ_{s-pdi}/k_BT increases to values greater than 1, the peak

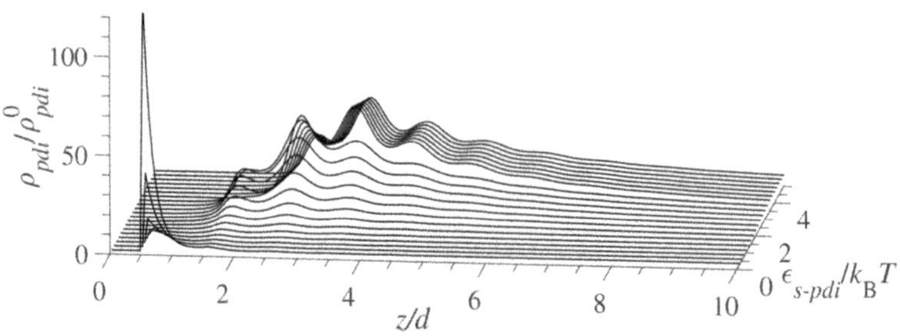

Figure 9.1. Normalized PDI density profiles for varying solvation LJ energy ϵ_{s-pdi}/k_BT between 0.25 and 5 in steps of $\Delta\epsilon_{s-pdi}/k_BT = 0.25$. The remaining parameters are the same as in figure 8.1.

near the wall disappears and more distant density peaks start forming at locations that correspond to roughly integer numbers of ionic diameters. Around $\epsilon_{s-pdi}/k_BT = 3.0$, the PDIs are mostly located in the approximately third molecular layer from the surface. After $\epsilon_{s-pdi}/k_BT = 3.25$, the highest peak moves farther away from the charged interface, and the PDIs density in the immediate vicinity of the reactive surface becomes quite low. These rearrangements of the density profiles are driven by the effect of the solvation interactions and the requirement for fixed chemical potentials for all species in the grand thermodynamic functional (8.1). The detailed density profile affects the surface charge by means of the charge regulation surface chemical equilibrium, which shifts according to the local concentration of PDIs.

Figure 9.2 demonstrates the change in the surface charge $\tilde{\sigma}$, and potential $\tilde{\psi}_s$, as the PDI solvation energy ϵ_{s-pdi} increases. The physical location of the surface charge and potential is at the reactive, charge regulating surface. It complements figure 9.1 and offers an illustration of the surface response to the rearrangement of the PDI density in the fluid. Both curves exhibit a sharp change at $\epsilon_{s-pdi}/k_BT = 3.25$, which is also where the PDI density peak moves abruptly farther into the solution, away from the charged surface.

The surface charge $\tilde{\sigma}$ (see figure 9.2(a)) is practically zero at low solvation energies, $\epsilon_{s-pdi}/k_BT \sim 1$, and approaches a fully dissociated state ($\tilde{\sigma} = -0.66$) for large energies, $\epsilon_{s-pdi}/k_BT \gtrsim 4$. The transition from a fully associated (neutral) to a fully dissociated (negatively charged) surface is not gradual, but has the steep change at $\epsilon_{s-pdi}/k_BT = 3.25$, discussed above. This drop correlates with the shift in the PDI density maximum shown in figure 9.1.

The surface potential $\tilde{\psi}_s$ in figure 9.2(b) shows a similar trend. It is very low at $\epsilon_{s-pdi}/k_BT \sim 1$, then exhibits an abrupt change to even more negative values at $\epsilon_{s-pdi}/k_BT = 3.25$, and levels off for $\epsilon_{s-pdi}/k_BT \gtrsim 4$ to $\tilde{\psi}_s = -22$. Such high potentials are physically unlikely for typical colloidal systems. Still, figure 9.2 is useful in demonstrating the qualitative trends that are the result of the solvation of the PDIs.

The increase of the surface charge and potential negativity leads to attraction of positive non-PDIs near the surface. Hence, the overall fluid positive charge

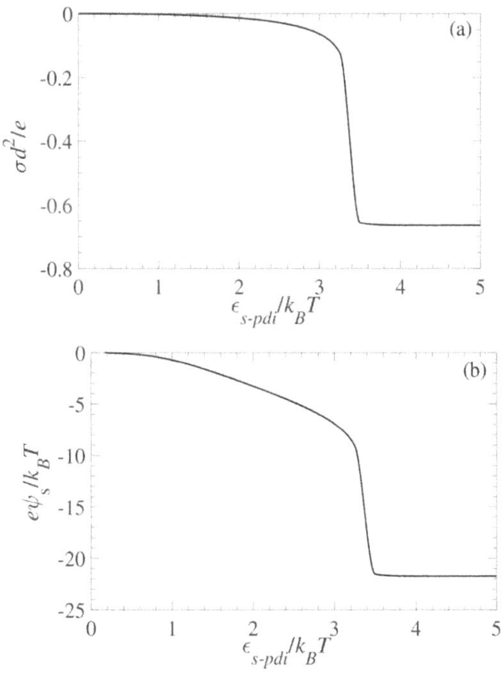

Figure 9.2. (a) Surface charge $\tilde{\sigma} = \sigma d^2/e$ and (b) potential $\tilde{\psi}_s = e\psi_s/k_BT$ at the reactive wall–solution charge regulating interface as functions of the PDI solvation energy ϵ_{s-pdi}/k_BT. (Adapted from [5], with the permission of AIP Publishing.)

density in the vicinity of the EDL interface increases with the solvation of the positive PDIs [5, 6].

The fluid charge distribution in the EDL, ρ_{el}, is shown in figure 9.3. As the PDI becomes more solvated, the surface negative charge increases and the positive counter charge density in the adjacent fluid increases (see figure 9.3(a)). The fluid charge density increase is relatively moderate up to $\epsilon_{s-pdi}/k_BT = 3.25$, and spikes to much greater values after that, following the trends in the surface charge and potential, as evident from the inset in figure 9.3, which provides a more detailed presentation of the density profiles. The fluid charge density profiles at low PDI solvation energies ($0 \leqslant \epsilon_{s-pdi}/k_BT \leqslant 1.2$) are shown in figure 9.3(b). The peaks next to the wall exhibit a minimum at $\epsilon_{s-pdi}/k_BT \simeq 0.5$. The reason of that non-monotonic behavior is the coupling between the PDIs solvation and the surface charge regulation. Poorly solvated PDIs tend to accumulate at the wall. They neutralize the surface and populate the fluid in the immediate vicinity with an excess of positive PDIs. As the solvation energy increases, the PDI density near the surface decreases, which leads to a reduction in the local positive fluid charge density. At the same time, the surface becomes more negative since there are less PDIs to participate in the charge regulating equilibrium (2.4)–(2.6). The negative surface charge, however, attracts the positive non-PDIs, and this effect becomes more pronounced above $\epsilon_{s-pdi}/k_BT \simeq 0.5$, leading to an increase in the fluid positive charge density at the wall.

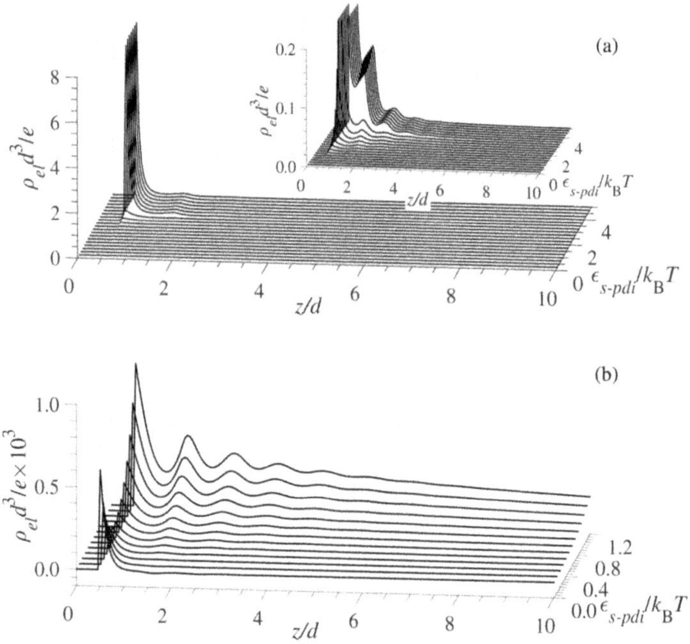

Figure 9.3. Fluid charge density ρ_{el} in the EDL as a function of the PDI solvation energy ϵ_{s-pdi}/k_BT. (a) The LJ solvation energy parameter varies between $\epsilon_{s-pdi}/k_BT = 0.25$ and 5 in increments $\Delta\epsilon_{s-pdi}/k_BT = 0.25$. The inset provides a more detailed illustration of the charge density variation at higher solvation energies. (b) The LJ solvation parameter varies between $\epsilon_{s-pdi}/k_BT = 0$ and 1.2 in increments $\Delta\epsilon_{s-pdi}/k_BT = 0.1$. (Reprinted from [5], with the permission of AIP Publishing.)

It is important to realize that some experimental methods (such as those based on electrokinetic phenomena) assess the charge and potential after the density peak [9–13], at the so-called plane of shear. The values, for both the charge and the potential at that location, are usually very different [4–6, 14].

9.2 Solvation of the positive non-potential determining ions

The non-PDIs are not involved in any form of a chemical interaction with the reactive surface groups. However, they indirectly affect the chemical equilibrium, and therefore, the resultant surface charge and potential (see equation (2.6)). The solvation interaction determines the local density of the ions next to the charged wall and in the fluid as illustrated in figure 9.4.

If the solvation energy of the positive ions is $\epsilon_{s-pos}/k_BT \lesssim 1$ (all other fluid components in the bulk interact with LJ energy $\epsilon_{ij}/k_BT = 1$), then the peaks in the density profile near the wall are high, and mostly due to the excluded volume effects, as well as the Coulombic attraction to the negatively charged surface. There are more distant peaks that decay with distance (see the inset in figure 9.4). As the solvation energy ϵ_{s-pos} increases, the first peak in the density profile sharply decreases, while the second and even third peaks grow to become more pronounced

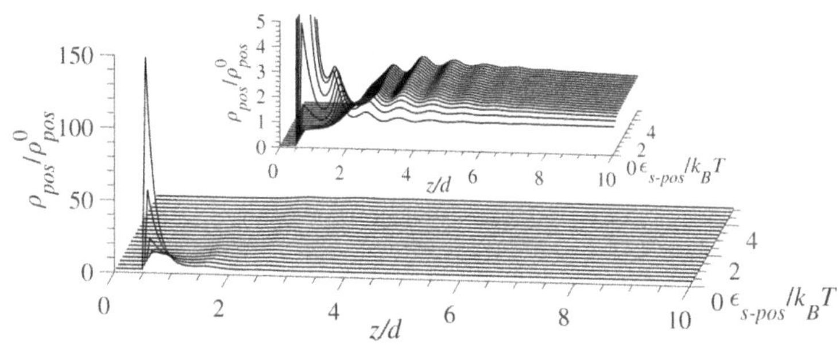

Figure 9.4. Normalized density profiles of the positive non-PDIs for varying solvation LJ energy ϵ_{s-pos}/k_BT between 0.25 and 5 in steps of $\Delta\epsilon_{s-pos}/k_BT = 0.25$. The remaining parameters are the same as in figure 8.1. The inset provides more detailed view of the lower density peaks, far from the surface.

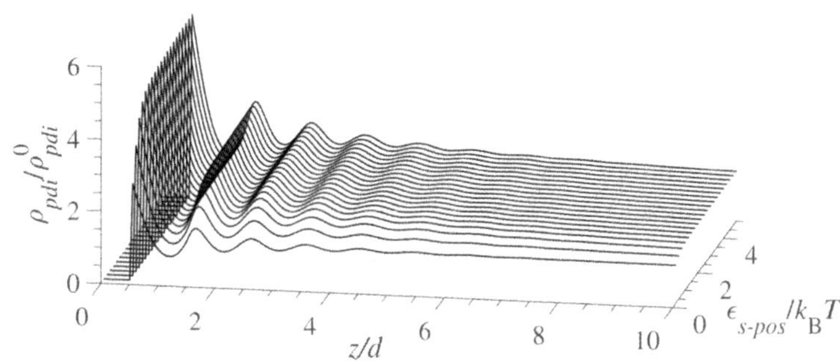

Figure 9.5. Normalized density profiles of the PDIs for varying solvation LJ energy for the positive non-PDIs. ϵ_{s-pos}/k_BT between 0.25 and 5 in steps of $\Delta\epsilon_{s-pos}/k_BT = 0.25$. The plot shows how the PDIs are indirectly affected by the non-Coulombic solvation of the positive non-PDIs.

(see the inset in figure 9.4). The local density values are ultimately determined by the equating the local chemical potentials to those predetermined in equation (8.1).

The presence of positive non-PDIs at the EDL-wall interface affects the local concentration of PDIs. Greater numbers of positive non-PDIs obstruct the inclusion of PDIs in the immediate vicinity of the charged surface, and vice versa, as seen from the comparison of the density distributions shown in figures 9.4 and 9.5. The density of the PDIs in the first layer, next to the wall, steadily increases and reaches a plateau at $\epsilon_{s-pos}/k_BT \gtrsim 1$, Hence, the reduction of the positive non-PDIs numbers at wall (figure 9.4) favors the increase of the PDI density (figure 9.5).

The resultant surface charge $\tilde{\sigma}$ and potential $\tilde{\psi}_s$ are shown in figure 9.6. The surface charge (see figure 9.6(a)) is defined at the reactive surface where the charge regulation reaction occurs. According to the discussion above, the non-reactive positive ions are in the first layer near the charged wall, as implied by figure 9.4 at $\epsilon_{s-pos}/k_BT = 1$. As a result, the density of the PDIs in the immediate vicinity of the reactive surface is lower (see figure 9.5), which in turn leads to less efficient neutralization of the reactive surface and more negative charge as shown in

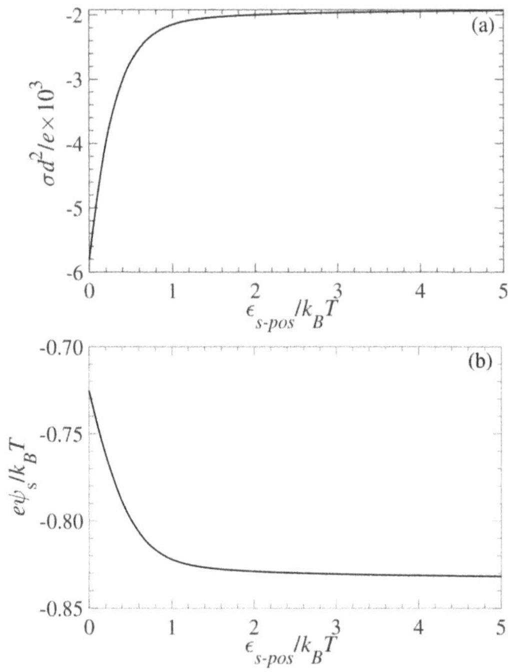

Figure 9.6. (a) Surface charge $\tilde{\sigma} = \sigma d^2/e$ and (b) potential $\tilde{\psi}_s = e\psi_s/k_BT$ at the reactive wall–solution charge regulating interface as functions of the positive non-PDI solvation energy ϵ_{s-pos}/k_BT.

figure 9.6(a). More importantly, however, the presence of positive ions at the wall brings the surface potential to values that are close to the Nernst potential (in this case, $\tilde{\psi}_N = -4.6$) and thus reduces the surface charge according to the charge regulation condition (2.6). As the solvation energy ϵ_{s-pos} increases, the density of the positive ions near the charged surface decreases, the potential becomes more negative, and the unobstructed PDIs approach the surface to further neutralize it.

The surface potential change is plotted in figure 9.6(b). It exhibits an opposite trend when compared to the surface charge. It is less negative in the case of weakly solvated positive non-PDIs and becomes more negative with increasing the ionic solvation.

The substantial change in both the surface charge and potential that occurs between $\epsilon_{s-pos}/k_BT = 0.25$ and 1 is in agreement with the results depicted in figure 9.4 and 9.5. These results indicate that the weakly solvated positive ions tend to accumulate at the surface and reduce the number of PDIs there through competition. This shifts the charge regulation equilibrium to greater surface dissociation and more negative charge. Stronger solvation of the positive non-PDIs leads to a reduction of the surface negative charge, and allows for the negative non-PDIs to populate the fluid layers in the immediate vicinity of the surface, which contributes to the increased negativity of the surface potential with the solvation of the positive non-PDIs.

The fluid charged density, ρ_{el}, is shown in figure 9.7. It peaks near the surface for $\epsilon_{s-pos}/k_BT < 1$. Interestingly, as the positive ions become more solvated and removed

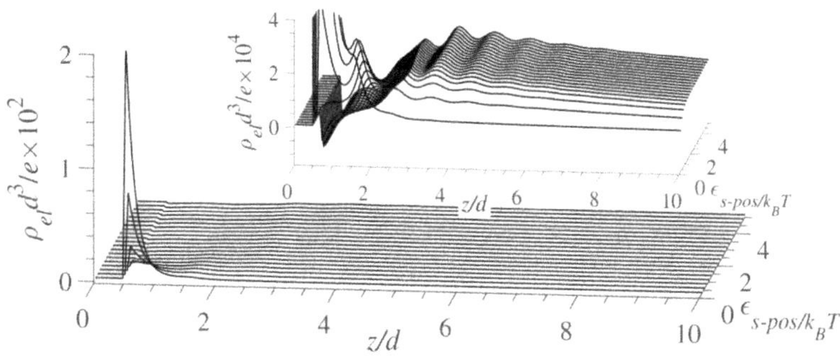

Figure 9.7. Fluid charge density ρ_{el} in the EDL as a function of the positive non-PDIs solvation energy ϵ_{s-pos}/k_BT. The inset provides a more detailed illustration of the charge density variation near the wall.

farther into the bulk, the fluid charge density exhibits a sign reversal and the overall fluid charge density near the wall becomes negative. This observation can be explained by realizing that as the positive non-PDIs are removed from the layer next to the charged wall, the PDIs do not experience competition from the positive non-PDIs, and have an unobstructed access to the reactive surface. Hence, the surface is neutralized through the charge regulation mechanism (2.4)–(2.6). The reduction of the negative surface charge (see figure 9.6(a)) allows for the negative ions to populate the layer next to the wall in numbers that exceed their counterparts (the positive PDIs and non-PDIs). The consequence of this ionic rearrangement is the increase of the overall negative surface potential shown in figure 9.6(b). The negative fluid charge turns positive at a distance of the order of 2-3 molecular diameters away from the wall. This positive charge density is mostly due to the positive non-PDIs, which are attracted by the negative ions near the wall. At the same time, the positive ions are far enough from the wall and are able to support a solvation shell.

Clearly, the charge and potential at the shear plane is different from those at the true substrate solution interface. Details on how c-DFT can be applied to account for the solvation effect on the electrokinetic charge and potential are available elsewhere [4–6, 14].

9.3 Solvation of the negative non-potential determining ions

The solvation of the negative non-PDIs also has an indirect effect on the surface charge, potential, and the overall EDL properties. The negative ion density near the charged interface of the EDL is lower because of the electrostatic repulsion from the negative wall. This is evident in figure 9.8, which indicates that the peaks near the wall are lower (in comparison with positive ions case shown in figure 9.4), and the slope, away from the surface, is generally positive. The density at the wall decreases with the solvation energy ϵ_{s-neg}. While this behavior is analogous to cases for the PDIs and the positive non-PDIs, its effect on the surface charge and potential is different.

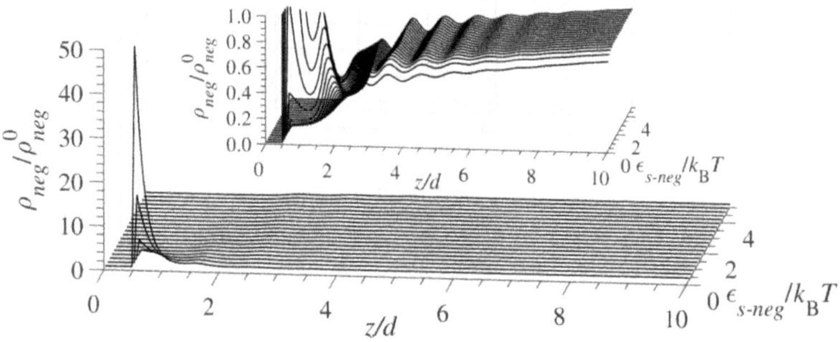

Figure 9.8. Normalized density profiles of the negative non-PDIs for varying solvation LJ energy ϵ_{s-neg}/k_BT. The step in the variation of the LJ parameter is $\Delta\epsilon_{s-neg}/k_BT = 0.25$. The remaining parameters are the same as in figure 8.1. The inset provides a more detailed illustration of the negative ions density variation in the EDL.

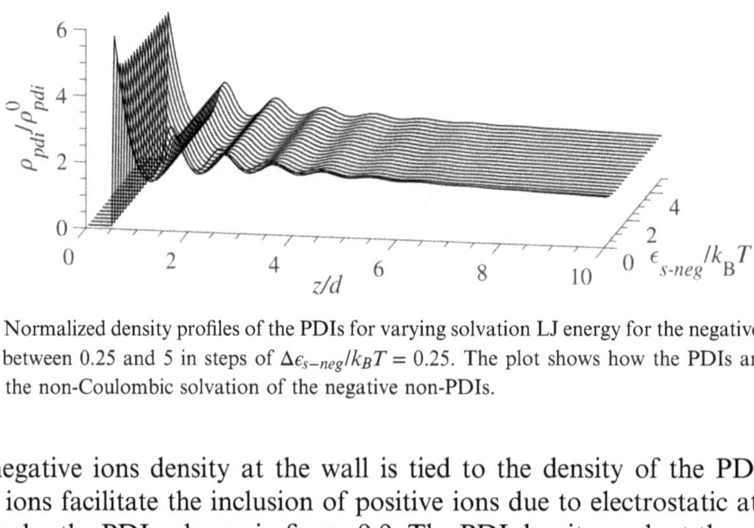

Figure 9.9. Normalized density profiles of the PDIs for varying solvation LJ energy for the negative non-PDIs. ϵ_{s-neg}/k_BT between 0.25 and 5 in steps of $\Delta\epsilon_{s-neg}/k_BT = 0.25$. The plot shows how the PDIs are indirectly affected by the non-Coulombic solvation of the negative non-PDIs.

The negative ions density at the wall is tied to the density of the PDIs. More negative ions facilitate the inclusion of positive ions due to electrostatic attraction. This includes the PDIs, shown in figure 9.9. The PDI density peak at the wall is the highest for poorly solvated negative ions. As the negative ions become more solvated and their density near the wall drops, the PDI numbers in the first layer decrease accordingly. This leads to a shift in the surface chemical equilibrium in agreement with the charge regulation condition (2.4)–(2.6).

Figure 9.10 depicts the effect of the negative ions solvation on the surface charge $\tilde{\sigma}$ and potential $\tilde{\psi}_s$. The surface charge (see figure 9.10(a)) becomes more negative with the increase of the negative ion solvation energy ϵ_{s-neg}. The reason is that the presence of negative ions near the surface favors the inclusion of positive ions from the bulk, including PDIs, which reduces the surface negative charge. The reduction of the negative ions density at the reactive interface is accompanied by a reduction of the PDIs density, which facilitates the dissociation of the surface groups. As a result, the surface becomes more negative. This scenario is in agreement with

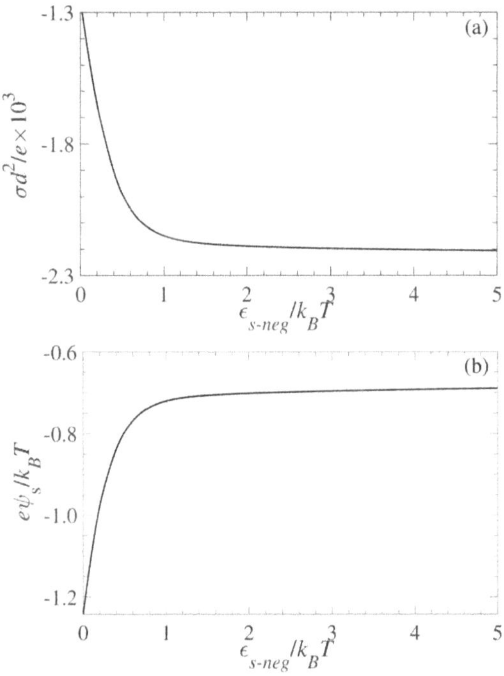

Figure 9.10. (a) Surface charge $\tilde{\sigma} = \sigma d^2/e$ and (b) potential $\tilde{\psi}_s = e\psi_s/k_BT$ as a function of the negative non-PDI solvation energy ϵ_{s-neg}/k_BT.

figure 9.9, which clearly shows a reduction the PDI density with the increase of the solvation energy ϵ_{s-neg}.

The surface potential dependence on the solvation energy of the negative non-PDIs, ϵ_{s-neg}, is presented in figure 9.10(b). The surface potential is $\tilde{\psi}_s = 1.225$ at $\epsilon_{s-neg}/k_BT = 0$. It then steeply changes and reaches a plateau. Most of the change in the potential occurs between $\epsilon_{s-neg}/k_BT = 0$ and 1, which again agrees with the PDI reduction shown in figure 9.9.

Figure 9.11 depicts the fluid charge density ρ_{el} in the EDL as a function of the negative ions solvation ϵ_{s-neg}/k_BT. The charge near the wall is negative for $\epsilon_{s-neg}/k_BT < 1$, very low for $\epsilon_{s-neg}/k_BT \sim 1$, and becomes increasingly positive for all values of $\epsilon_{s-neg}/k_BT > 1$. It follows the trend in the PDIs density, shown in figure 9.9, as well in the surface charge $\tilde{\sigma}$ and potential $\tilde{\psi}_s$ presented in figure 9.10. The low solvation energy favors the accumulation of negative ions near the charged surface. This leads to a simultaneous increase of the local PDI density, as the oppositely charged ions are attracted into the subsurface layer. The PDIs reduce the negative surface charge, which in turn allows for a further increase of the negative ions density. Increasing the solvation energy leads to a decrease in the number of negative ions at the wall, as well as of that for the PDIs. The surface equilibrium shifts towards more negative values of the surface charge, which attracts positive non-PDIs. The surface potential becomes less negative, partially due to the local increase of the positive charge.

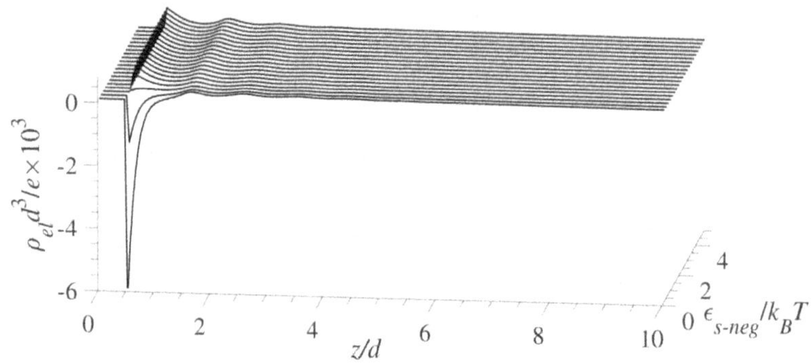

Figure 9.11. Fluid charge density ρ_{el} in the EDL as a function of the negative non-PDIs solvation energy ϵ_{s-neg}/k_BT.

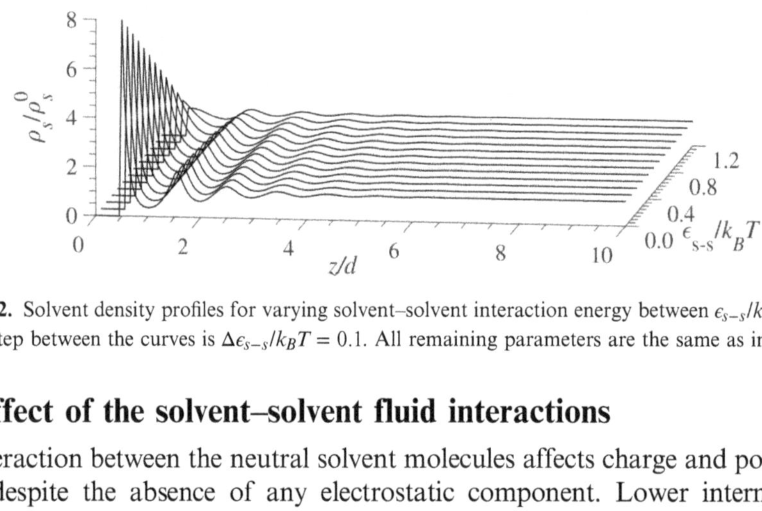

Figure 9.12. Solvent density profiles for varying solvent–solvent interaction energy between $\epsilon_{s-s}/k_BT = 0$ and 1.2. The step between the curves is $\Delta\epsilon_{s-s}/k_BT = 0.1$. All remaining parameters are the same as in figure 8.1.

9.4 Effect of the solvent–solvent fluid interactions

The interaction between the neutral solvent molecules affects charge and potential of EDLs despite the absence of any electrostatic component. Lower intermolecular attraction increases the solvent density in the layer adjacent to the wall (see figure 9.12), and therefore reduces the number of ions through an excluded volume competition. The solvent–solvent LJ energy is varied between 0 and 1.2 k_BT. The upper limit is chosen to prevent the system from getting too close to the LJ fluid triple point [15]. The ionic solvation also is affected, as the solvent molecules may preferentially be attracted, or avoid the proximity of the ions, depending on the relative strength of the solvent–solvent LJ energy ϵ_{s-s} to that of the ionic solvation interactions.

Figure 9.13 presents the density profiles of the PDIs in the EDL as a function of the solvent–solvent LJ interaction energy ϵ_{s-s}. The PDI density at the surface is low for weak solvent interactions. In fact, the first peaks in the density are lower than the second peaks for $\epsilon_{s-s}/k_BT \lesssim 0.7$. For $\epsilon_{s-s}/k_BT > 0.7$, the trend reverses and the highest density peaks are the ones nearest to the charged surface. These variations in the PDI density at the reactive interface are reflected in the resultant surface charge $\tilde{\sigma}$ and potential $\tilde{\psi}_s$ values as evident in figure 9.14.

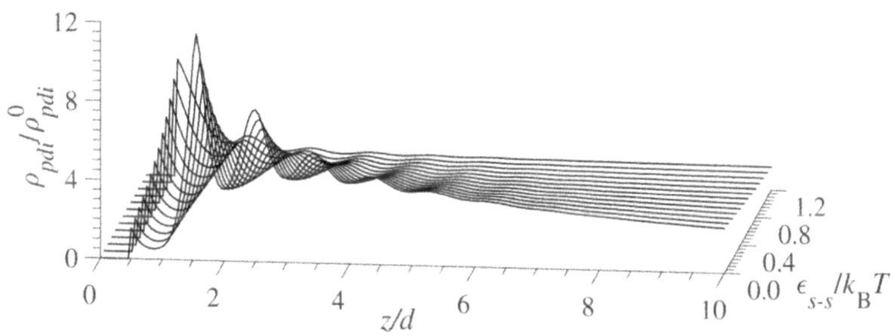

Figure 9.13. Normalized density profiles of the PDIs for varying interaction LJ energy between the solvent molecules. ϵ_{s-s}/k_BT between 0 and 1.2 in steps of $\Delta\epsilon_{s-s}/k_BT = 0.1$.

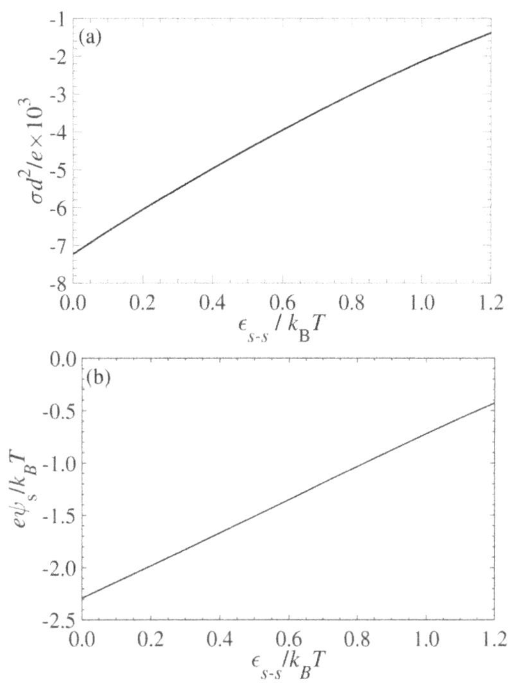

Figure 9.14. (a) Surface charge $\tilde{\sigma} = \sigma d^2/e$ and (b) potential $\tilde{\psi}_s = e\psi_s/k_BT$ as a function of the solvent–solvent interaction energy ϵ_{s-s}/k_BT.

The surface charge magnitude is the greatest for a hard sphere solvent ($\epsilon_{s-s}/k_BT = 0$), and decreases (i.e., becomes more positive) as the LJ energy becomes more attractive (see figure 9.14(a)). The same is true for the surface potential magnitude, shown in figure 9.14(b). Remarkably, the potential dependence on the solvent attractive energy ϵ_{s-s}/k_BT is almost linear.

The variation of the surface charge and potential is quite significant, which again (see also figure 8.4) reinforces the notion of the importance of the liquid structure and its coupling to the charge regulation surface reactions.

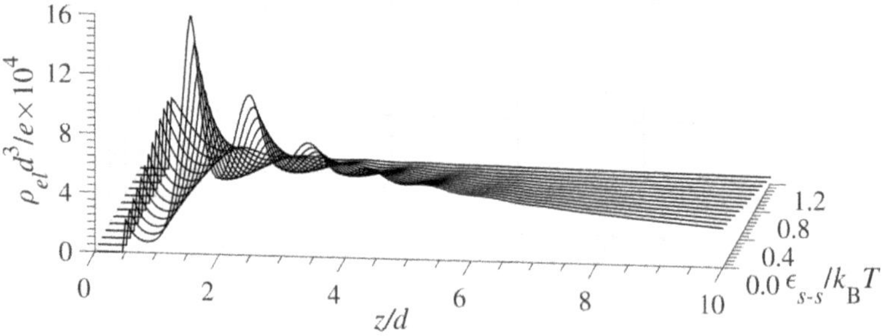

Figure 9.15. Fluid charge density ρ_{el} as a function of the solvent–solvent LJ energy ϵ_{s-s}/k_BT.

Figure 9.15 illustrates the effect of solvent interactions ϵ_{s-s} on the fluid charge density in the EDL. The shape of the curves is very similar to the density profiles for the PDIs in figure 9.13. This is not surprising since the fluid charge is predominantly due to the positive ions that are attracted to the negatively charged surface. In addition, the PDIs have the same diameter as the positive non-PDIs, which contributes to the similarity between the two figures.

9.5 Conclusions

The neutral solvent molecules in the electrolyte solution interact among themselves and with the dissolved ions. These interactions affect the local density of all solution components. For example, the solvation of the potential determining ions directly affects their number density in the vicinity of the reactive charge regulating interface. Hence, the surface chemical equilibrium (2.1) may shift, depending on the mass action law. Poor solvation increases the PDI density at the surface and facilitates the neutralization of the negative charge. Strong solvation reduces the PDI density at the wall, which leads to an increase of the surface negative charge magnitude. The latter determines the fluid charge and potential distribution in the EDL.

The solvation of the positive and negative non-PDIs has an indirect effect of the surface charge regulation. The variation of their local densities at the wall has an impact on the positive PDIs, which are attracted or repelled by the Coulombic forces that act between the ions, thus affecting the surface equilibrium (2.1).

The interactions between the neutral solvent molecules affects the surface charge, potential and overall structure of the EDL. Weak interactions saturate the fluid layer at the surface with solvent molecules, and the PDIs are expelled by excluded volume effects. Strong solvent–solvent interactions have the opposite effect, which facilitates the surface partial neutralization by the PDIs.

References

[1] Oleksy A and Hansen J-P 2006 Towards a microscopic theory of wetting by ionic solutions. i. surface properties of the semi-primitive model *Mol. Phys.* **104** 2871–83

[2] Oleksy A and Hansen J-P 2009 Microscopic density functional theory of wetting and drying of solid substrate by an explicit solvent model of ionic solutions *Mol. Phys.* **107** 2609–24

[3] Gillespie D, Petsev D N and van Swol F 2020 Electric double layers with surface charge regulation using density functional theory *Entropy* **22** 132

[4] Vangara R, Brown D C R, van Swol F and Petsev D N 2017 Electrolyte solution structure and its effect on the properties of electric double layers with surface charge regulation *J. Colloid Interface Sci.* **488** 180–9

[5] Vangara R, van Swol F and Petsev D N 2017 Solvation effects on the potential and charge distributions in electric double layers *J. Chem. Phys.* **147** 214704

[6] van Swol F and Petsev D N 2018 Solution structure effects on the properties electric double layers with surface charge regulation assessed by density functional theory *Langmuir* **34** 13808–20

[7] Lee S S, Fenter P, Park C, Sturchio N C and Nagy K L 2010 Hydrated cation speciation at the muscovite (001)-water interface *Langmuir* **26** 16647–51

[8] Lee S S, Fenter P, Park C and Sturchio N C 2012 Monovalent ion adsorption at the muscovite (001)solution interface: Relationships among ion coverage and speciation, interfacial water structure, and substrate relaxation *Langmuir* **28** 8637–50

[9] Thompson A 2003 Nonequilibrium molecular dynamics simulation of electro-osmotic flow in a charged nanopore *J. Chem. Phys.* **119** 7503–11

[10] Qiao R and Aluru N R 2003 Ion concentrations and velocity profiles in nanochannel electroosmotic flows *J. Chem. Phys.* **118** 4692–701

[11] Qiao R and Aluru N R 2004 Charge inversion and flow reversal in a nanochannel electro-osmotic flow *Phys. Rev. Lett.* **92** 198301

[12] Qiao R and Aluru N R 2005 Scaling of electrokinetic transport in nanometer channels *Langmuir* **21** 8972–7

[13] Wu P and Qiao R 2011 Physical origins of apparently enhanced viscosity of interfacial fluids in electrokinetic transport *Phys. Fluids* **23** 072005

[14] Vangara R, van Swol F and Petsev D N 2018 Solvophilic and solvophobic surfaces and non-coulombic surface interactions in charge regulating electric double layers *J. Chem. Phys.* **148** 044702

[15] Mastny E A and de Pablo J J 2007 Melting line of the Lennard-Jones system, infinite size, and full potential *J. Chem. Phys.* **127** 104504

IOP Publishing

Molecular Theory of Electric Double Layers

Dimiter N Petsev, Frank van Swol and Laura J D Frink

Chapter 10

Surface solvation and non-Coulombic ion–surface interactions

The substrate–solution non-Coulombic interactions significantly affect the charge and potential at the EDL interface. The solvent–substrate interaction defines the surface properties in terms of solvophilicity or solvophobicity. It is central to all capillary and wetting phenomena [1], and contributes to the liquid structuring in confined spaces, as well as to the overall interactions between surfaces in liquids [2–8].

The non-Coulombic interactions between ions and the surface determine their affinity to physically adsorb at the surface (see chapter 2). Below, we present a brief overview of the effects of these surface interactions, using the c-DFT approach that includes all solution species. The interactions between all species in the fluid are $\epsilon_{ij}/k_BT = 1$. The interactions with the charged wall are also set to $\epsilon_{i-w}/k_BT = 1$ except for one component (i.e., solvent, PDIs, positive non-PDIs, or negative ions), which is examined. Additional details are available in [9–11].

10.1 Solvent–surface interactions. Solvophilic and solvophobic surfaces

The interactions between the solvent molecules and the charged wall determine the local density of the former near the charged interface. Examples of the density profiles for varying interactions with the wall ϵ_{s-w} are shown in figure 10.1. The surface is considered solvophobic if the interaction with the solvent molecules ϵ_{s-w}/k_BT is less than the interactions between the fluid components in the solution, which in this case are $\epsilon_{ij}/k_BT = 1$. Consequently, for $\epsilon_{s-w}/k_BT > 1$, the surface is considered solvophilic. The profiles show that the solvent density at the wall increases with the LJ attraction as all peaks in the distributions exhibit an increase with ϵ_{s-w}/k_BT.

The effect of the surface solvation on the local density of the PDIs is demonstrated in figure 10.2. It shows that the PDIs become displaced as the surface

doi:10.1088/978-0-7503-2276-8ch10

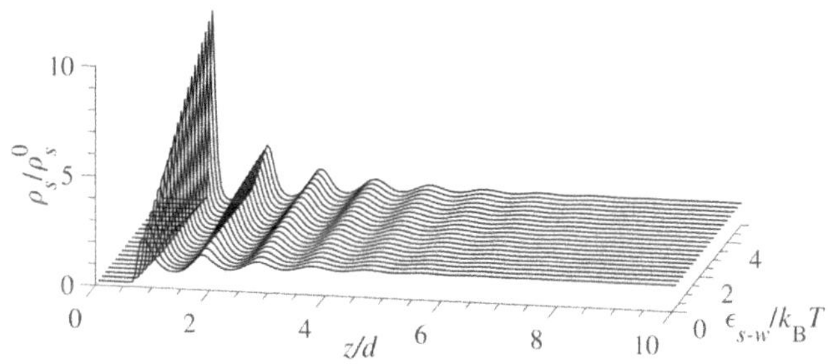

Figure 10.1. Normalized density profiles of the solvent molecules for varying surface solvation LJ energy ϵ_{s-w}/k_BT. The step in the variation of the LJ parameter is $\Delta\epsilon_{s-w}/k_BT = 0.25$. The remaining parameters are the same as in figure 8.1. (Adapted from [11], Copyright (2018) American Chemical Society.)

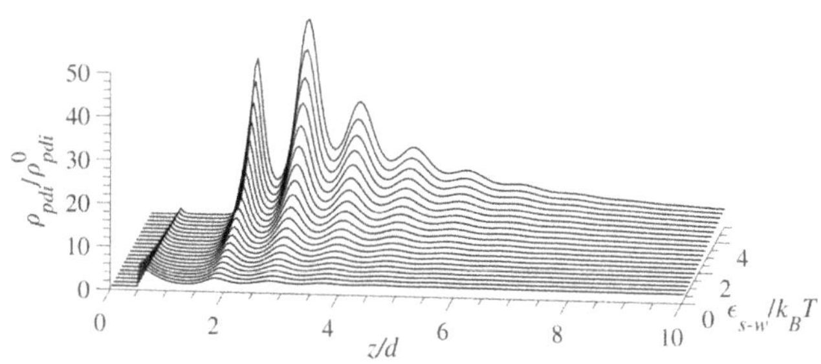

Figure 10.2. Normalized density profiles of the PDIs for varying surface solvation LJ energy ϵ_{s-w}/k_BT. The step in the variation of the LJ parameter is $\Delta\epsilon_{s-w}/k_BT = 0.25$. (Adapted from [11], Copyright (2018) American Chemical Society.)

becomes less solvophobic and more solvophilic. This is accompanied by an increase in the local PDI density around two diameters away from the surface. This is driven by the increases of the surface negative charge magnitude and potential as seen in figure 10.3, which attracts the positive PDIs closer to the surface.

The increase of the surface charge, $\tilde{\sigma}$, and potential, $\tilde{\psi}_s$, magnitudes, displayed in figure 10.3, follows the change of the PDI density at charge regulating interface. The PDI concentration for the solvophobic walls, in this example, is greater, which leads to a shift in the charge regulation equilibrium towards less negative surface charge in accordance with equations (2.4)–(2.6). However, as the solvent–wall attractions increases, the local PDI density decreases and the surface charge and potential become more negative.

The fluid charge in the EDL is overall positive and is determined by the surface charge as illustrated in figure 10.4. The shape of the curves resembles the PDI density profiles because the total fluid charge is mostly due to the positive PDIs and non-PDIs, which are attracted to the negatively charged wall. While the negative ions also contribute to the fluid charge density, they are mostly expelled because of the electrostatic repulsion, and therefore, their effect is hard to notice at this scale.

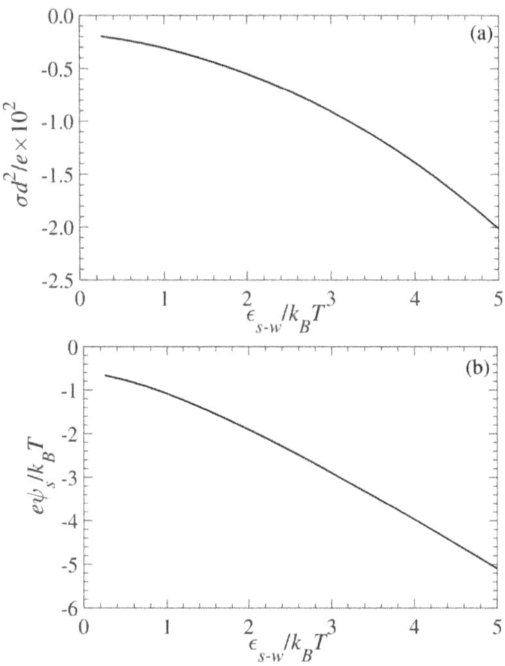

Figure 10.3. (a) Surface charge $\tilde{\sigma} = \sigma d^2/e$ and (b) potential $\tilde{\psi}_s = e\psi_s/k_BT$ as a function of the solvent–wall interaction energy ϵ_{s-w}/k_BT.

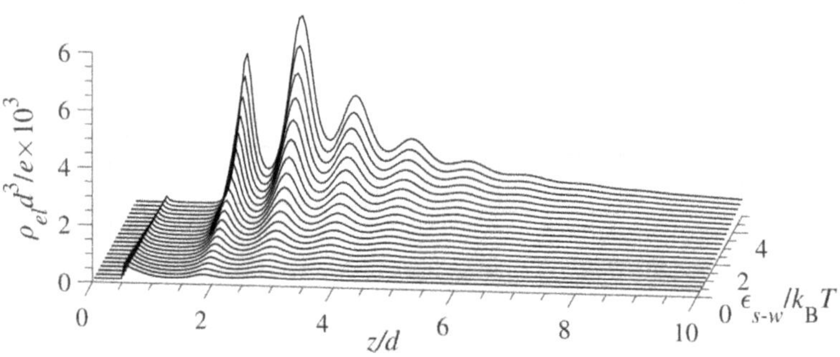

Figure 10.4. Fluid charge density ρ_{el} as a function of the solvent–surface interaction energy ϵ_{s-w}/k_BT. The step in the variation of the LJ parameter is $\Delta\epsilon_{s-w}/k_BT = 0.25$.

10.2 Effect of the non-Coulombic interactions between the potential determining ions and the charged wall

The specific non-Coulombic LJ interaction between the PDIs and the wall substrate has a major effect on the surface charge by manipulating the chemical equilibrium (2.4)–(2.6). An increase of the local concentrations of PDIs neutralizes the surface, while a lower PDI density at the wall facilitates an increase of the negative charge.

Figure 10.5 shows the PDI density profiles for different values of the ion-surface non-Coulombic, LJ interaction energies ϵ_{pdi-w}. The densities at the wall (represented by the peaks at $z/d \sim 1$) sharply increase with the LJ energy. As ϵ_{pdi-w}/k_BT increases, the charge and potential and the EDL interface becomes less negative, which is shown in figure 10.6. At $\epsilon_{pdi-w}/k_BT \sim 3$, both the surface charge, $\tilde{\sigma}$, and potential, $\tilde{\psi}_s$ become essentially equal to zero. The further increase of the LJ energy ϵ_{pdi-w}/k_BT

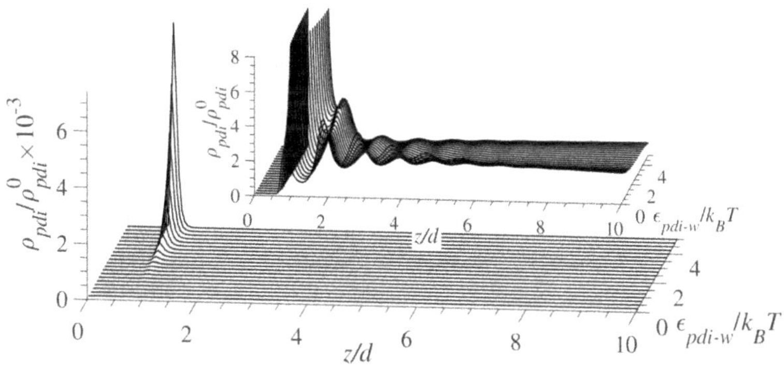

Figure 10.5. Density profiles for the PDIs at different values for the surface interaction energy ϵ_{pdi-w}/k_BT. The step in the variation of the LJ parameter is $\Delta\epsilon_{pdi-w}/k_BT = 0.25$. The inset provides a more detailed illustration of the PDIs density variation in the EDL.

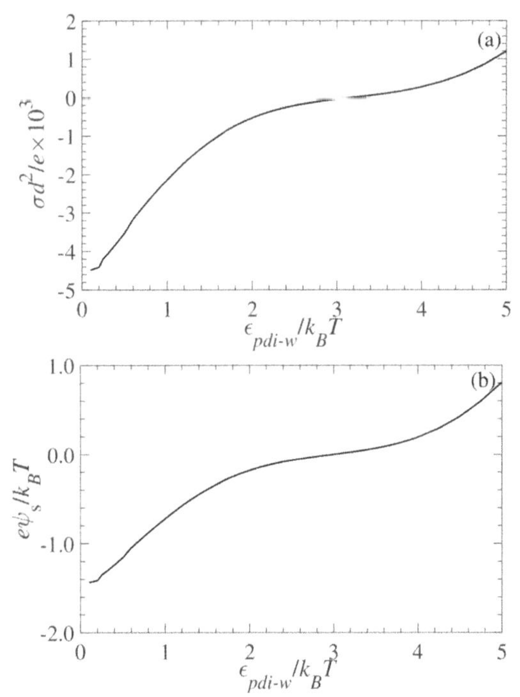

Figure 10.6. (a) Surface charge $\tilde{\sigma} = \sigma d^2/e$ and (b) potential $\tilde{\psi}_s = e\psi_s/k_BT$ as a function of the solvent–wall interaction energy ϵ_{pdi-w}/k_BT.

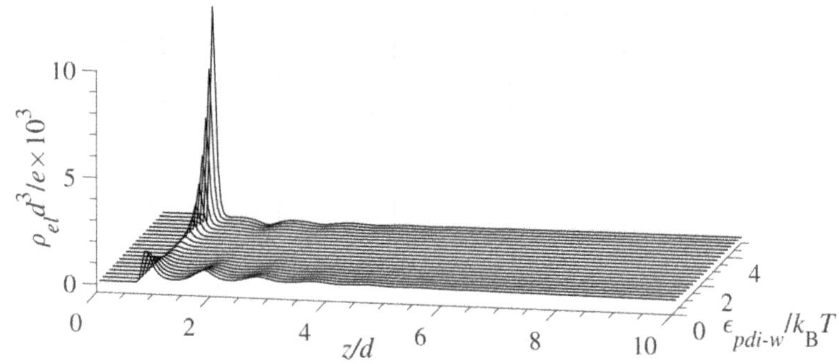

Figure 10.7. Fluid charge density ρ_{el} as a function of the PDI-surface interaction energy ϵ_{pdi-w}/k_BT. The step in the variation of the LJ parameter is $\Delta\epsilon_{pdi-w}/k_BT = 0.25$.

reverts the surface charge and potential to positive values. These changes are entirely driven by the surface charge regulation condition (2.4)–(2.6). The role of the non-Coulombic LJ interactions ϵ_{pdi-w} is that they control the density of the surface-reactive PDIs in the immediate vicinity of the EDL interface, thus shifting the equilibrium in accordance with the mass action law.

The fluid charge density profiles, at different values for the LJ energies ϵ_{pdi-w}/k_BT, are plotted in figure 10.7. The density peaks at the wall exhibit a non-monotonic behavior, showing a well-defined minimum around $\epsilon_{pdi-w}/k_BT \simeq 2.2$. The surface charge and potential are very negative for $\epsilon_{pdi-w}/k_BT = 0$, because the PDI's density is low and the charge regulating equilibrium is skewed towards greater surface dissociation. This leads to Coulombic attraction of positive ions. This includes the non-PDI positive ions, which are in greater numbers overall. The accumulation of positive ions near the wall increases the local positive charge density as indicated by the peaks and low ϵ_{pdi-w}/k_BT. Note that this Coulombic attraction leads to the appearance of more distant peaks in the positive fluid charge density ρ_{el} centered around $z/d \simeq 2, 3, 4, \ldots$. The increased attraction of PDIs, neutralizes the surface charge, and thus, reduces the Coulombic attraction of positive counterions. As shown in figure 10.6 above, the surface charge and potential exhibit and charge polarity reversal, which will turn the Coulombic attraction into repulsion. However, the LJ wall attraction for $\epsilon_{pdi-w}/k_BT > 2.2$ overcomes the Coulombic repulsion and leads to a high density of the positive PDIs (see also figure 10.5). Interestingly, the overall charge densities for $\epsilon_{pdi-w}/k_BT \gtrsim 3$ are negative beyond the first positive peak. They also exhibit minima (instead of maxima peaks) around $z/d \simeq 2, 3, 4, \ldots$. This implies that the overpopulation of positive PDIs in the layer next to the EDL–substrate interface generates positive electric field in the fluid, which attracts mostly negative ions at distances greater that $z/d \simeq 1$.

10.3 Effect of the non-Coulombic positive ions—surface interactions

The positive non-PDIs interaction with the substrate affects the local density of the positive ions in the immediate vicinity of the charged interface, and thus, it influences the charge regulation process by changing the density of the PDIs.

Figure 10.8 presents the density profiles for the positive non-PDIs for different values of the ion-surface non-Coulombic attractive energy ϵ_{pos-w}. The density of positive ions at the wall steeply increases with ϵ_{pos-w}/k_BT. The maxima of the density peaks near the wall fall on an increasing exponential curve. The great relative number of positive ions near the charged surface creates an excluded volume and electrostatic barrier for the PDIs, which have to compete in order to reach the reactive, charge-regulating interface. The effect of the non-Coulombic LJ energy of interaction between the positive non-PDIs and the surface on the density profiles of the PDIs is shown in figure 10.9. The PDIs' peaks height, near the wall-EDL interface, decreases with the increase of positive non-PDIs density (see figure 10.8). The more distant peaks at $z/d \simeq 2, 3, 4, \ldots$, follow a similar height decreasing behavior as the positive non-PDIs, become more attracted to the wall due to the non-Coulombic LJ interaction ϵ_{pos-w}.

The effects of the positive non-PDIs' interactions with the substrate wall on the surface charge and potential are shown in figure 10.10.

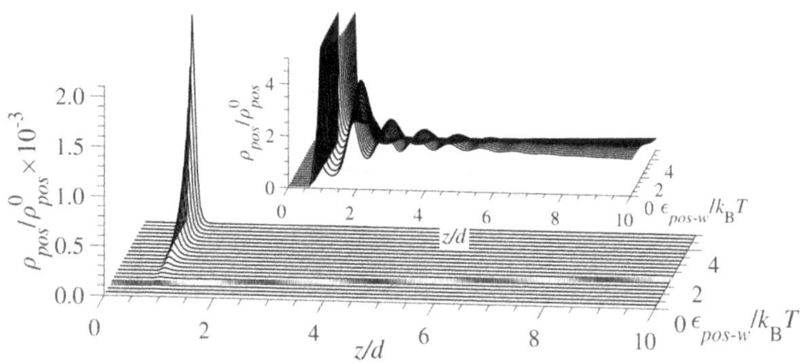

Figure 10.8. Density profiles for the positive non-PDIs at different values for the surface interaction energy ϵ_{pos-w}/k_BT. The step in the variation of the LJ parameter is $\Delta\epsilon_{pos-w}/k_BT = 0.25$.

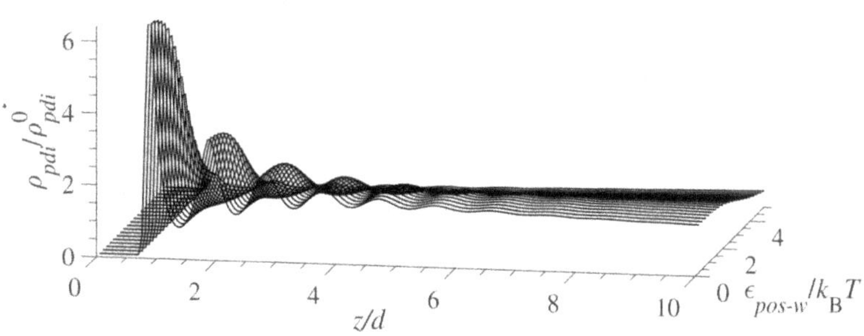

Figure 10.9. Density profiles for the PDIs at different values for the non-PDI positive ions-wall interaction energy ϵ_{pos-w}/k_BT. The step in the variation of the LJ parameter is $\Delta\epsilon_{pos-w}/k_BT = 0.25$. The LJ energy of non-Coulombic surface attraction for the PDIs is $\epsilon_{pdi-w}/k_BT = 1$ for all curves.

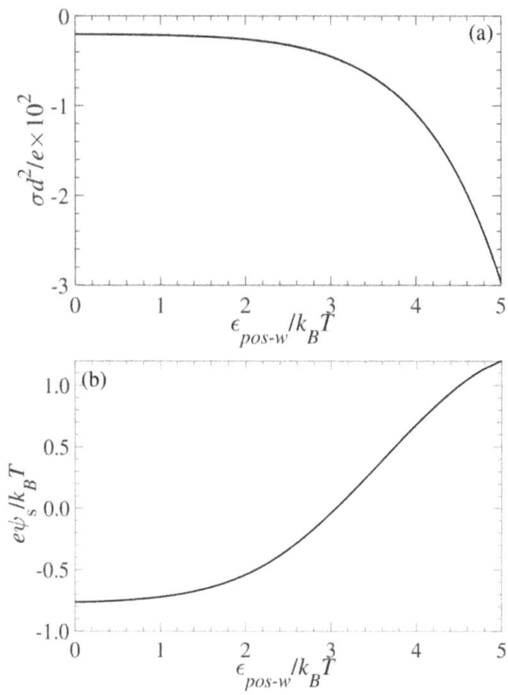

Figure 10.10. (a) Surface charge $\tilde{\sigma} = \sigma d^2/e$ and (b) potential $\tilde{\psi}_s = e\psi_s/k_BT$ as a function of the positive ions-wall interaction energy $\epsilon_{pos-w}/k_\text{B}T$.

The surface (see figure 10.10(a)) becomes more negative with the increased attraction of the positive ions to the wall. This is driven by the concurrent reduction of the PDI density at the wall, which favors surface dissociation according to the charge regulation mechanism (2.4)–(2.6).

The surface potential is shown in figure 10.10(b). The potential negativity decreases with ϵ_{pos-w}, and turns positive at $\epsilon_{pos-w}/k_\text{B}T \gtrsim 3$. This is due to the accumulation of great numbers of positive ions, which leads to a positive shift in the local potential [10].

The fluid charge distribution ρ_{el} in the EDL is depicted in figure 10.11. The main figure shows a growing peak near the wall with ϵ_{pos-w}. Most of the fluid charge is due to the positive non-PDIs. Hence, the peak's maxima increases exponentially with the non-Coulombic surface attraction, similarly to the positive ions densities plotted in figure 10.8. The inset in figure 10.11 offers greater detail of the peak structure, particularly at greater distances from the charged interface. It also shows that as the surface charge turns positive, the fluid charge is mostly negative after the first positive peak near the wall. The charge profile exhibits well-defined minima at $z/d \gtrsim 2$, which implies that layers of negative charges are formed next to the first positive peak.

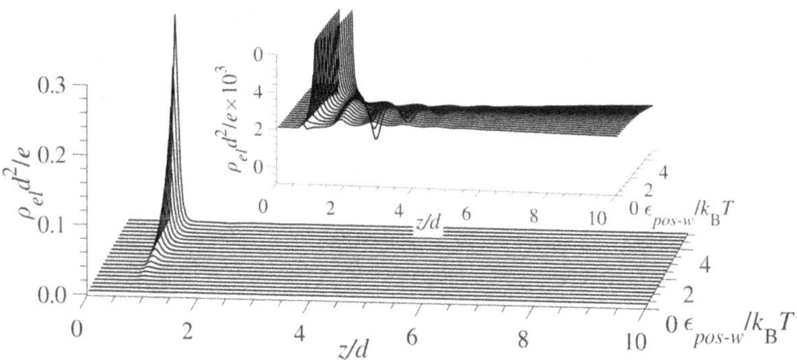

Figure 10.11. Fluid charge density ρ_{el} as a function of the positive non-PDI-surface interaction energy ϵ_{pos-w}/k_BT. The step in the variation of the LJ parameter is $\Delta\epsilon_{pos-w}/k_BT = 0.25$. Inset: rescaling the vertical axis provides more details of the peak structure.

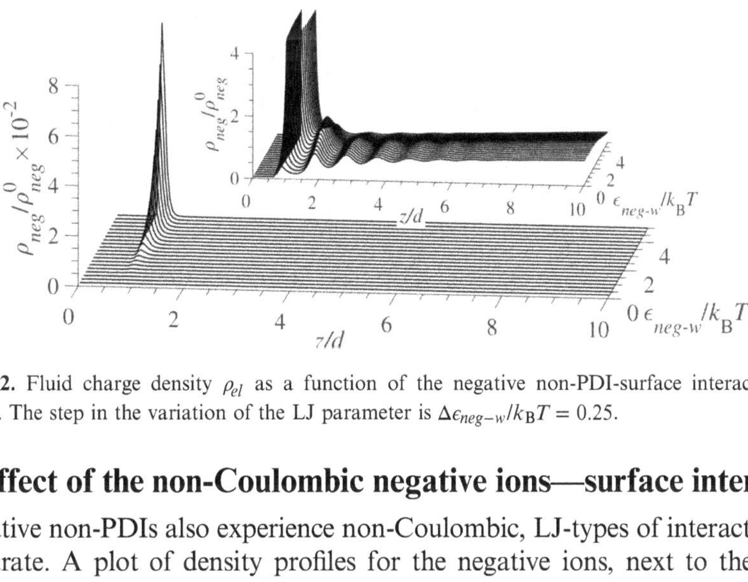

Figure 10.12. Fluid charge density ρ_{el} as a function of the negative non-PDI-surface interaction energy ϵ_{neg-w}/k_BT. The step in the variation of the LJ parameter is $\Delta\epsilon_{neg-w}/k_BT = 0.25$.

10.4 Effect of the non-Coulombic negative ions—surface interactions

The negative non-PDIs also experience non-Coulombic, LJ-types of interactions with the substrate. A plot of density profiles for the negative ions, next to the charged surface, and for different values for the interaction energy ϵ_{neg-w}/k_BT, is shown in figure 10.12. Unsurprisingly, the negative ions density increases with the LJ attraction, which is evident by the growing peak near the EDL–substrate interface. Refining the vertical axis shows multiple peaks in the density profiles that form at greater distances away from the surface (see inset). The first and highest peaks exponentially increase in height with ϵ_{neg-w}/k_BT. The second row of peaks, however, pass through a maximum at $\epsilon_{neg-w}/k_BT \simeq 2.5$. The more distant peaks exhibit a decreasing trend with ϵ_{neg-w}/k_BT. This is due to the significant accumulation of negative ions in the first peaks, which repels ions with the same charge polarity at distances $z/d \gtrsim 2$. At the same time, the positive ions in the solution are attracted to the layer of negative ions. This includes the PDIs, whose density profiles are depicted in figure 10.13.

The PDI density at the wall increases proportionally to that of the negative ions. Hence, the PDI density at the charged wall increases with the interaction energy

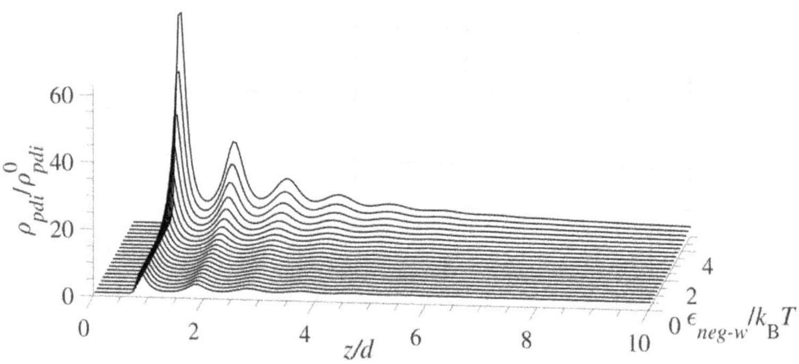

Figure 10.13. Density profiles for the PDIs at different values for the non-PDI positive ions–wall interaction energy $\epsilon_{neg-w}/k_\mathrm{B}T$. The step in the variation of the LJ parameter is $\Delta\epsilon_{neg-w}/k_\mathrm{B}T = 0.25$. The LJ energy of non-Coulombic surface attraction for the PDIs is $\epsilon_{pdi-w}/k_\mathrm{B}T = 1$ for all curves.

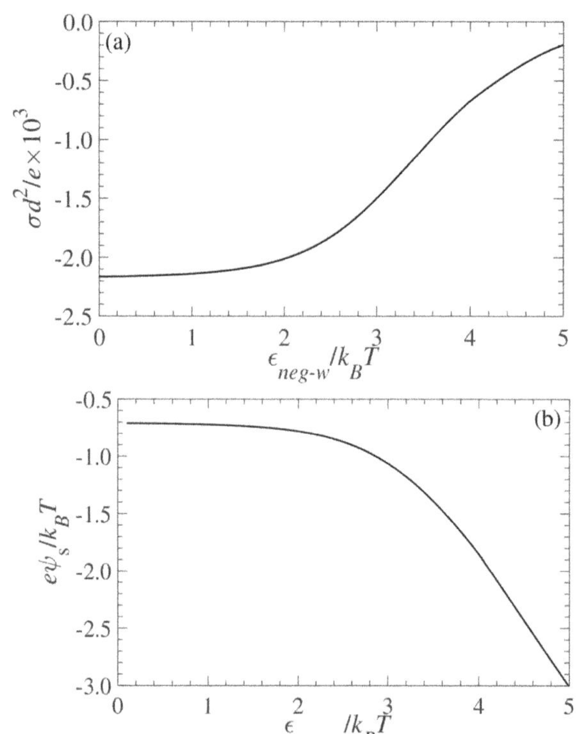

Figure 10.14. (a) Surface charge $\tilde{\sigma} = \sigma d^2/e$ and (b) potential $\tilde{\psi}_s = e\psi_s/k_\mathrm{B}T$ as a function of the negative ions-wall interaction energy $\epsilon_{neg-w}/k_\mathrm{B}T$.

$\epsilon_{neg-w}/k_\mathrm{B}T$. This is true also for distances of the order of $z/d \simeq 2, 3, 4, \ldots$, where well-defined density peaks are evident. The increase in the PDI density at the wall affect the charge regulation equilibrium (2.4)–(2.6), and reduces the surface charge σ—see figure 10.14(a).

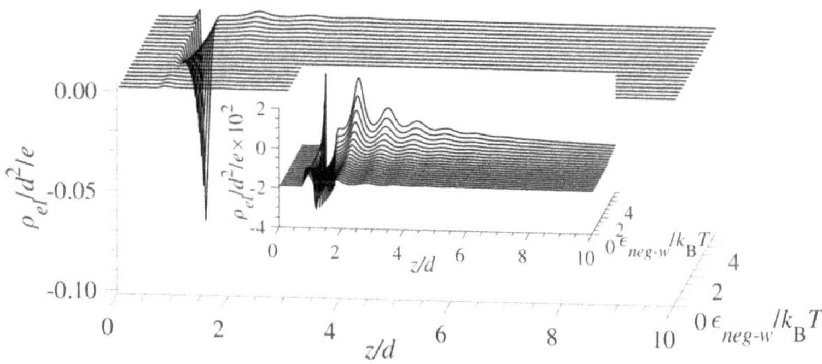

Figure 10.15. Fluid charge density ρ_{el} as a function of the negative non-PDI-surface interaction energy $\epsilon_{neg-w}/k_{\mathrm{B}}T$. The step in the variation of the LJ parameter is $\Delta\epsilon_{pos-w}/k_{\mathrm{B}}T = 0.25$. Inset: rescaling the vertical axis provides more details of the peak structure.

The surface potential, $\tilde{\psi}_s$, shows the opposite trend—see figure 10.14. It becomes more negative with the accumulation of negative ions near the wall. The greater potential negativity is in agreement with the greater concentrations of positive PDIs, shown in figure 10.13.

The fluid charge density in the EDL, ρ_{el}, is plotted in figure 10.15. The charge distribution profiles are rather complex. In the case of weak attraction between the negative ions and the wall, the fluid charge densities show relatively short, positive peaks near the charged surface. The peaks decrease and reach a minimum around $\epsilon_{neg-w}/k_{\mathrm{B}}T \simeq 1.5$. This is also the energy value at which a negative charge density minimum forms and becomes deeper with the increasing $\epsilon_{neg-w}/k_{\mathrm{B}}T$. The positive charge density peaks also start to increase beyond $\epsilon_{neg-w}/k_{\mathrm{B}}T \simeq 1.5$. They are centered slightly closer to the charged wall than the negative minima. This implies that as the negative ions adsorption at the wall increases, there are always some positive ions that would be attracted in the vicinity due to the increasingly negative potential at the surface (see figure 10.14). This positive charge density is located closer to the negatively charged surface than the negative ions, which are repelled a little farther into the fluid. The inset in figure 10.15 offers a more detailed picture of the charged density structure inside the EDL, near the wall. The overall fluid charge density for $z/d \gtrsim 2$ is positive, which is likely due to the significant negative charge in the first layer next to the wall.

10.5 Conclusions

The non-Coulombic interactions (e.g. LJ, and excluded volume) between the solution components and the wall are no less important than they are in the fluid. These interactions involve the neutral solvent and the dissolved ionic species, and contribute to the solution composition near the charge regulating surface. The PDIs compete with the rest of the solution species for the finite space at the wall, and that determines the surface charge through the reactions (2.1).

The interaction energy between a solid substrate and a liquid is an important material property. It depends on the particular choice of materials and determines properties such as the surface wetting and contact angles at the three-phase line [1]. The surface properties may range from solvophilic (the liquid experiences strong molecular attraction to the solid) to solvophobic (the interactions are weakly attractive, or even repulsive). Note that the surface solvophilicity and/or solvophobicity always depends on the particular combination of a substrate and a liquid. For example, glass is solvophilic (or hydrophilic) with respect to water, but is strongly solvophobic (or hydrophobic) if the contact liquid is mercury.

The non-Coulombic interactions between the ions and the surface contribute to the adsorption, and thus, to the composition of the Stern layer [see equation (2.13)]. Hence, these interactions have an impact on the charge and potential at the surface and in the fluid.

References

[1] de Gennes P-G, Brochart-Wyart F and Quere D 2003 *Capillarity and Wetting Phenomena* (New York: Springer)

[2] Marcelja S and Radic N 1976 Repulsion of interfaces due to boundary water *Chem. Phys. Lett.* **15** 129–30

[3] Jonsson B and Wennerstrom H 1983 Image-charge forces in phospholipid bilayer systems *J. Chem. Soc. Faraday Trans. 2* **79** 19–35

[4] Kjellander R 1984 On the image-charge model for the hydration force *J. Chem. Soc. Faraday Trans. 2* **80** 1323–48

[5] Attard P and Patey G N 1991 Continuum electrostatic interactions between planar lattices of dipoles and the possible relevance to the hydration force *Phys. Rev.* A **43** 2953–62

[6] Paunov V N, Kaler E W, Sandler S I and Petsev D N 2001 A model for hydration interactions between apoferritin molecules in solution *J. Colloid Interface Sci.* **240** 640–3

[7] Valle-Delgado J J, Molina-Bolivar J A, Galisteo-Gonzalez F, Galvez-Ruiz M J, Feiler A and Rutland M W 2005 Hydration forces between silica surfaces: Experimental data and predictions from different theories *J. Chem. Phys.* **123** 034708

[8] Israelachvili J N 2011 *Intermolecular and Surface Forces* 3rd edn (New York: Academic)

[9] Vangara R, Brown D C R, van Swol F and Petsev D N 2017 Electrolyte solution structure and its effect on the properties of electric double layers with surface charge regulation *J. Colloid Interface Sci.* **488** 180–9

[10] Vangara R, van Swol F and Petsev D N 2018 Solvophilic and solvophobic surfaces and non-coulombic surface interactions in charge regulating electric double layers *J. Chem. Phys.* **148** 044702

[11] van Swol F and Petsev D N 2018 Solution structure effects on the properties electric double layers with surface charge regulation assessed by density functional theory *Langmuir* **34** 13808–20

IOP Publishing

Molecular Theory of Electric Double Layers

Dimiter N Petsev, Frank van Swol and Laura J D Frink

Chapter 11

The potential distribution in the electric double layer and its relationship to the fluid charge

11.1 The Poisson equation for structured electrolyte solutions

The electrostatic potential ψ in the fluidic part of the electric double layer (EDL) is related to the local charge density ρ_{el} via the Poisson equation (see equation (1.8))

$$\nabla^2\psi(\mathbf{r}) = -\frac{\rho_{el}(\mathbf{r})}{\varepsilon\varepsilon_0}.$$

The effects of all Coulombic and non-Coulombic interactions are included the fluid charge density, $\rho_{el}(\mathbf{r})$, and therefore reflect on the potential, ψ. Hence, combining the Poisson equation with the c-DFT approach allows to determine the effects of the detailed molecular and ionic interactions on the electrostatic potential distribution, $\psi(\mathbf{r})$, in the EDL.

Below, we consider an example of the relationship between the fluid charge and potential in a single, flat EDL, while accounting for the fluid interactions within the framework of the c-DFT as described above. The Poisson equation in this case reads:

$$\frac{d^2\psi(z)}{dz^2} = -\frac{\rho_{el}(z)}{\varepsilon\varepsilon_0} \tag{11.1}$$

where

$$\rho_{el}(z) = \sum_i q_i\rho_i(z). \tag{11.2}$$

The summation in (11.2) is over all charged species in the solution.

The fluid charge density profiles exhibit multiple maxima and minima, which are due to the non-Coulombic energy contribution, and particularly, to the excluded volume contributions (see figures 9.3, 9.7, 9.11, 9.15, 10.4, 10.7, 10.11, and 10.15).

doi:10.1088/978-0-7503-2276-8ch11

Still, the corresponding electrostatic potential distributions in the fluid, in the vicinity of the charged interface, are very often represented by smooth, monotonic functions [1–3].

Figure 11.1 presents examples of the fluid charge profiles (figure 11.1(a)) and electrostatic potential distribution (figure 11.1(b)) in an EDL. The charge profile exhibits well defined peaks that reflect the local solution structure near the wall. All species have the same diameter, which leads to the peaks being separated by distance of the order of $z/d \simeq 1$, except for the first peak, which is located at $z/d \simeq 0.5$. In contrast, the electrostatic potential changes monotonically with the distance from the wall and shows no maxima and/or minima. This can be explained by noticing that the derivative $d^2\psi/dz$ is related to the curvature of the function $\psi(z)$. Hence, equation (11.1) relates the curvature of $\psi(z)$ to the sign and value of the fluid charged density ρ_{el}. As long as the overall fluid charge density in the EDL does not change sign, neither does the curvature. Note that the Poisson equation states that the derivative $d^2\psi/dz^2$ is proportional to $-\rho_{el}$. Therefore, for $\rho_{el} > 0$, the curvature of $\psi(z)$ is negative. This is evident from figure 11.1(b), where the potential plot has a negative curvature everywhere in the EDL. The actual magnitude of the local curvature of $\psi(z)$ in figure 11.1(b) varies in accordance with the charge density oscillations, shown in figure 11.1(a), but always remains positive. These variations, however, are practically not discernible on the potential plot.

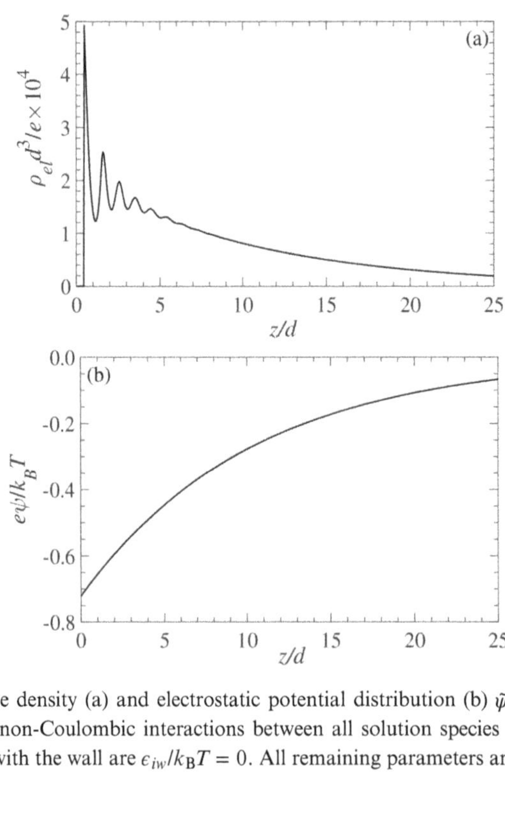

Figure 11.1. Fluid charge density (a) and electrostatic potential distribution (b) $\tilde{\psi} = e\psi/k_BT$ In the electric double layer. The fluid, non-Coulombic interactions between all solution species are $\epsilon_{ij}/k_BT = 1$. The non-Coulombic interactions with the wall are $\epsilon_{iw}/k_BT = 0$. All remaining parameters are the same as in figure 8.1.

The fluid charge density, $\rho_{el}(z)$ can assume negative values as seen in figures 9.7, 9.11, 10.11, and 10.15. This means that, in some cases, the curvature of the potential distribution, $\psi(z)$, may change its sign in addition to the magnitude. An example for such a case is shown in figure 11.2, which depicts the fluid charge (figure 11.2(a)) and electrostatic potential (figure 11.2(b)) distributions in the EDL. The fluid charge exhibits a negative dip next to the charged surface, then steeply increases into the positive territory. There are a number of structural peaks, which disappear for $z/d \gtrsim 7$. At larger distances the fluid charge density profile is positive, gradually decaying to zero at infinity.

The electrostatic potential distribution (figure 11.2(b)) curve is still monotonically increasing and any variations in the sign or magnitude of the functional curvature are hardly visible. Therefore, it is helpful to examine the relationship [1]

$$\Delta\tilde{\psi}(z) = \tilde{\psi}(z) - \left[\tilde{\psi}_s + \left(\frac{\partial\tilde{\psi}}{\partial z}\right)_{z=0} z\right], \tag{11.3}$$

which is the nonlinear residual of the potential $\tilde{\psi}(z)$.

The function $\Delta\tilde{\psi}(z)$ is plotted in figure 11.3. The linear component of the potential change with distance is subtracted, which allows to demonstrate the variation of the curvature sign. Figure 11.3 indicates that the curvature in

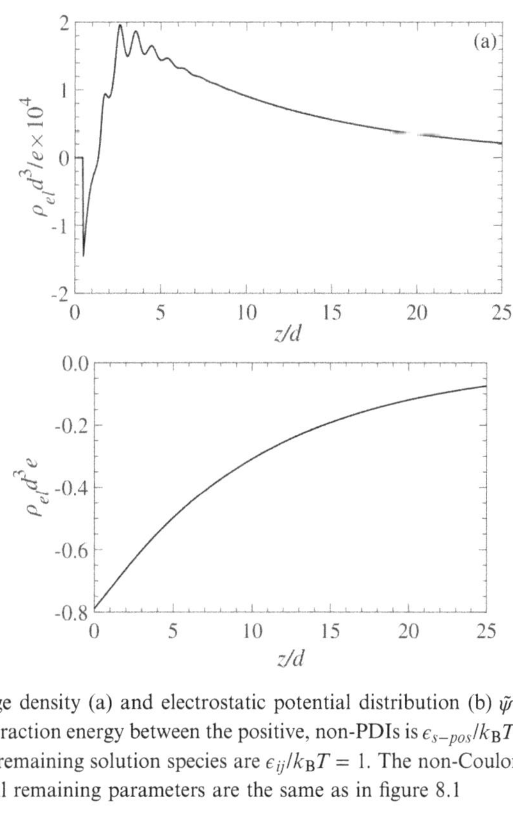

Figure 11.2. Fluid charge density (a) and electrostatic potential distribution (b) $\tilde{\psi} = e\psi/k_BT$ in the electric double layer. The LJ interaction energy between the positive, non-PDIs is $\epsilon_{s-pos}/k_BT = 2$. The non-Coulombic interactions between all remaining solution species are $\epsilon_{ij}/k_BT = 1$. The non-Coulombic interactions with the wall are $\epsilon_{iw}/k_BT = 0$. All remaining parameters are the same as in figure 8.1

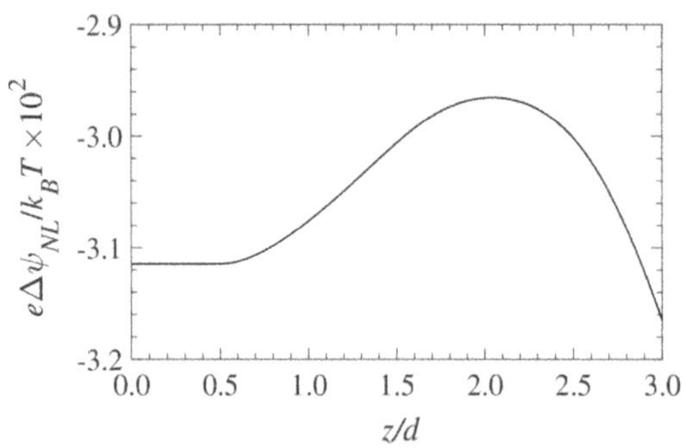

Figure 11.3. Nonlinear residual of the potential distribution function in figure 11.2. (Reprinted with permission from [2], Copyright (2018) American Chemical Society.)

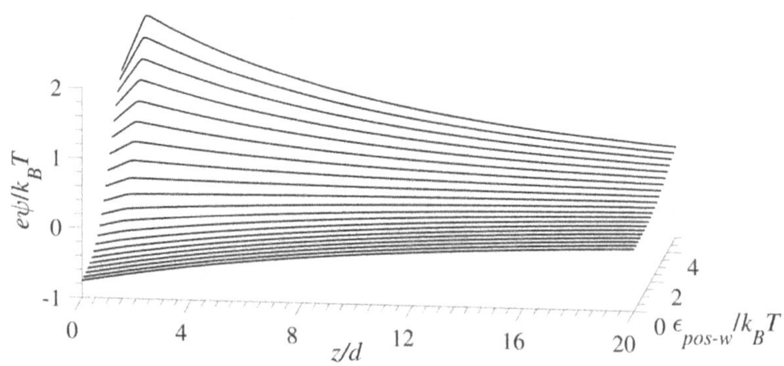

Figure 11.4. Electrostatic potential distribution profiles $\tilde{\psi}(z)$ for varying LJ attractive energy between the positive non-PDIs and the wall ϵ_{pos-w}/k_BT between 0 and 5 in steps $\Delta\epsilon_{pos-w}/k_BT = 0.25$.

the function $\tilde{\psi}(z)$ (see figure 11.2(b)) is zero for $z/d < 0.5$. It is positive for $0.5 < z/d \lesssim 1.25$, and turns negative for $z/d \gtrsim 1.25$.

There are cases where the variations in the fluid charge densities over distance are so extreme that the electrostatic potential may exhibit a non-monotonic behavior. Such an example is shown in figure 11.4, which shows the dependence of the electrostatic potential, $\tilde{\psi}(z)$, on the LJ attractive energy, $\epsilon_{pos-w}k_BT$, between the positive non-PDIs and the EDL interface. All curves exhibit a linear portion with a positive slope between $z/d = 0$ and 0.5. The slope of the remainder of the curves is also positive for $\epsilon_{pos-w}/k_BT \lesssim 3$. As the LJ wall attraction exceeds 3 k_BT, the density of positive ions in the layer next to the negative surface becomes so high that the local potential value changes from negative to positive. This is accompanied by a slope and curvature change for the function $\tilde{\psi}(z)$ at $z/d = 0.5$. The slope becomes negative and the curvature turns positive.

11.2 Molecular interpretation of the Helmholtz planes, the Stern–Grahame layer, and the electrokinetic shear plane

The early theories of charged interfaces between substrates and electrolyte solutions were understandably very simple by today's standards and lacked much detail (see chapter 1 for more details). Still, they provided initial insights and identified some of the important elements and length-scales. Helmholtz [4] considered such an interface to be a capacitor with plates that represent the charges on the substrate and on a virtual plane in the solution, next to the interface. The substrate charges correspond the inner Helmholtz plane (IHP), while the charges in the solution form the outer Helmholtz plane (OHP).

The Helmholtz model is not a particularly realistic, since at finite temperatures, the ions in the solution are in constant random motion, and therefore unlikely to assume the plane ordering as presumed by the model. This shortcoming was avoided by Gouy [5, 6] and Chapman [7], who proposed the point-charge analysis based on the Poisson–Boltzmann equation. Still, Stern [8], and later Grahame [9] realized the solution's molecular layer, in direct contact with the substrate, may organize in a structure not unlike the OHP. In their model, the EDL is composed by a Stern–Grahame layer with a thickness of the order of the ion or solvent molecular diameter, followed by a structureless solution of point-sized ions dissolved in a continuum solvent. The location of the Stern–Grahame virtual plane runs through the centers of the ions and solvent molecules in the wall-adjacent solution layer. The Grahame model focuses on the role of the solvent molecules and according to it, the IHP corresponds to the layer of ions, covalently bound to the charged surface, while the OHP is located at the first layer of solvated ions next to the wall.

The Stern–Grahame plane is not to be confused with the electrokinetic shear plane. The latter is a concept that is introduced in an attempt to better define and justify important experimental quantities such as the electrokinetic ζ-potential [10–12]. The shear plane location is determined by the hydrodynamics of relative motion of the solution, relative to the substrate, or vice versa. In contrast, the Stern–Grahame layer structure is governed by the thermodynamic energies of interaction between ions and solvent molecules with the substrate wall.

It is obvious that the IHP, OHP, Stern–Grahame layer, and shear plane are not very well defined. A real-world charged interface would be composed of ions and/or solvent molecules that differ in size, which makes it difficult to precisely locate any of the planes. In addition, most materials have an intrinsic surface roughness, which further aggravates the problem.

An EDL model with molecular resolution, based on modern statistical mechanics and computational methods, allows one to obtain deeper insights in the detailed structure of the liquid near the charged surface and shed additional light on the concepts of the IHP, OHP, Stern–Grahame layer, and the shear plane. A brief overview of the problem is outlined below. Additional details are available in [2].

The c-DFT molecular analysis of EDLs, described above, offers a better understanding of the structure and the various locations (i.e., planes) where the charged species and solvent molecules organize. For example the reactive surface at the substrate solution interface is the key contributor to the charge origination and regulation process. It reacts with the PDIs and the process typically involves electron transfers and forming (or breaking) of chemical bonds. On the other hand, ions and solvent molecules that are in the immediate vicinity of the substrate wall can be considered constituents of the Stern–Grahame layer. They are not involved in a chemical reaction with the surface, but are subjected to a physical adsorption governed by Coulombic and non-Coulombic forces. The Stern–Grahame surface can then be defined by centers of charges for the ionic participants in the layer. If the ions and solvent molecules have different sizes, the location of the Stern–Grahame planes is problematic and can be considered as an average quantity. This picture differs from the Stern–Grahame model where the chemical and physical bonding occurs at the same location. Since the PDIs have finite size, the reaction of charge exchange (e.g., a proton—see equation (2.1)) occurs at a different location than the geometric center of the PDI. The detailed analysis of the solution structure in the EDL indicates that the are additional layers of ions and solvent represented by peaks in the density profiles. These, of course, are not evident from the Gouy–Chapman, and Stern–Grahame analyses. Hence, the true physical situation is considerably more complex than these early models, and a more rigorous approach is preferable.

The location of electrokinetic shear plane depends on a combination of thermodynamic and hydrodynamic forces. It represents the surface where the fluid starts moving relative to the substrate. Molecular dynamics analysis [13–17] shows that this typically happens at the plane located after the first layer of ions and solvent molecules, or at about one diameter away from the reactive surface. This observation is in agreement with the non-slip boundary condition that is commonly used in fluid mechanics [18].

11.3 Conclusions

The electrostatic potential distribution in the EDL can be related to the fluid charge density profiles by means of the Poisson equation (1.8). c-DFT offers an approach that provides a lot of specific molecular detail for the fluid charge density. Using the c-DFT derived fluid charge density in conjunction with equation (1.8) offers a much better insight into the properties of the EDL potential distribution in comparison to the Poisson–Boltzmann approximation (3.2). Using the fluid charge density profiles, obtained by c-DFT, as an input for equation (1.8) allows one to assess the effect of the solvent molecules and/or the non-Coulombic interactions on the precise potential distribution in the EDL.

A better and more detailed representation of the fluid charge density and potential distributions in the EDL offers a molecular interpretation of historically established quantities such as the Helmholtz planes, the Stern and Grahame layer, and the plane of electrokinetic shear.

References

[1] Vangara R, van Swol F and Petsev D N 2018 Solvophilic and solvophobic surfaces and non-coulombic surface interactions in charge regulating electric double layers *J. Chem. Phys.* **148** 044702

[2] van Swol F and Petsev D N 2018 Solution structure effects on the properties electric double layers with surface charge regulation assessed by density functional theory *Langmuir* **34** 13808–20

[3] Gillespie D, Petsev D N and van Swol F 2020 Electric double layers with surface charge regulation using density functional theory *Entropy* **22** 132
[4] Helmholtz H 1853 Ueber einige gesetze der vertheilung elektrischer strme in krperlichen leitern mit anwendung auf die thierisch–elektrischen versuche *Ann. Phys. Chem.* **165** 211–33
[5] Gouy G 1910 Sur la constitution de la charge electrique 'a la surface dun electrolyte *J. Physique* **9** 457–68
[6] Gouy G 1917 Sur la fonction electrocapillaire *Ann. Phys.* **7** 129–84
[7] Chapman D L 1913 A contribution to the theory of electrocapillarity *Phil. Mag.* **25** 475–81
[8] Stern O 1924 Zur theorie der elektrolytischen doppelschicht *Z. Electrochem* **30** 508–16
[9] Grahame D C 1947 The electrical double layer and the theory of electrocapillarity *Chem. Rev.* **41** 441–501
[10] Dukhin S S and Derjaguin B V 1974 Equilibrium double layer and electrokinetic phenomena *Surface and Colloid Science* ed E Matijevic vol 7 (New York: Wiley Interscience) ch 2 pp 50–272
[11] Hunter R J 1981 *Zeta Potential in Colloid Science* (New York: Academic)
[12] Russel W B, Saville D A and Schowalter W R 1989 *Colloidal Dispersions* (New York: Cambridge Univsersity Press)
[13] Thompson A 2003 Nonequilibrium molecular dynamics simulation of electro-osmotic flow in a charged nanopore *J. Chem. Phys.* **119** 7503–11
[14] Qiao R and Aluru N R 2003 Ion concentrations and velocity profiles in nanochannel electroosmotic flows *J. Chem. Phys.* **118** 4692–701
[15] Qiao R and Aluru N R 2004 Charge inversion and flow reversal in a nanochannel electro-osmotic flow *Phys. Rev. Lett.* **92** 198301
[16] Qiao R and Aluru N R 2005 Scaling of electrokinetic transport in nanometer channels *Langmuir* **21** 8972–7
[17] Wu P and Qiao R 2011 Physical origins of apparently enhanced viscosity of interfacial fluids in electrokinetic transport *Phys. Fluids* **23** 072005
[18] Landau L D and Lifshitz E M 1987 *Fluid Mechanics* (Amsterdam: Pergamon)

IOP Publishing

Dimiter N Petsev, Frank van Swol and Laura J D Frink

Chapter 12

Electric double layers containing multivalent ions

The charge number, or valency, of the ions present in the electric double layer (EDL) is important. This is not surprising since they are subjected to the electrostatic forces exerted by charged interface into the solution. Hence, the ion valency has a profound effect on the properties of charged colloidal systems. An example is the Schulze–Hardy rule [1, 2], which states that the critical electrolyte concentration, necessary to precipitate a suspension of charged colloids, is proportional to q_i^{-6}, where q_i is the charge of counterions to the particles. Below, we present an overview of the ion valency effects coupled to the explicit molecular structure of the solution driven by the presence of the solvent and various non-Coulombic interactions. The focus is on the non-potential determining ions (non-PDIs). The PDIs are assumed to be monovalent in the examples discussed below. Additional details are available in [3].

12.1 Multivalent ion density profiles in the electric double layer

The density profiles for a variety of positive and negative non-PDIs are shown in figure 12.1. The ion valency for both positive and negative ions is varied between 1 and 3. The curves present all possible combinations. Note that ionic numbers for a combination of positive and negative ions is adjusted in such a way that the total ionic strength (including the PDIs) is always $I = 0.01$ M. This choice is made to assure that the screening parameter κ remains the same in all cases. All remaining parameters are the same as those used in chapter 8.

Figure 12.1(a) presents density profile results for the positive counterions next to the negatively charged surface. The curves are split into three distinct groups, depending on the counterion valency. The top three curves correspond to binary electrolytes that have a trivalent positive ion. The middle triplet of curves is for binary electrolytes that contain a divalent positive ion. Finally, the bottom group of

doi:10.1088/978-0-7503-2276-8ch12

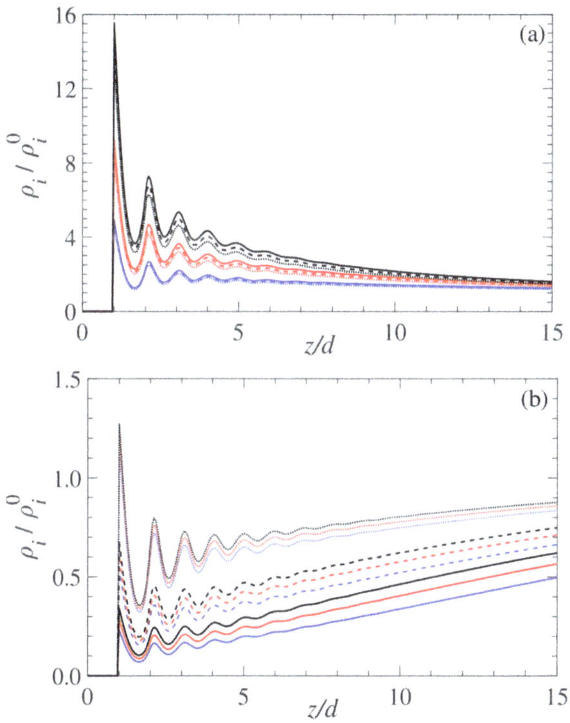

Figure 12.1. Normalized density profiles for non-PDIs. (a) Positive ions. Top curves (black): 3:3 (solid), 3:2 (dashed), 3:1 (dotted); middle curves (red): 2:3 (solid), 2:2 (dashed), 2:1 (dotted); bottom curves (blue): 1:3 (solid), 1:2 (dashed), 1:1 (dotted). (b) Negative ions. Top curves (dotted): 3:1 (black), 2:1 (red), 1:1 (blue); middle curves (dashed): 3:2 (black), 2:2 (red), 1:2 (blue); bottom curves (solid): 3:3 (black), 2:3 (red), 1:3 (blue). The remaining parameters are the same as in figure 8.1. (Reprinted from [3] ©2019 IOP Publishing. Reproduced with permission. All rights reserved.)

curves shows the monovalent positive ions densities. The finer distinction between the curves, within each triplet, demonstrates the effect of the negative ion valency. The solid lines correspond to trivalent negative ions. The dashed lines present the divalent negative ions. The monovalent negative ions are depicted by the dotted lines in the figure. The effect of the negative ions valency on the density profiles for the positive ions is more subtle than that for the positive ions valency. In comparing the density plots, one should remember that all curves are normalized by the respective bulk densities away from the charged surface, which explains the slight variations. The solid curves are scaled (divided) by the smallest factor, the dashed by a greater one and the dotted by the greatest. Hence, the solid curves are always the highest, and the dotted—the lowest in the plots.

The Coulombic interaction's effect of the ions valency on the density profiles is also evident from the slopes of the curves at distances $z/d \gtrsim 6$. It is greatest for the trivalent positive ions, less for the divalent ions, and least for the monovalent ions.

The density profiles near the negative, charge-regulating surface are plotted in figure 12.1(b). All the data are normalized with the respective densities at infinity.

The curves are again bundled in groups of three, this time based on the valency of the negative ions. The curves at the top are monovalent negative ions since they experience the least repulsion by the negatively charged wall. The curves at the middle are for the different electrolytes that have a divalent negative ion. The curves at the bottom are for the trivalent negative ions, which are subjected to the strongest repulsion by the surface. The smaller differences between the curves in each triplet are due to the positive ions and the variations are due to similar scaling reasons as described above.

12.2 Effect of the non-potential-determining ions valency on the density profiles of the potential determining ions in the electric double layer

Varying the valency of the non-PDIs indirectly affects the density profiles for the PDIs in the EDL. The results are shown in figure 12.2. The curves for various combinations of positive and negative ion valencies are relatively close to each other, but are still distinguishable. The density variations of the PDIs near the EDL interface, coupled with the charge regulation conditions (2.4)–(2.6), determine the surface charge and potential—see table 12.1.

The surface charge is governed by the number of PDIs in the Stern–Grahame layer. This number decreases with the valency of the positive non-PDIs, which leads to an increase in the surface charge magnitude, $\tilde{\sigma}$. The trend exhibited by the surface potential, $\tilde{\psi}_s$, is the opposite. It becomes less negative as the valency of the positive ion increases (see also [3]).

The effect of the negative ions valency is less pronounced. Greater negative ion charge number leads to a lower surface charge (for a fixed valency of the respective positive ion). The surface potential becomes more negative with the negative ions valency.

The fluid charge density profiles in the EDL, ρ_{el}, are shown in figure 12.3. It evident from the plots that ions that have a greater positive charge experience

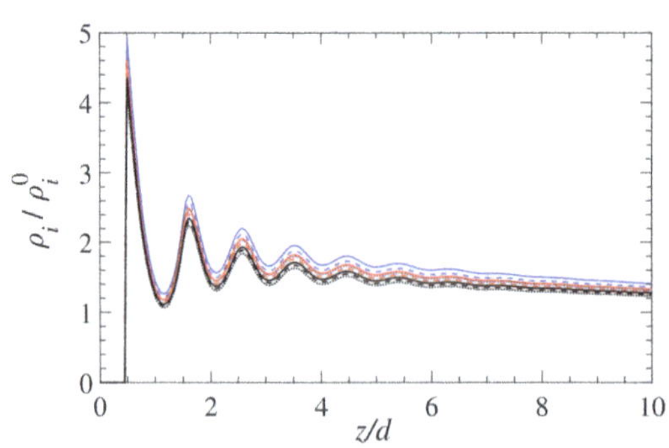

Figure 12.2. Normalized density profile of the PDIs in the presence of different electrolytes. the different curves are (black): 3:3 (solid), 3:2 (dashed), 3:1 (dotted); (red): 2:3 (solid), 2:2 (dashed), 2:1 (dotted); (blue): 1:3 (solid), 1:2 (dashed), 1:1 (dotted). The remaining parameters are the same as in figure 8.1. (Reprinted from [3] ©2019 IOP Publishing. Reproduced with permission. All rights reserved.)

Table 12.1. Surface charge $\tilde{\sigma} = \sigma d^2/e$, and surface potential $\tilde{\psi}_s = e\psi_s/k_BT$. The overall ionic strength is adjusted to a maintain a fixed Debye length $\kappa^{-1} = 10.6d$. $|q_1|$ and $|q_2|$ are the positive and negative ions charge numbers for the background, non-PDI electrolyte.

$\lvert q_1\rvert : \lvert q_2\rvert$	$\sigma d^2/e$	$e\psi_s/k_BT$
1:1	-2.150×10^{-3}	−0.720 8
1:2	-2.068×10^{-3}	−0.758 1
1:3	-1.990×10^{-3}	−0.794 6
2:1	-2.263×10^{-3}	−0.671 4
2:2	-2.196×10^{-3}	−0.700 2
2:3	-2.133×10^{-3}	−0.728 0
3:1	-2.369×10^{-3}	−0.627 8
3:2	-2.314×10^{-3}	−0.650 3
3:3	-2.262×10^{-3}	−0.671 9

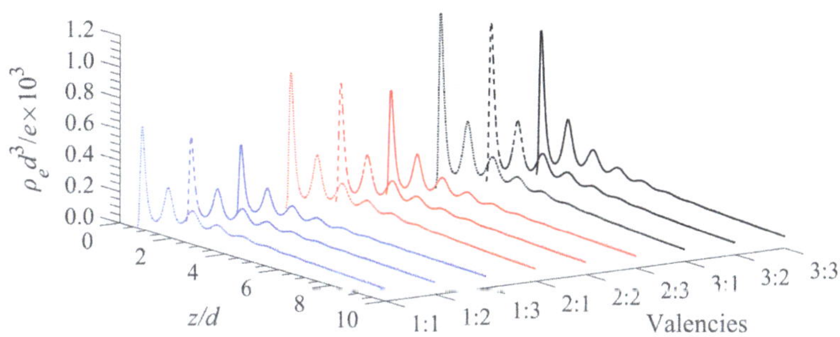

Figure 12.3. Fluid charge density profiles near a charge-regulating interface. The different curves correspond to: (blue): 1:3 (solid), 1:2 (dashed), 1:1 (dotted); (red): 2:3 (solid), 2:2 (dashed), 2:1 (dotted); (black) 3:3 (solid), 3:2 (dashed), 3:1 (dotted). The remaining parameters are the same as in figure 8.1. (Reprinted from [3] ©2019 IOP Publishing. Reproduced with permission. All rights reserved.)

stronger attraction towards the negative surface than those with lower valency. The effect is purely Coulombic in nature. The negative ions valency also affects the fluid charged density profiles, albeit less than the valency of the positive ones. The ions with greater negative charge numbers are more strongly repelled by the negatively charged wall and are expelled farther into the fluid. This forces some of the positive ions to move away into the fluid as well in order to maintain the externally fixed chemical potential.

12.3 Non-Coulombic surface interactions, charge and potential distributions in the Stern–Grahame layer and beyond

The increased presence of multivalent ions in the Stern–Grahame layer has an effect on the properties of EDLs. It can influence the charge regulation equilibrium (2.1), which governs the surface charge and potential. Another

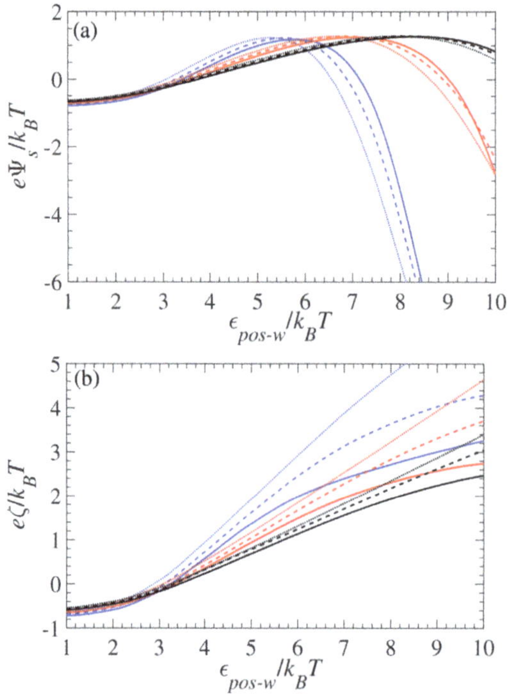

Figure 12.4. Potential vs wall interaction energy ϵ_{pos-w}/k_BT energy for the positive non-PDIs. (a) Surface potential $\tilde{\psi}_S = e\psi_S/k_BT$ at the IHP. (b) Electrokinetic $\tilde{\zeta} = e\zeta/k_BT$ potential at the plane of shear. The different curves correspond to (black): 3:3 (solid), 3:2 (dashed), 3:1 (dotted); (red): 2:3 (solid), 2:2 (dashed), 2:1 (dotted); (blue): 1:3 (solid), 1:2 (dashed), 1:1 (dotted). All remaining parameters are the same as in figure 8.1. (Reprinted from [3] ©2019 IOP Publishing. Reproduced with permission. All rights reserved.)

consequence is that the multivalent ions in sufficient numbers in the Stern–Grahame layer may exceed the charge at the inner Helmholtz plane (IHP). Hence, a strong adsorption of multivalent counterions can change the apparent surface charge and potential, if assessed by an electrokinetic measurement, which probes the shear plane, outside the Stern–Grahame layer [4–8].

Figure 12.4 demonstrates the effect of surface adsorption of the positive non-PDIs on the surface potential. The plots in figure 12.4(a) show the surface potential at the IHP for different LJ interactions energies with the charged wall ϵ_{pos-w}/k_BT. The particular curves are for various combinations of the positive and negative non-PDI valencies. At lower LJ surface attraction, the ion attachment at the wall is predominantly driven by Coulombic forces, and hence, the ion valency is important. As the LJ energy ϵ_{pos-w}/k_BT increases, the positive ions accumulate in great numbers and change the surface potential from negative to positive. However, the further increase of both the LJ attractions and density of positive ions in the Stern–Grahame layer prevents that PDIs from approaching and reacting with the charge-regulating surface. As a result, surface equilibrium shifts towards greater surface negativity. This brings the surface potential back into the negative range. The major differences between the curves are due to the that valencies of the positive ions. The minor variations within each set of curves are dependent on the valencies of the negative ions.

The electrokinetic ζ-potential data are plotted in figure 12.4(b). The behavior of the potentials at the shear plane is very different than the those at the IHP. The potential is negative at low LJ energies $\epsilon_{pos-w}/k_BT \lesssim 3$, and turns positive beyond that. The ζ-potential monotonically increases as ϵ_{pos-w}/k_BT varies between 1 and 10. The potential has the highest positive value for the 1:1 type of electrolyte, and the lowest for the 3:3 electrolyte. This counter intuitive results is explained by the fact that the overall ionic density is the highest for 1:1 and is the lowest for the 3:3 electrolyte, which is a consequence of the imposed requirement that the screening parameter κ = const. Hence, despite the fact that the trivalent positive counterions carry a greater charge, their number is lower, while the monovalent counterions are able to populate the Stern–Grahame layer more efficiently and hence, lead to a more positive ζ-potential.

The non-Coulombic adsorption of negative ions at the negatively charged EDL interface mostly affects the surface charge regulation equilibrium. The number of negative ions in the Stern–Grahame layer increases with the attractive LJ energy ϵ_{neg-w}/k_BT, which attracts the positive PDIs and hence, facilitates the reduction of the negative surface charge according to the reactions (2.1). The effect of negative ions adsorption on the ζ-potential is less dramatic in comparison with the that of the positive ions in the sense that there is no difference in the signs of the surface potential ψ_s at the IHP and the ζ-potential at the shear plane. More specifics are available in [3].

12.4 Conclusions

The charge and potential distributions in EDLs, depend on the type of dissolved ions. The ions are involved in a variety of interactions with the solvent, the charged EDL interface, and with each other. Hence, the properties of EDLs depend on the specific ions that are present in the solution. The ionic charge is an important property as it affects the Coulombic interactions, and their interplay with other non-Coulombic forces, that are present, between the solution species

References

[1] Verwey E J W and Overbeek J T G 1948 *Theory of the Stability of Lyophobic Colloids* (Amsterdam: Elsevier)

[2] Derjaguin B V, Churaev N V and Muller V M 1987 *Surface Forces* (New York: Plenum)

[3] Vangara R, Stoltzfus K, F van Swol, York M and Petsev D N 2019 Coulombic and non-coulombic effects in charge-regulating electric double layers *Mater. Res. Express* **6** 086331

[4] Dukhin S S 1974 *Surface and Colloid Science* ed E Matijevic vol 7 (New York: Wiley) ch 1 pp 1–49

[5] Tadros T F and Lyklema J 1968 Adsorption of potential-determining ions at the silica-aqueous electrolyte interface and the role of some cations *J. Electroanal. Chem.* **17** 265–75

[6] Thompson A 2003 Nonequilibrium molecular dynamics simulation of electro-osmotic flow in a charged nanopore *J. Chem. Phys.* **119** 7503–11
[7] Qiao R and Aluru N R 2003 Ion concentrations and velocity profiles in nanochannel electroosmotic flows *J. Chem. Phys.* **118** 4692–701
[8] Qiao R and Aluru N R 2004 Charge inversion and flow reversal in a nanochannel electro-osmotic flow *Phys. Rev. Lett.* **92** 198301

IOP Publishing

Molecular Theory of Electric Double Layers

Dimiter N Petsev, Frank van Swol and Laura J D Frink

Chapter 13

Ionic size effects

Most of the results, discussed in this study, are for solutions in which all species have the same size. In reality, the size of solutes and solvent molecules may differ [1], which affects the properties of the electric double layer (EDL) [2–4]. This chapter presents an overview of the species size variation effects on the EDLs, when coupled to a surface charge regulation and in the explicit presence of a solvent. The analysis below follows [5].

13.1 Ionic size variations and solution density

In order to illustrate the size effects we use a model system that is similar to the one defined in chapter 8. The analysis is limited to the size variation of the positive or negative non-potential determining ions (non-PDIs) relative to that of the solvent. The size of the PDIs is tied to that of the solvent through equation (2.3), and hence is not a subject to any variations.

To study the ion-size effects on the double layer requires that we give some thought to the bulk fluid that is in equilibrium with the double layer. Ideally, one would follow the experimental lead and keep the pressure constant, and find the bulk density of the mixture. This is not always straightforward. The next best option is to keep the overall reduced mixture density $\bar{\rho}$ (or, alternatively, the overall mixture packing fraction $\xi = \frac{\pi}{6}\bar{\rho}$) constant, where

$$\bar{\rho} = \sum_{i=1}^{M} \rho_i d_i^3. \tag{13.1}$$

If we choose component $i = 1$ to be the solvent, we can also rewrite this as

$$\bar{\rho} = \bar{\rho}_1 + \sum_{i=2}^{M} \rho_i d_i^3. \tag{13.2}$$

The idea then is to adjust the solvent density, $\bar{\rho}_1$, such that the overall mixture density, $\bar{\rho}$ stays constant. In other words,

$$\bar{\rho}_1 = \bar{\rho} - \sum_{i=2}^{M} \rho_i d_i^3. \tag{13.3}$$

The size analyses, presented below, are for fixed average density $\bar{\rho} = 0.8$. This allows to separate the effects that are due to varying the diameter of a given ionic species from others that might results from density variations.

The LJ energy parameter of fluid interaction is $\epsilon_{ij}/k_B T = 1$ for all species. The LJ parameter for all interactions with the charged wall are also set to $\epsilon_{iw}/k_B T = 1$. The solution is composed of solvent, positive PDIs, positive non-PDIs, and a common negative counterion for both. The concentration of the potential determining ions corresponds to pH = 2. The bulk dielectric permittivity is $\varepsilon = 78.5$ at temperature $T = 298$ K. The surface charge is determined by the charge regulation conditions (2.4)–(2.6) with parameters $\Gamma d^2/e = 0.66$ (Γ being the number of surface reactive groups per unit area), $\mathrm{pK}_- = 6$, and $\mathrm{pK}_+ = -2$.

13.1.1 Positive non-PDI size variation effects

The size variation of the positive non-PDIs impacts the solution structure near the charged interface through a combination of the local packing geometry and the long-ranged interactions (LJ and Coulombic), which are size-dependent. Figure 13.1

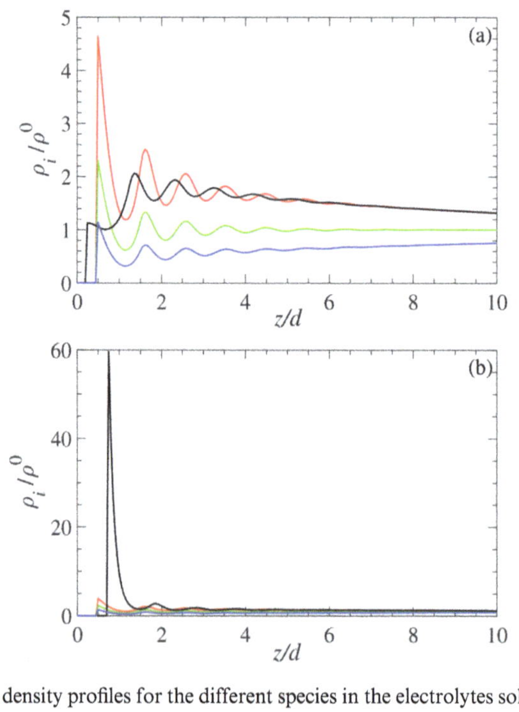

Figure 13.1. Normalized density profiles for the different species in the electrolytes solution: positive non-PDIs (black), PDIs (red), negative non-PDIs (blue), and solvent (green). (a) $d_{pos}/d = 0.5$, and (b) $d_{pos}/d = 1.5$. The remaining parameters are listed in the text. (Reproduced with permission from [5], Copyright (2021) Taylor and Francis Ltd. http://tandfonline.com.)

shows an example of the density distributions of the different solution components for $d_{pos}/d = 0.5$ (figure 13.1(a)) and $d_{pos}/d = 1.5$ (figure 13.1(b)). The density peaks for the positive ions are out of phase in relation to the rest of the solution components. In the case of $d_{pos}/d = 0.5$, the first peak in the density distribution is unsurprisingly around $z/d = 0.5$. If the positive ions are larger (i.e., $d_{pos}/d = 1.5$), then the density at the interface peaks at $z/d = 1.5$. The density profiles for the rest of the solution species seem, at least qualitatively, unaffected and exhibit peaks that correspond to their size, which is d. This is because correlation effects of higher order are ignored in the mean-field c-DFT approach, used to obtain figure 13.1.

Another illustration of size variation on the density profile of the positive non-PDIs is shown in figure 13.2. The density peak, near the charged wall, moves away as the ionic diameter increases, while it height increases. This variation has an impact on the local density of PDIs (see figure 13.3), which decreases as the positive non-PDI diameter increases. This shifts the surface charge regulating equilibrium (2.1) toward greater surface charge, while reducing the surface potential (see figure 13.4). The reason is because the increase in the positive non-PDIs size creates a positive excluded volume barrier, which prevent the PDIs from approaching the surface, while reducing the potential negativity near the EDL interface.

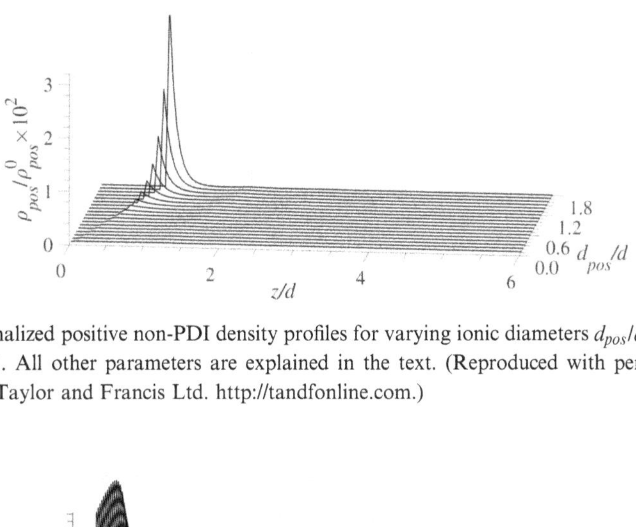

Figure 13.2. Normalized positive non-PDI density profiles for varying ionic diameters d_{pos}/d between 0.1 and 1.8 in steps of 0.1. All other parameters are explained in the text. (Reproduced with permission from [5], Copyright (2021) Taylor and Francis Ltd. http://tandfonline.com.)

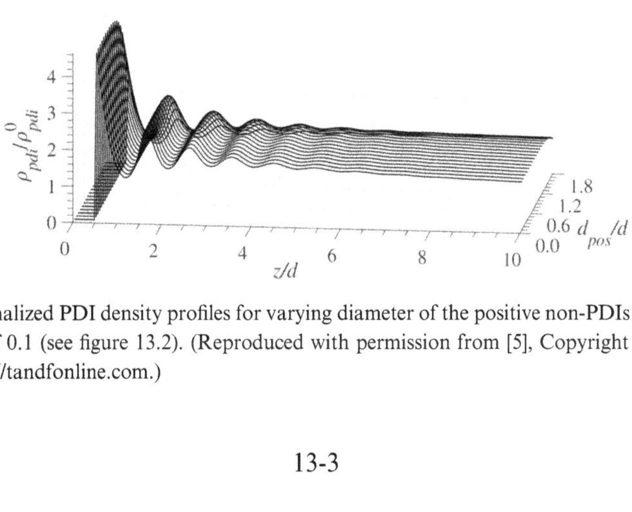

Figure 13.3. Normalized PDI density profiles for varying diameter of the positive non-PDIs d_{pos}/d between 0.1 and 1.8 in steps of 0.1 (see figure 13.2). (Reproduced with permission from [5], Copyright (2021) Taylor and Francis Ltd. http://tandfonline.com.)

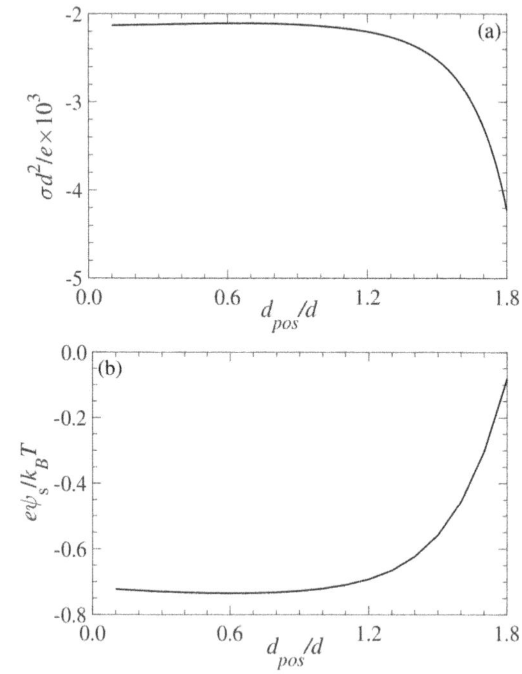

Figure 13.4. Surface charge (a), and potential (b) as a function of the positive non-PDI diameter d_{pos}/d. The remaining parameters are listed in the text. (Reproduced with permission from [5], Copyright (2021) Taylor and Francis Ltd. http://tandfonline.com.)

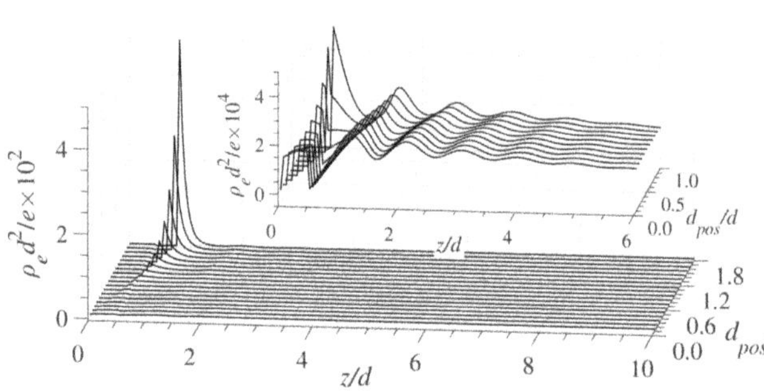

Figure 13.5. Fluid charge density distribution for varying positive non-PDI diameter d_{pos}/d. The inset provides more detailed features of the charge density distributions. (Reproduced with permission from [5], Copyright (2021) Taylor and Francis Ltd. http://tandfonline.com.)

The fluid charge profile ρ_e in the EDL is plotted in figure 13.5. It consists predominantly of positive non-PDIs, and therefore the shape of the density curves resembles those in figure 13.2. However, zooming into the plot (see the inset in figure 13.5) reveals some interesting details. The peaks near the charged wall appear to have their tops 'cut off'. This shape is due to the fact that planes of positive and negative charges are shifted with respect to each other. This is evident also in figure 13.1, which shows the density distributions of all solutions species for the particular cases of $d_{pos}/d = 0.5$ and $d_{pos}/d = 1.5$. The fluid charge density is equal to the difference between the local values of the positive and negative ionic densities. This leads to the particular shape of the charge density curves between $z = d_{pos}$ and $z = d_{neg}$, shown in figure 13.5.

13.1.2 Negative non-PDI size variation effects

Figure 13.6 shows the normalized density profiles for all solution components at two different sizes of the negative ions. The plots in figure 13.6(a) are for $d_{neg}/d = 0.5$, while those in figure 13.6(b) are for $d_{neg}/d = 1.5$. The curves for the positive PDIs and non-PDIs lie on top of each other and are therefore indistinguishable. The negative ions density oscillations are out of phase in comparison with the rest of the same-sized solution species. The densities near the wall are lower in comparison with the positive ions case, discussed above, because of the electrostatic repulsion from the negatively charged EDL interface.

A more detailed illustration of the negative ions diameter on their density near the charged wall is offered in figure 13.7. Unsurprisingly, the density peak near the surface shifts away in accordance with the size increase, while increasing in height. This is very similar to the positive ions size variation effect discussed above, and implies that it is mostly driven by excluded volume effects.

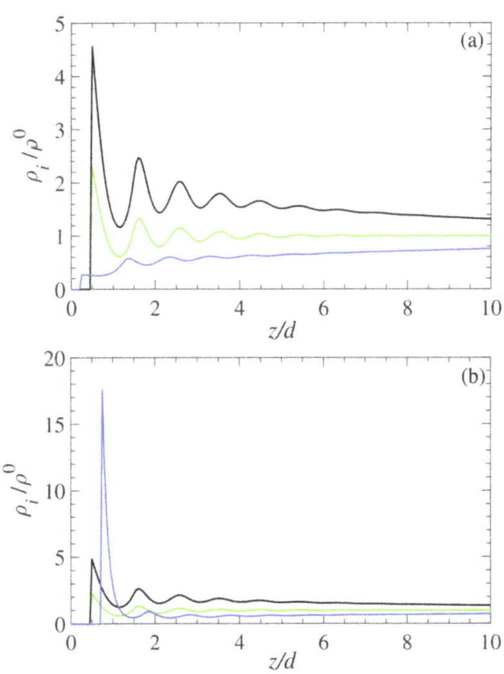

Figure 13.6. Normalized density profiles for the different species in the electrolytes solution: positive non-PDIs (black), PDIs (red), negative non-PDIs (blue), and solvent (green). The positive PDIs and non-PDIs curves lie on top of each other. (a) $d_{pos}/d = 0.5$, and (b) $d_{pos}/d = 1.5$. The remaining parameters are listed in the text. (Reproduced with permission from [5], Copyright (2021) Taylor and Francis Ltd. http://tandfonline.com.)

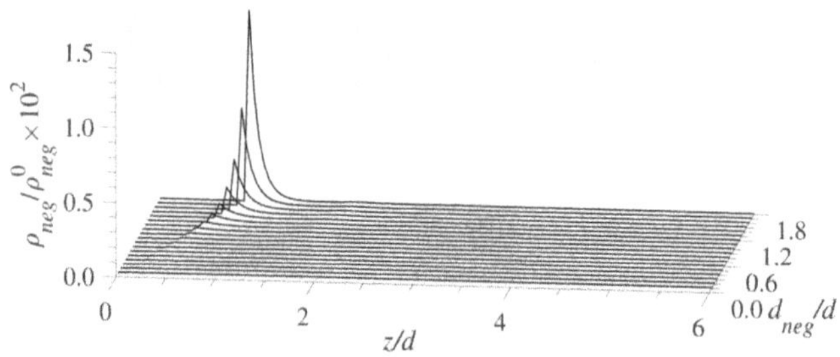

Figure 13.7. Normalized negative ionic density profiles for varying ionic diameters d_{neg}/d between 0.0 and 1.8 in steps of 0.1. All other parameters are explained in the text. (Reproduced with permission from [5], Copyright (2021) Taylor and Francis Ltd. http://tandfonline.com.)

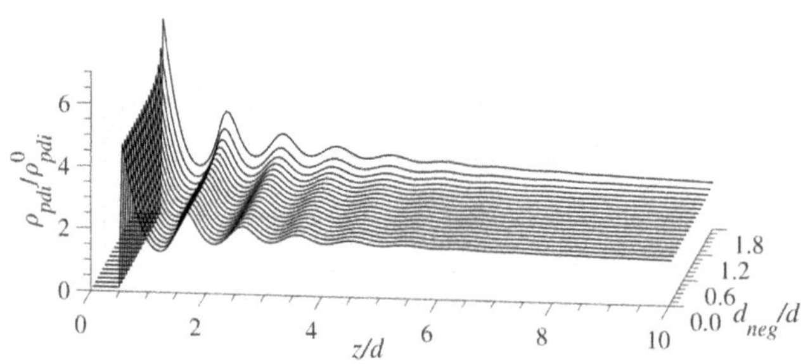

Figure 13.8. Normalized PDI density profiles for varying diameter of the positive non-PDIs d_{pos}/d between 0.1 and 1.8 in steps of 0.1 (see figure 13.2). (Reproduced with permission from [5], Copyright (2021) Taylor and Francis Ltd. http://tandfonline.com.)

The density of the PDIs does not significantly change for $d_{neg}/d \lesssim 1$, but shows an increasing trend for larger negative ions (see figure 13.8). This leads to a decrease of the surface charge negativity (see figure 13.9(a)) because of the surface charge regulation (2.1). The surface potential, however, becomes more negative as the size of the negative ions increases as seen in figure 13.9(b).

The corresponding fluid charge distribution in the EDL is depicted in figure 13.10. The overall fluid charged in the EDL is positive for smaller negative ions. This is observation is in agreement with figure 13.8, which shows that lower densities of the PDIs at the same conditions. Hence, the surface charge is less neutralized and more negative (see also figure 9.10(a)), which attracts positive non-PDIs in the immediate vicinity of the wall. As the negative ions increase in diameter, their local density also increases (see figure 13.7). This leads to the formation of a negative charge dip in figure 13.10 larger negative ions. However, the smaller positive ions are able to fit in the spaces that form between the larger negative ions and the negative surface, attracted there by the negative electrostatic surface potential (see figure 9.10(b)). The peaks tops are cut off again, because the planes of localization of positive and negative charges are shifted with respect to each other.

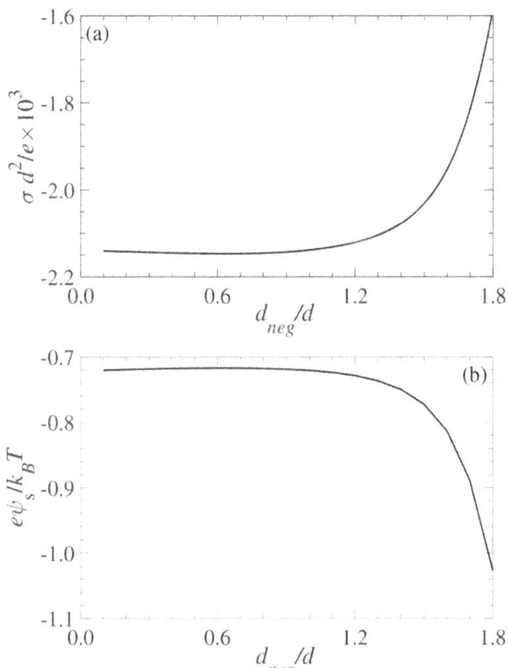

Figure 13.9. Surface charge (a), and potential (b) as a function of the negative non-PDI diameter d_{neg}/d. The remaining parameters are listed in the text. (Reproduced with permission from [5], Copyright (2021) Taylor and Francis Ltd. http://tandfonline.com.)

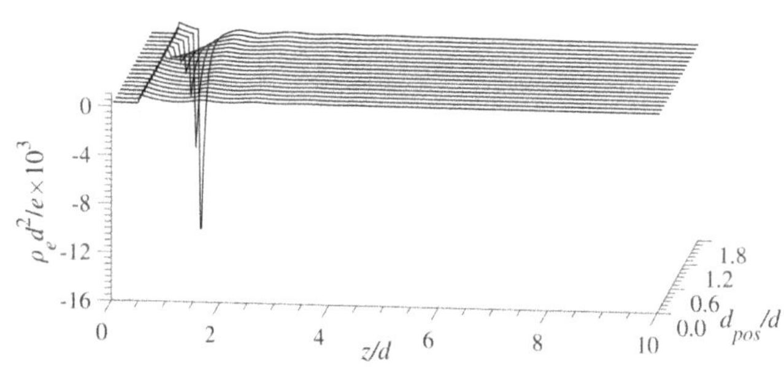

Figure 13.10. Fluid charge density distribution for varying negative non-PDI diameter d_{neg}/d. The inset provides more detailed features of the charge density distributions. (Reproduced with permission from [5], Copyright (2021) Taylor and Francis Ltd. http://tandfonline.com.)

13.2 Conclusions

The size impact of the dissolved ions in solutions offer another example of the importance of the non-Coulombic interactions for the properties of EDLs. Our analysis suggests, that the main effects are due to excluded volume and packing constraints near the charged EDL interface. These couple to the charge regulation surface chemical reaction, by affecting the local density of the PDIs.

References

[1] Marcus Y 1988 Ionic radii in aqueous solutions *Chem. Rev.* **88** 1475–98

[2] López-García J J and Horno J 2011 Poisson–Boltzmann description of the electrical double layer including ion size effects *Langmuir* **27** 13970–4

[3] Valisko M, Kristof T, Gillespie D and Boda D 2018 A systematic monte carlo simulation study of the primitive model planar electrical double layer over an extended range of concentrations, electrode charges, cation diameters and valences *AIP Adv.* **8** 025320

[4] Valisko M, Boda D and Gillespie D 2007 Selective adsorption of ions with different diameter and valence at highly charged interfaces *J. Phys. Chem.* C **111** 15575–85

[5] Prakash D J, Denoyer L, Vangara R, Baca J M, van Swol F and Petsev D N 2021 Classical density functional analysis of the ionic size effects on the properties of charge regulating electric double layers *Molecular Phys.* e1937737

Part III

Numerical methods

Chapter 14

Molecular simulation: methods

14.1 Background

Molecular simulation (or sometimes computer simulation) is a still rapidly developing field that is highly successful in providing information on bulk and interfacial fluids. There exist two basic methods of classical molecular simulation: molecular dynamics (MD) and Monte Carlo (MC). MC can be performed for lattice models as well as off-lattice models. Both methods start with defining a Hamiltonian, an energy function that depends on the positions of the system particles (e.g., atoms, molecules, spins, etc). Each seeks to generate a sequence of relevant configurations from a starting configuration. The configurations then exist of a list of N particle positions and, in the case of MD, velocities of all the particles. Properties of the system are then obtained as averages over these configurations.

We point out that prior to the advent of computer generated liquid structures there were many attempts to build physical models in the laboratory. John Finney [1], a student of J D Bernal, has published a very illuminating and entertaining account of Bernal's (and others') contributions to build 'homogenous, coherent and essentially irregular assemblages containing no crystalline regions'. Ground breaking work was performed using random packings of spheres (steel ball bearings)[1] to generate the close packing density and the radial distribution function, $g(r)$, see figure 14.1.

14.2 Molecular dynamics methods

Of the two methods, MD is perhaps the most intuitive. From an initial set of N positions and N velocities, i.e. at time $t = 0$, one solves Newton's equations of motion and generates a set of new positions and velocities. This is identical to calculating, say, the trajectories of planets around the Sun. Each model system is defined by specifying the Hamiltonian, that expresses the interaction (typically by

[1] A ball-and-spoke model of a random close-packed structure is on display in the British Science Museum in London, UK.

doi:10.1088/978-0-7503-2276-8ch14

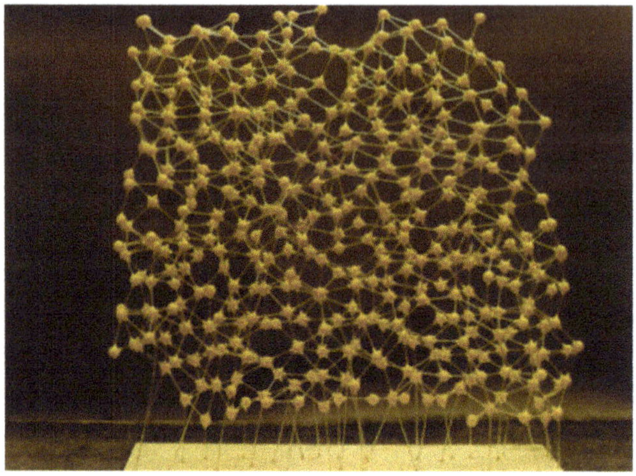

Figure 14.1. A ball-and-spoke model displayed in the British Science Museum due to J D Bernal. The model was built using the sphere coordinates of a paint-fixed random close-packed sphere assembly. (Reproduced from [1] Taylor & Francis Ltd. http://tandfonline.com, CC BY 3.0.)

pair-potentials) acting between the different particles. For hard particles these are the simplest of all, namely the particles experience an infinite potential energy when they overlap, and zero potential energy otherwise. Thus for two hard spheres of diameters σ_i and σ_j and we have

$$\phi(r_{ij}) = \begin{cases} 0; & r_{ij} > \sigma_{ij} \\ \infty; & r_{ij} < \sigma_{ij} \end{cases} \tag{14.1}$$

where $\sigma_{ij} = \alpha_{ij}(\sigma_i + \sigma_j)/2$ with α_{ij} a coefficient that controls the additivity of the interaction. If $\alpha_{ij} > 1$ then the two particles appear larger than their arithmetic mean and hence experience an enhanced repulsion over the simple additive spheres. On the other hand, if $\alpha_{ij} < 1$ a reduced repulsion results. The parameter α allows us one to promote or discourage mixing of various species in a mixture.

This simple pair-potential leads to significant simplifications in the statistical mechanics of hard particle systems. Foremost amongst these is the fact that the phase diagram is a function of the density (or packing fraction) only, and *not* of the temperature (T). This follows immediately from the fact that for hard particles the Boltzmann factor ($\exp(-\phi/kT)$) reduces to either 0 or 1, and is hence independent of T. As a consequence, hard particle systems only have one single isotherm as all the properties scale linearly with the kinetic energy, i.e. T.

14.2.1 Hard sphere dynamics

Next, we will discuss the first molecular dynamics simulation ever performed, by Alder and Wainwright in 1957 [2]. This paper was jointly published with another hard sphere simulation, MC this time, by Wood and Jacobson [3]. Both papers came out of US National laboratories (Livermore and Los Alamos), where there was access to 'electronic computers' [2]. The topic was the hard sphere fluid–solid transition, manifested by two distinct branches in the pressure–volume plane.

For more detailed descriptions the reader is referred to the literature (see e.g. [4–7]).

Shared by both techniques is the concept of the simulation space, or simulation box as it is commonly known. Usually one performs the simulations in a rectangular box with one of the vertices coinciding with the origin of the coordinate system. Thus, the coordinates run from 0 to 1 in the x-direction and from 0 to y_c, and 0 to z_c in the y- and z-directions, respectively. Hence the particle coordinates can be thought of as fractional coordinates, similar to the description of a crystallographic unit cell. Typically, periodic boundary conditions (PBCs) are applied in order to minimize finite size effects. Thus, the simulation box is considered to be the unit cell of an infinite lattice of periodic images. For short-range interactions, the PBCs are applied with the so-called minimum image (MI) convention. This means that a particle i only interacts with the nearest image of all possible images of particle j.

It is important to realize that the MI is not in any way implied by the presence of PBCs. Indeed, when the interactions are long-ranged, such as in charged systems, one applies and Ewald summation [8] and (e.g., see also [5]) or Lekner's method [9–11] to be discussed below.

Given the positions and velocities at time $t = 0$, the time to the next collision between two spheres, i and j is calculated from

$$\left|\mathbf{r}_i(t) - \mathbf{r}_j(t)\right| = \sigma_{ij} \tag{14.2}$$

or, equivalently

$$(\mathbf{r}_i(0) + \mathbf{v}_i(0)t - \mathbf{r}_j(0) - \mathbf{v}_j(0)t)^2 = \sigma_{ij}^2. \tag{14.3}$$

After reorganizing, and introducing some short hand notation (i.e. $\mathbf{r}_{ij} \equiv \mathbf{r}_i - \mathbf{r}_j$ and $\mathbf{v}_{ij} \equiv \mathbf{v}_i - \mathbf{v}_j$) we arrive at a simple quadratic equation for the collision time t:

$$v_{ij}^2(0)t^2 + 2\mathbf{r}_{ij}(0) \cdot \mathbf{v}_{ij}(0)t + (r_{ij}^2(0) - \sigma_{ij}^2) = 0. \tag{14.4}$$

Of the two possible positive roots of this equation the short time root is the physically relevant root t_c, when the centers are first separated by σ. The long time root simply describes the next time at which the sphere centers are separated by σ namely, the time at which spheres have finished moving through each other.

Note that in order for two spheres to possibly collide their relative velocity component along their vector of separation, i.e. $\mathbf{r}_{ij}(0) \cdot \mathbf{v}_{ij}(0)$ has to be negative. That is, they have to be approaching.

The new velocities at the collision can be derived from the equations expressing conservation of momentum and conservation of energy. Only the velocity components along the $\mathbf{r}_{ij}(0)$-direction are affected. That is, at the collision the normal components of the velocity vectors (the velocity components that lie along the direction of $\mathbf{r}_{ij}$ given by $\mathbf{v}_i^\perp \equiv (\mathbf{v}_i \cdot \mathbf{r}_{ij})\mathbf{r}_{ij}/r_{ij}^2$ and $\mathbf{v}_j^\perp \equiv (\mathbf{v}_j \cdot \mathbf{r}_{ij})\mathbf{r}_{ij}/r_{ij}^2$) are interchanged,

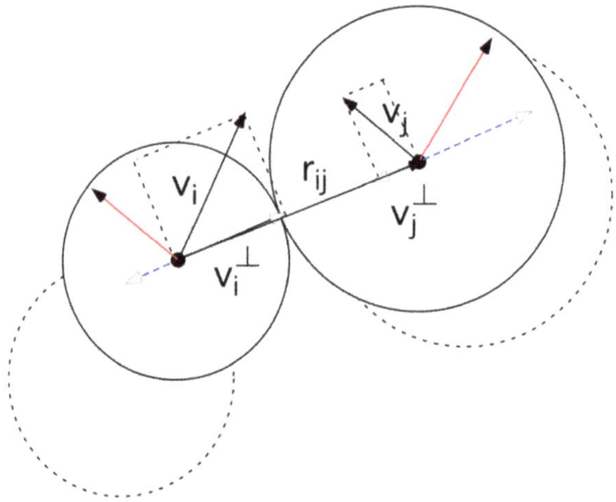

Figure 14.2. An illustration of the collision between two spheres of equal mass, i and j. The dashed circles show the positions at $t = 0$ while the solid circles denote the positions at $t = t_c$, when the spheres are in contact. To obtain the new velocities (red vectors) from the original vectors $\mathbf{v}_i$ and $\mathbf{v}_j$, one merely exchanges the normal component of the velocities between particle i and j (denoted by $\mathbf{v}_i^\perp$ and $\mathbf{v}_j^\perp$, respectively) to give the dashed blue vectors. Upon adding the blue vectors to the unchanged tangential components of $\mathbf{v}_i$ and $\mathbf{v}_j$ the red vectors result, i.e. $\mathbf{v}'_i = \mathbf{v}_j^\perp + \mathbf{v}_i^\parallel$ and $\mathbf{v}'_j = \mathbf{v}_i^\perp + \mathbf{v}_j^\parallel$. This construction is the 3D generalization of the 1D collision, where only normal components exist. It is straightforward to verify that this construction conserves both the total momentum and the total energy.

while the tangential components remain unchanged, see figure 14.2. Thus, writing the interchange in terms of $\mathbf{v}_{ij} \equiv \mathbf{v}_i - \mathbf{v}_j$ we obtain,

$$\begin{aligned} \mathbf{v}'_i &= \mathbf{v}_i - (\mathbf{v}_{ij} \cdot \mathbf{r}_{ij})\mathbf{r}_{ij}/r_{ij}^2 \\ \mathbf{v}'_j &= \mathbf{v}_j + (\mathbf{v}_{ij} \cdot \mathbf{r}_{ij})\mathbf{r}_{ij}/r_{ij}^2 \end{aligned} \tag{14.5}$$

which is equivalent to the alternative presented in the caption of figure 14.2. Notice that given the initial velocities $\mathbf{v}_i$ and positions $\mathbf{r}_i$ for all the particles i, the essence of a simple event-driven MD algorithm can now be summarized as follows:

1. For each possible particle i determine the earliest collision time (exact within machine precision) and collision partner.
2. Search the table of N collision times for the earliest collision time interval, t_c, and identify the corresponding collision partners k and l.
3. Update the positions of all the N particles to those at t_c, using $\mathbf{r}_i(t_c) = \mathbf{r}_i(0) + \mathbf{v}_i(0)t_c$ (i.e. straight line movement).
4. Perform the collision for the pair (k, l), i.e. update $\mathbf{r}_k$, $\mathbf{r}_k$, $\mathbf{v}_l$, $\mathbf{v}_l$ according to equations (14.5).
5. Update the collision time table for all those particles affected by the collision just performed between k and l,. This includes k and l as well as all those particles that initially had k or l scheduled as a collision partner.
6. Repeat from 2.

A key observation is that the collision times are irregular, i.e., not equally spaced. Rather, the configuration moves through time with unequal steps. Typically, any configuration in an event-driven simulation has one pair of touching spheres. This is a small but important detail to keep in mind.[2] The generalization to colliding spheres of different masses is straightforward.[3] The collision times between any two particles i and j, is found as the root of a quadratic equation. And it is this simple property that makes the trajectories exact. We note in passing that adding a gravitational field or letting the particle diameters grow linearly with time [12] does not affect the quadratic nature of the equation for the collision time, and hence such modifications do not change the algorithm significantly.

Averages obtained from MD simulations are time averages

$$\bar{A} = \frac{1}{\tau}\int_0^{\tau} dt\ A(t). \tag{14.7}$$

The ergodic hypothesis states that for an equilibrium system the time average (in the limit $\tau \to \infty$) will equal the ensemble average, that is $\bar{A} = \langle A \rangle$. In other words, the MC and MD averages should be equal for sufficiently long simulation runs.

14.2.2 Continuous potentials

The dynamics for hard core particles is easy, as the particles move along straight lines between collisions. This is generally true for a potential that consists of a number of discontinuities, such as the square-well potential or a hard-shoulder potential. Things are decidedly different for continuous intermolecular potentials, because the force is continuously changing as the particle moves. To solve Newton's equations of motion one needs the force, F_i on each particle i. Typically the force is a function of the position of all the particles, i.e., $F_i = F_i(r_1, \ldots r_N)$. Given that the particles are propelled to a later time, and the positions all change, the forces will have to be recalculated at the new time. Depending on the details of the potential the sequence of configurations will be obtained by time stepping or an event driven method. For smooth continuous potentials one uses time stepping, generating new configurations at times $t = 0, \Delta t, 2\Delta t, \ldots$, i.e., a time series that is equally spaced. The far more common continuous potential MD method, which would not be introduced until six years later [13], solves Newton's equations in an approximate stepwise manner, with timesteps of the order of 10^{-14} s. Rahman used the so-called

[2] It implies that if radial distribution data is collected with event-driven MD, there will be a a slight (positive) bias near $r = \sigma_{ij}$ It is not difficult to arrange an equally-spaced list of configurations, but this comes at the price of performing an operation that scales like N.

[3] The resulting expressions are (cf, equations (14.5)),

$$\begin{aligned} \mathbf{v}'_i &= \mathbf{v}_i - \frac{2m_j}{m_i + m_j}(\mathbf{v}_{ij} \cdot \mathbf{r}_{ij})\mathbf{r}_{ij}/r_{ij}^2 \\ \mathbf{v}'_j &= \mathbf{v}_j + \frac{2m_i}{m_i + m_j}(\mathbf{v}_{ij} \cdot \mathbf{r}_{ij})\mathbf{r}_{ij}/r_{ij}^2. \end{aligned} \tag{14.6}$$

predictor–corrector algorithm. Shortly thereafter Loup Verlet [14] produced an algorithm that was slightly superior. It is now known as the Stormer–Verlet algorithm. In the preferred 'velocity-Verlet' [5, 15] form it reads

$$\begin{aligned} \mathbf{r}(t+\Delta t) &= \mathbf{r}(t) + \mathbf{v}(t)\Delta t + \frac{1}{2}\mathbf{a}(t)\Delta t^2 \\ \mathbf{v}(t+\Delta t) &= \mathbf{v}(t) + \frac{1}{2}[\mathbf{a}(t) + \mathbf{a}(t+\Delta t)]\Delta t \end{aligned} \tag{14.8}$$

where the acceleration, $m\mathbf{a}$ is the force, the negative gradient of the potential. That is,

$$\mathbf{a} = \mathbf{F} = m\mathbf{a} = -\nabla\phi(r) = -\frac{\mathbf{r}}{r}\frac{\partial\phi}{\partial r}. \tag{14.9}$$

In the form presented, equation (14.8) is would appear as if one needs two force calculations per time step. That is not true, which can be seen from the standard implementation:

1. form intermediate half-step velocity: $\mathbf{v}(t + \frac{1}{2}\Delta t) = \mathbf{v}(t) + \frac{1}{2}\mathbf{a}(t)\Delta t$.
2. calculate full-step position: $\mathbf{r}(t + \Delta t) = \mathbf{r}(t) + \mathbf{v}(t)\Delta t$.
3. use the new positions to calculate the acceleration: $\mathbf{a}(t + \Delta t) = \mathbf{F}(t + \Delta t)/m$.
4. complete the full-step velocity: $\mathbf{v}(t + \Delta t) = \mathbf{v}(t + \frac{1}{2}\Delta t) + \frac{1}{2}\mathbf{a}(t\frac{1}{2}\Delta t)\Delta t$.

The time step is chosen such that the algorithm conserves energy: the sum of the potential energy and kinetic energy has to remain constant. That is, it can fluctuate in the 4th decimal place, say, but it should not drift (downward or, more typically, upward) In reduced time units a typical time step for a LJ simulation is $\Delta t \approx 0.005\sqrt{m\sigma^2/\epsilon}$, or about 10^{-14} s for argon.

The velocity-Verlet algorithm (e.g., [15]) has convenient features that the predictor–corrector algorithm does not, including that it is time-reversible.

Newton's equations of motion have the total energy as a conserved property of the motion. In other words, the natural MD ensemble is *NVE*, or micro canonical ensemble. There are many situations where the canonical, or *NVT*, ensemble is preferred. Similarly, and in addition, many experimental investigations (e.g., concerning fluid mixtures) proceed under the condition of constant pressure as well, and hence the isothermal–isobaric (*NpT*) ensemble is the more natural choice. These considerations stimulated an explosion of algorithmic research to accomplish *NVT* and *NpT* dynamics. Details can be found in the books by Allen and Tildesley [5] as well as by Smit and Frenkel [7].

Time savings techniques

From the beginning of the field of molecular simulation there have been efforts to speed up the computer codes that executed the MD algorithm. A number of the tried and proven approaches are equally applicable to the MC techniques described in the next section.

Figure 14.3. A two-dimensional illustration of the periodic boundary condition and the minimum image convention. The central box contains the primary particles (solid colors) that are actually followed in the simulation. The neighboring boxes (with lighter colored image particles) illustrate the various periodic images. If a primary particle crosses a boundary, a replacement image particle enters through the opposite boundary. The same happens when a particle exists through two boundaries simultaneously. When calculating the intermolecular interactions (e.g., energy, force, virial) involving the particle marked blue, we draw a circle around its center. The interacting neighbors to be included are in the primary box as well as in the image boxes. Here (three) image particles are indicated with the marking *I*. Note that there is no absolute origin in a PBC system, i.e., one can move the back rectangular grid and find that nothing (e.g., energy, force, virial, total momentum) changes.

As mentioned simulators applied so-called periodic boundary conditions (pbc),[4] a commonly encountered boundary in many physics problems. This implied that the computational simulation box (of dimensions L_x by L_y) of interest was surrounded by identical copies, see figure 14.3. This means that the system has no surfaces, and can be considered 'infinite but periodic'. Clearly, the largest fluctuation in density is of a wave length $\min[L_x, L_y]$.

Typically, the calculation intermolecular interactions are limited to the neighbors that fall into a circle (or sphere in 3D) of radius smaller than one half box length (i.e., $(1/2)\min[L_x, L_y]$). It is common to consider interaction potentials that are explicitly cut at a certain distance. For example, many LJ studies use a LJ potential that is cut at a distance 2.5σ. This leaves a small discontinuity at $r = 2.5\sigma$. In MD it is not uncommon to simply 'ignore' this discontinuity.

[4] These boundary conditions (bc) are also known as a toroidal bc. The initial 2D, flat simulation box can be distorted and folded back on itself to become the surface of a torus.

14.2.3 Monte Carlo

In Monte Carlo simulations, introduced by Metropolis *et al* in 1953 [16], one uses random numbers and an acceptance criterion to generate a Markov chain (e.g., [5]) of configurations ('states') that eventually will have each configuration appear with the correct Boltzmann probability, namely $\exp(-U/kT)$. Here $U = \sum_{i<j}\phi(r_{ij})$ is the total potential energy of a configuration, T is the temperature and k denotes Boltzmann's constant. Ensemble averages for a property A may then be obtained simply as an average over the chain of N_{steps} configurations, i.e.

$$<A> = \frac{1}{N_{\text{steps}}} \sum_{k=1}^{N_{\text{steps}}} A_k. \tag{14.10}$$

To randomly generate relevant configurations with true Boltzmann weights is an impossible task, as most configurations have overlaps and associated large potential energies (or indeed infinite energies for hard care particles). This is discussed, for instance, in the book by Allen and Tildesley [5]

A Markov chain is a chain of states, where each state is generated from the previous one in a stochastic fashion (by 'trials') using a transition matrix, $\boldsymbol{\pi}$. The generation of the chain is performed such that the outcome of a trial is only dependent upon the preceding state and not on any of the previous states (or trials). Metropolis and the couples Rosenbluth and Teller [16] achieved the 'impossible feat' of generating configurations with the correct Boltzmann probability by only controlling the transition probability of going from configuration A_k to A_{k+1} (and vice versa). This transition probability is proportional to $\exp(-(U_{k+1} - U_k)/kT)$, the ratio of the Boltzmann factors of the two states k and $k + 1$. That is, the transition probability is only a function of the energy difference. Within Markov chain theory one can then show that in the limit $N_{\text{steps}} \to \infty$ each configuration will be visited with the correct probability. Starting from any initial vector $\boldsymbol{\rho}(1)$ one obtains the correct limiting vector of the probabilities of each state by letting the transition matrix operate on in it many times. In other words,

$$\boldsymbol{\rho}_{\text{limit}} = \lim_{N_{\text{steps}}\to\infty} \boldsymbol{\rho}(1)\boldsymbol{\pi}^{N_{\text{steps}}}.$$

The first algorithm to accomplish this feat is now appropriately referred to as the Metropolis algorithm. It can be derived by imposing the microscopic reversibility condition, and it proceeds as follows: starting from a low energy configuration, such as a regular crystal lattice, attempted MC moves are performed by giving a randomly selected particle a small random displacement. The potential energy difference, $\Delta U \equiv U_{k+1} - U_k$, is calculated and a decision is made concerning the acceptance or rejection of the new configuration. If the new energy is lower ($\Delta U < 0$) one accepts the new configuration outright. Otherwise one calculates $\exp(-\Delta U/kT)$, and compares this to a random number, ξ from the interval (0,1]. If the $\xi < \exp(-\Delta U/kT)$ one accepts the new configuration, otherwise the old configuration is kept and counted again. Schemes different from the Metropolis algorithm are also possible and have been described in the literature.

A few comments are in order at this point. Firstly, for hard particles (without additional repulsive or attractive forces) the total potential energy, U, is a binary function. It is either zero (no overlaps) or infinity (at least one overlapping pair), and hence the Boltzmann factor, $\exp(-U/kT)$, is either one or zero, respectively. Consequently, systems of hard particles only visit configurations that are free of overlaps, and each of these configurations is equally probable. Thus, when we attempt an MC move for a hard particle i, it suffices to evaluate the acceptability of the new configuration by checking overlaps between i and neighboring j, by searching for the *first* overlap. Once any overlap is detected the attempted move can be aborted immediately. This makes MC of hard particles a very cheap and straightforward technique.

MC can be performed in various ensembles. The most common is the NVT ensemble of fixed volume and fixed number of particles, but two other ensembles also find wide application and are both relevant to the discussion here. The first of these, the isobaric or NpT ensemble has a variable volume, and the other, the so-called grand or μVT ensemble has a fixed volume but a variable number of particles. The isobaric ensemble was first implemented for hard disks by Wood [17]. It has all the features of $NVT - MC$ but, in addition, it possesses volume moves, small random volume changes, $\Delta V = V_{k+1} - V_k$, that are accepted with a probability $\exp(-(\Delta U + \Delta V)/kT)$, where the pressure p is simply a pre-set parameter for the MC simulation. Although an infinite number of implementation schemes are possible, in practice one performs about an attempted volume change every N particle moves. In principle, the volume can be changed in various ways. However, the most convenient way to change the volume is by a simple scaling operation, or equivalently by scaling the particle size. Note that simple scaling leaves the relative particles positions in the simulation box untouched, thus avoiding a need to update the particle positions (an operation of order N).

The details of Monte Carlo moves

As mentioned, the MC technique consists of generating the correct Markov Chain through the application of a transition matrix, the Metropolis scheme [16] being equivalent to defining one such matrix. Underlying this scheme is the concept of selecting the particular change of state k to $k + 1$, which has so far been left unspecified. Given state n how do we select a proposed state m? This is what constitutes a proposed or attempted 'Monte Carlo move', and it is formally controlled by the matrix element α_{nm} of a so-called stochastic (or underlying) matrix. The matrix $\boldsymbol{\alpha}$ is typically symmetric and specifies the probability with which states can be selected from other states.

There is great freedom in specifying $\boldsymbol{\alpha}$. For simulations of spherical particles the accepted method for attempting a MC move boils down to

1. select a particle at random, and
2. select three random numbers, ξ_k; $k = 1, 3$ from the interval $[0, 1)$ and perform a translation according to

$$
\begin{aligned}
x' &= x + \Delta r\ (2\xi_1 - 1)\\
y' &= y + \Delta r\ (2\xi_2 - 1)\\
z' &= z + \Delta r\ (2\xi_3 - 1)
\end{aligned}
\tag{14.11}
$$

where Δr is an arbitrary maximum displacement parameter that is set by the user. This scheme is computationally efficient and certainly corresponds to a symmetric matrix $\boldsymbol{\alpha}$. The parameter, Δ_r, is used to get a decent acceptance ratio (the ratio of accepted moves and attempted moves). If it is set too large, a particle overlap is likely to be created and the acceptance ratio is too close to zero. There is formally no problem with this, but practically there is, since it makes the Markov chain sample phase space too inefficiently. In fact, very large values of Δr amount to something close to a particle 'creation move' common to the μVT ensemble. Similarly, if $\Delta r \to \infty$ the acceptance ratio approaches 1, but the gain is zero too, as we barely move through phase space, and neighboring states are too similar. Clearly, the optimum value of Δr should also vary with density (decreasing with density). In addition, for mixtures there should be a different value of Δr for every species.

The 'best' possible value for Δr can be based on various criteria. One of which is to optimize with respect to the computational time and size of the error in the property of interest, e.g. pressure, $g(r)$, etc. Intuitively, we want to move through phase space in the most efficient way. Since for hard spheres, it is the first overlap detection that matters, it serves to aim for a slightly lower acceptance ratio, somewhere between a low and high of roughly 10 and 50%.

For nonspherical particles the MC move must be expanded to include rotational moves. One often uses a method originally put forward by Barker and Watts [18]. In this procedure one first randomly selects one of the three space-fixed axes of the Cartesian coordinates system of the MD box, and then performs a change in the orientation of the particle. Thus, the MC scheme outlined above is augmented with the following steps:

1. select an axis at random
2. randomly select an angle γ from the interval $[-\Delta\gamma, \Delta\gamma]$
3. form the appropriate rotation matrix. For example, a rotation around the $x-$ axis employs the following matrix

$$
\begin{pmatrix}
1 & 0 & 0\\
0 & \cos\gamma & \sin\gamma\\
0 & -\sin\gamma & \cos\gamma
\end{pmatrix}
$$

4. let the rotation matrix operate on the three body vectors **a**, **b** and **c**.

As for the translation, $0 < \Delta\gamma < 2\pi$ is a free parameter for the rotation. However, unlike Δr, $\Delta\gamma$ is always fixed and cannot be automatically adjusted at this point. This is because, although variations are certainly possible, for nonspherical particles users often combine translation and rotation into one single compound MC move.

Long-range interactions, bulk systems

The very long-range nature of the Coulombic interaction requires a special approach. Whereas, for dispersion interactions one uses the minimum image convention, this is not sufficient for charge interactions (and dipolar interactions). The potential decays as $r^{-}1$, but the number of particles, and thus the number of interactions, growths as r^2, so, naively, the total energy, say, might be thought not to converge. This is encountered in molecular simulations as well as integral equation theory. Of course, the solution to this 'problem' is saved by the fact that there are both positive as well as negative interactions, providing the convergence of the total energy.

The slow decay of the interaction between particles i and j implies that many images of particle j have to be included. For bulk systems, this was first accomplished by employing the so-called Ewald summation [8], first introduced to calculate crystal energies. The method hinges on the fact that the images are periodic. Hence, the summation of interactions between i and *all* the images of j

$$v_{ij} = \frac{q_i q_j}{4\pi\epsilon_0} \sum_{\mathbf{n}} \frac{1}{|\mathbf{r}_i - \mathbf{r}_j + \mathbf{n}L|} \tag{14.12}$$

where $\mathbf{r}_j$ denotes the position of j in the primary box; $\mathbf{n}L$ is a lattice vector, and the collection of images of j have positions $\mathbf{r}_j + \mathbf{n}L$, which highlights the periodic nature of a crystal. Here, we have assumed that the basic cell (simulation box) is cubic. The total potential energy results when v_{ij} is summed over all pairs,

$$E = \frac{1}{2}\sum_{i=1}^{N}\sum_{j=1}^{N}{}' v_{ij} = \frac{1}{4\pi\epsilon_0}\frac{1}{2}\sum_{\mathbf{n}}\sum_{i=1}^{N}\sum_{j=1}^{N}{}' \frac{q_i q_j}{|\mathbf{r}_i - \mathbf{r}_j + \mathbf{n}L|}. \tag{14.13}$$

The prime on the summation over j indicates that we *exclude* the term $j = i$ but *only* when we are in the primary cell, $\mathbf{n} = 0$. The factor of 1/2 corrects for the fact that the summations 'double count' by visiting every pair i and j twice.

Unfortunately, the triple summation of equation (14.13) is not of practical use as is. That is because the expression is only conditionally convergent, as well as slowly convergent provided an elegant practical solution, by splitting the summation into two different summations, one in real space and one in Fourier space. Both these series converge rapidly and absolutely.

The final result is

$$\begin{aligned} E = &\frac{1}{8\pi\epsilon_0}\sum_{\mathbf{n}}\sum_{i=1}^{N}\sum_{j=1}^{N}{}' \frac{q_i q_j}{|\mathbf{r}_i - \mathbf{r}_j + \mathbf{n}L|}\,\mathrm{erfc}\left(\frac{|\mathbf{r}_i - \mathbf{r}_j + \mathbf{n}L|}{\sqrt{2}\sigma}\right) \\ &+ \frac{1}{2V\epsilon_0}\sum_{\mathbf{k}\neq 0}|S(\mathbf{k})|^2 k^{-2} e^{-\sigma^2 k^2/2} \\ &- \frac{1}{4\pi\epsilon_0\sqrt{2\pi}\sigma}\sum_i q_i^2. \end{aligned} \tag{14.14}$$

The first term is the a real space summation, tempered by the erfc function. The second term is a summation in Fourier space, also tempered by the exponential function. The last term is the self-energy term. The Ewald technique ends with an expression that has a free parameter, σ, the width of the Gaussian charge distribution. In practice, one selects a value for σ that produces an optimal convergence for both the real space and the reciprocal series.

Long-range interactions, interfacial systems

A fluid against a wall presents a case that is fundamentally different from a bulk electrolyte in that the periodicity is limited to the two dimensions that are parallel to the x and y directions, say, while the z direction is taken perpendicular to the wall. Note that the 2D periodicity does not imply that the wall has to be flat, although it often will be. In 1989 John Lekner [9, 10] presented a novel method using mathematical transformations to speed up the convergence. Tildesley [11] and, later, Moreira and Netz [19] provide excellent reviews of Lekner's elegant method. Here we will closely follow Tildesley's write-up [11].

$$\mathbf{f}_{ij} = q_i q_j \sum_{\mathbf{n}} \frac{\mathbf{r}_i - \mathbf{r}_j - \mathbf{n}}{|\mathbf{r}_i - \mathbf{r}_j - \mathbf{n}|^3}. \tag{14.15}$$

Here $\mathbf{n}$ is a two-dimensional lattice vector, and the factor of $4\pi\epsilon_0$ has been included in the charges. We introduce reduced distances,

$$\begin{aligned} x &\equiv L^{-1}(x_i - x_j) \\ y &\equiv L^{-1}(y_i - y_j) \\ z &\equiv L^{-1}(z_i - z_j) \end{aligned} \tag{14.16}$$

and write the force on atom i due to j as $\mathbf{f}_{ij} = q_i q_j L^{-1}(F_x, F_y, F_z)$

$$\begin{aligned} F_x(x, y, z) &= \sum_{l=-\infty}^{\infty} \sum_{m=-\infty}^{\infty} (x + l)[(x + l)^2 + (y + m)^2 + z^2]^{-3/2} \\ F_y(x, y, z) &= \sum_{l=-\infty}^{\infty} \sum_{m=-\infty}^{\infty} (y + m)[(x + l)^2 + (y + m)^2 + z^2]^{-3/2} \\ F_z(x, y, z) &= \sum_{l=-\infty}^{\infty} \sum_{m=-\infty}^{\infty} z[(x + l)^2 + (y + m)^2 + z^2]^{-3/2}. \end{aligned} \tag{14.17}$$

To make these summations rapidly convergent required Lekner [9, 10] to introduce three transformations. the details are collected in appendix D.[5] The final results are

[5] We note that $F_y(x, y, z) = F_x(y, x, z)$, and so only one of these has to be evaluated.

$$F_x(x, y, z) = 8\pi \sum_{l=1}^{\infty} l \sin 2\pi lx \sum_{m=-\infty}^{\infty} K_0\left(2\pi l\sqrt{(y+m)^2 + z^2}\right)$$
$$F_y(x, y, z) = 8\pi \sum_{l=1}^{\infty} l \sin 2\pi ly \sum_{m=-\infty}^{\infty} K_0\left(2\pi l\sqrt{(x+m)^2 + z^2}\right)$$
$$\begin{aligned} F_z(x, y, z) = &\frac{2\pi \sinh 2\pi z}{\cosh 2\pi z - \cos 2\pi y} \\ &+ 8\pi z \sum_{l=1}^{\infty} l \cos 2\pi lx \sum_{m=-\infty}^{\infty} [(y+m)^2 + z^2]^{-1/2} \\ &\times K_1\left(2\pi l\sqrt{(y+m)^2 + z^2}\right). \end{aligned} \tag{14.18}$$

Here, K_0 and K_1 denote the zeroth-order and first-order modified Bessel functions respectively. The electrostatic energy between two ions i and j is obtained by integration of the force. Writing the potential energy as $U_{ij} = q_i q_j v_{ij}/L$, we have

$$\begin{aligned} v_{ij}(x, y, z) = &4 \sum_{l=1}^{\infty} \cos 2\pi lx \sum_{m=-\infty}^{\infty} K_0\left(2\pi l\sqrt{(y+m)^2 + z^2}\right) \\ &- \ln(\cosh 2\pi z - \cos 2\pi y) + C. \end{aligned} \tag{14.19}$$

Here C is an integration constant. It is expressed in terms of the Madelung constant M, as $C = (1 + \sqrt{2} M - \ln 2 = 3.207\,11\ldots$. For a square lattice of alternating charges $M = 1.615\,54\ldots$. Note that Tildesley uses the ambiguous symbol log in equation (14.19) and the expression for C, but the intended logarithm is the natural logarithm, i.e., ln.

The sums over l and m have a large (i.e., infinite) number of terms that, in practice, requires a maximum index l_m to limit the summation over l and m. Tildesley [11] provides a very helpful comparison of Lekner's method and the approaches of Heyes [20] as well as those of Hautman and Klein [21].

Following Tildesley we performed similar calculations for a two-particle system and report on our calculations here in table 14.1. We considered the reduced pair

Table 14.1. Lekner's method results for a two-particle system. One charge of +1 is located at the origin, a second charge of −1 is located at (0.25, 0.10, 0.15). The primary box is a $L \times L$ square. We report the reduced energy (equation (14.19)) and reduced force components (see equation (14.18)).

l_m	v_{ij}	F_x	F_y	F_z
4	3.303 851	−7.174 029	−3.093 232	−6.642 854
6	3.301 738	−7.399 036	−3.092 673	−6.571 854
10	3.301 911	−7.374 532	−3.092 487	−6.579 291
14	3.301 912	−7.374 198	−3.092 487	−6.579 388
20	3.301 912	−7.374 194	−3.092 487	−6.579 389
100	3.301 912	−7.374 194	−3.092 487	−6.579 389

energy, v_{ij} (cf, equation (14.19)), as well as the reduced force components F_x, F_y and F_z (cf, equation (14.18)). For comparison, the energy for the *isolated* pair (ignoring all images) is 3.244 428. From the table we see that $l_m = 10$ provides sufficient accuracy.

The mathematical manipulations underlying the Ewald and Lekner methods are not restricted to Coulombic interactions. Wells and Chaffee have considered a general Ewald approach that covers inverse power potentials, r^{-n}. In addition, these authors provided methods for optimal techniques in the case of MC implementations.

14.3 The potential distribution theorem (PDT)

Widom, in 1963, derived the first method by which the chemical potential could be obtained for an *homogeneous* fluid as a direct ensemble average,

$$\beta\mu = \ln \rho\Lambda^3 - \ln\langle e^{-\beta U_t}\rangle . \tag{14.20}$$

Its derivation is deceptively simple and straightforward. From the thermodynamic definition of chemical potential in the canonical (i.e. NVT) ensemble, we have

$$\begin{aligned}\mu &= \left(\frac{\partial F}{\partial N}\right)_{V,T} \\ &= -kT \ln\left(\frac{Z_{N+1}}{Z_N}\right)\end{aligned} \tag{14.21}$$

using $F = -kT \ln Z_N$.

Widom showed that, the ratio of partition functions, Z, can be usefully rewritten [22] as follows,

$$\begin{aligned}\beta\mu = -\ln\left(\frac{Z_{N+1}}{Z_N}\right) &= -\ln\left(\frac{V}{\Lambda^3(N+1)}\right) - \ln\left(\frac{\int d\mathbf{r}^{N+1}e^{-\beta U(\mathbf{r}^{N+1})}}{\int d\mathbf{r}^N e^{-\beta U(\mathbf{r}^N)}}\right) \\ &= \ln(\rho\Lambda^3) - \ln\left(\frac{\int d\mathbf{r}^N \int d\mathbf{r}_{N+1}e^{-\beta\Delta U}e^{-\beta U(\mathbf{r}^N)}}{\int d\mathbf{r}^N e^{-\beta U(\mathbf{r}^N)}}\right) \\ &= \ln(\rho\Lambda^3) - \ln\langle e^{-\beta\Delta U}\rangle_N \\ &= \beta\mu_{id} + \beta\mu_{ex}\end{aligned} \tag{14.22}$$

where the first term is the ideal gas term and where $\Delta U \equiv U(\mathbf{r}^{N+1}) - U(\mathbf{r}^N)$, i.e., it is the difference in the potential energy between the $N+1$ and the N-particle system. This quantity can also be thought of as the energy of particle $N+1$ in the $(N+1)$-particle system.

The next key observation is that the particle $N+1$ is, in essence, an extra particle or test particle, a probe, that is virtually added to the actual N-particle system. Thus, imagine a MC simulation of N particles. Periodically the simulation is interrupted to select a random position to insert a test particle. The energy, $U_t = \Delta U$, of the test particle is calculated, and its Boltzmann factor is collected and averaged. This is nowadays referred to as Widom's insertion method [22]. Here, we derived it for the NVT-ensemble, but we point out that there are also similar expressions for other ensembles such as the isobaric ensemble (e.g., see the book by Frenkel and Smit [7]).

14.3.1 The PDT for an inhomogeneous fluid

We will start by considering the operational definition for the chemical potential in a single component inhomogeneous fluid, such as a hard sphere fluid in an external field. Widom [22, 23] showed that (1) a local chemical potential, $\mu(\mathbf{r})$, can be defined, and (2) can evaluated through the use of the so-called potential distribution theorem, which provides a description of the configurational contribution to the chemical potential. The PDT links the chemical potential at a position **r** to the local number density, $\rho(\mathbf{r})$, and the local potential energy of test particle, $U_t(\mathbf{r}) + V_{\text{ext}}(\mathbf{r})$ as follows,

$$\begin{aligned}\beta\mu(\mathbf{r}) &= \ln\rho(\mathbf{r}) - \ln\langle e^{-\beta(U_t(\mathbf{r})+V_{\text{ext}}(\mathbf{r}))}\rangle \\ &= \ln\rho(\mathbf{r}) - \ln\langle e^{-\beta(U_t(\mathbf{r}))}\rangle + \beta V_{\text{ext}}(\mathbf{r}).\end{aligned} \tag{14.23}$$

The angled brackets denote an ensemble average of the test particle insertion at **r**. Note that the test-particle is identical to the actual fluid particles, interacting through the same interaction potential. The difference lies in the fact that the test particle is a spectator that can probe the local energy, but does not in any way affect the fluid particles. In practical terms, say, if one performs a Monte Carlo or molecular dynamics simulation, then periodically one stops the simulation and inserts a test particle at position **r** and measures its potential energy. The latter comes in two contributions. The first is simply the local value of a one-body external field, $V_{\text{ext}}(\mathbf{r})$. This contribution does not depend on the position of the fluid particles but solely on the test-particle's location and hence can be taken outside the ensemble average. The second is due to the interactions with the particles that make up the actual fluid.

At equilibrium, the chemical potential is constant throughout the entire volume, independent of position. Now, for inhomogeneous fluids, each term on the right hand side of (14.23) varies with position. However, the PDT states that the sum of the two terms must be a constant, independent of position!

For a fluid between two planar hard walls we rewrite the above equation as

$$\beta\mu(z) = \ln\rho(z) - \ln\langle e^{-\beta U_t(z)}\rangle + \beta V_{\text{ext}}(z). \tag{14.24}$$

Specializing to a fluid between hard walls located at $z = 0$ and $z = L$ we have $V_{\text{ext}}(z) = 0$ for $0 < z < L$ and the writing of the PDT simplifies to

$$\beta\mu(z) = \ln\rho(z) - \ln\langle e^{-\beta U_t(z)}\rangle\, ;\, 0 < z < L \tag{14.25}$$

$$\equiv \ln\rho(z) + \beta\mu^{\text{ex}}(z) \tag{14.26}$$

where the second line defines the local excess chemical potential.

Further specializing to a hard sphere fluid leads to a simple geometrical interpretation of the second term. Namely, for a hard sphere fluid the Boltzmann factor $\exp(-\beta U_t(z))$ can only take on two values: unity when the test-particle does not overlap with any of the fluid particles and zero when there is an overlap. In other words, the ensemble average $<\exp(-\beta U_t(z))>$ is simply a probability, namely the probability that a hard sphere can be inserted at a distance z, without overlap.

Although it may seem unlikely at first, it also follows from equation (14.26) that the location of the peaks in $\rho(z)$ are the most likely positions for inserting a hard sphere. In fact, right at the wall is the most likely place. This is so because for a hard sphere fluid against a hard wall the the contact density is equal to the pressure ($\beta p = \rho(z = 0)$). For a hard sphere fluid, the latter always exceeds the density.

The spatial constancy of the chemical potential implies that when a system is in equilibrium, there is no *net* drift of matter from one part of the system to another. Rewriting the constancy of the chemical potential in differential terms, we have,

$$\frac{\partial \ln \rho(z)}{\partial z} = -\frac{\partial \mu^{ex}(z)}{\partial z} = \frac{\partial \ln <e^{-\beta U_t(z)}>}{\partial z}; \quad 0 < z < L. \tag{14.27}$$

Now, consider the net force of a particle in the fluid. In a bulk fluid this force is zero, because of symmetry. However, in the inhomogeneous fluid that force is not zero everywhere, but rather depends on position. For a fluid at a planar wall that the normal component of the force depends on the distance, z, form the wall. The tangential components are zero on the average.

The intrinsic chemical potential is defined as:

$$\mu_{int}(z) \equiv \mu - V_{ext}(z). \tag{14.28}$$

The intrinsic chemical potential is a useful concept in the presence of a *slowly* varying field. For example, the atmosphere has slowly varying properties with elevation from the Earth's surface. Specifically, in gravity the intrinsic chemical potential decreases with elevation, according to

$$\mu_{int}(z) = \mu - mg(z - z_0) \tag{14.29}$$

where m denotes the mass and g is the gravitational acceleration. The position z_0 is a reference position. Thus, a fluid in a gravitational field has an intrinsic chemical potential that varies linearly with distance from the reference point. Combined with a bulk equation of state that links the chemical potential with the bulk density (i.e., some relation $\mu(\rho)$), equation (14.29) constitutes an implicit equation for the local density, $\rho(z)$.

14.3.2 Consequences of the uniformity the chemical potential

In 1976 Leng, Rowlinson and Thompson [24] published the results of a test of the PDT. They found the liquid–vapor profile of the penetrable-sphere mode in the mean field approximation and calculated the local chemical potential using PDT. They found that the chemical potential was constant throughout the interface and the value equaled that of the coexisting bulk phases.

In 1987 Widom [23] published an clarifying paper on the uniformity of the chemical potential through an interfacial region. He discusses the Leng *et al* paper

[24] and addresses how insisting on the uniformity of the chemical potential can be used as a means to generate solutions. Some of the repercussions are further discussed in the book by Rowlinson and Widom [25].

14.4 Simulation routes to the grand potential

In general, in the various molecular simulation ensembles there are no simple routes to the entropy (S) or free energy (F or G). Instead, one is faced with the same job and experimentalist is: performing a thermodynamic integration of accessible observables like the pressure of the internal energy.

The (surface) grand potential of a planar system is an exception. For an inhomogeneous system with an interface perpendicular to the z-direction, we can obtain Ω and Ω_s as,

$$\begin{aligned}\Omega/A &= -\int_0^{\infty} dz\, p_T(z),\\ \Omega_s/A &= \int_0^{\infty} dz\, [p - p_T(z)]\\ &= \underbrace{\int_0^{\infty} dz\, [p_N(z) - p_T(z)]}_{\gamma} - \int_0^{\infty} dz\, z\, \rho(z) V'_{\text{ext}}(z)\end{aligned} \tag{14.30}$$

where the prime denotes a differentiation with repeat to z, and p is the bulk pressure, far away from a wall. The first term of the last line, denoted γ, the surface tension represents the contributions to Ω_s that are solely due to the inter-particle interactions (figure 14.4). The second term reflects the interactions due to the external field, $V_{\text{ext}}(z)$, figure 14.4. Here, we have introduced the normal and tangential(or lateral) components, p_N and p_T of the pressure tensor. Whereas the pressure is a scalar quantity in a bulk fluid, in an inhomogeneous fluid of a planar symmetry the pressure is a tensor

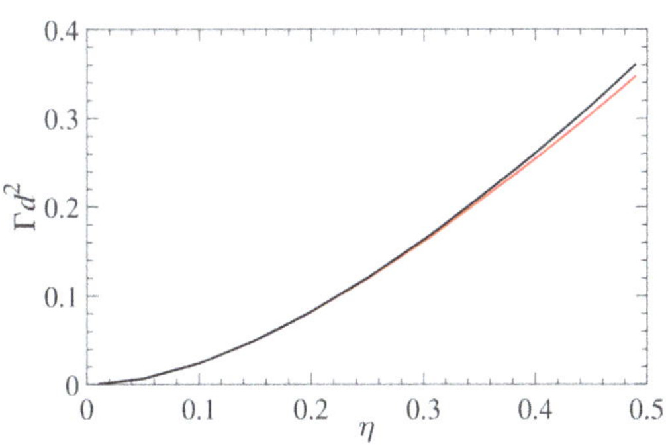

Figure 14.4. The surface adsorption for a system of hard spheres against a hard wall. The red curve is the SPT result, the black line is the simulation result of [26]

$$\mathbf{P}(z) = \begin{pmatrix} P_{xx} & P_{xy} & P_{xz} \\ P_{yx} & P_{yy} & P_{yz} \\ P_{zx} & P_{zy} & P_{zz} \end{pmatrix} = \begin{pmatrix} p_T(z) & 0 & 0 \\ 0 & p_T(z) & 0 \\ 0 & 0 & p_N(z) \end{pmatrix} \tag{14.31}$$

which in planar symmetry has only nonzero components along the diagonal, the off-diagonal (or shear) components all vanish. We have highlighted that, in general, $p_N(z)$ and $p_T(z)$ vary with position z. Hydrostatic equilibrium implies (for an introduction see [26])

$$p_N'(z) = -\rho(z)V_{\text{ext}}'(z), \tag{14.32}$$

where the prime indicates differentiation with respect to z.

In the case of a simple hard wall, or the absence of a wall (e.g., a liquid–vapor profile), $p_N'(z) = 0$, i.e, $p_N(z)$ is a constant for all z.

The challenge then is derive an expression that can be readily implemented in a molecular simulation [25] (figure 14.5). Schofield and Henderson [28] demonstrated in 1982 that although the normal component has a unique expression, $p_T(z)$ does not. Although the integral $\int dz\, p_T(z)$ has to be the same for all choices as this amounts to an observable (i.e., the surface tension), an infinity of choices still exist.[6] Two common examples of those are due to Irving and Kirkwood [29] (IK) in 1950 and Harasima [30] in 1958.

A useful expression for the IK convention for simulations was introduced in 1983 by Walton *et al* [31] is,

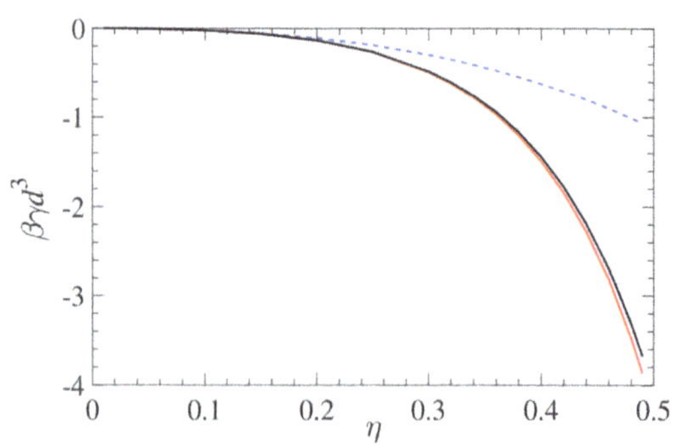

Figure 14.5. The surface tension for a system of hard spheres against a hard wall. The red curve is the SPT result, the black line is the simulation result [26], while the dashed blue line is the two-term virial equation of state due to Bellemans [27]. Note that the surface tension for a hard sphere/hard wall system is always negative. Finally, it is interesting to see that the SPT result (see chapter 5) is remarkably close to what was obtained from the simulations.

[6] The so-called 'surface of tension' is the first moment of the difference $p(z) - p_T(z)$. This quantity is *not* the same for all choices. This indicates that it is not an observable.

$$\beta p_N(z) = \rho(z) - \frac{1}{2A}\left\langle \sum_{i\neq j} \frac{z_{ij}^2}{r_{ij}} \frac{\beta\phi'(r_{ij})}{|z_{ij}|} \Theta\left(\frac{z - z_i}{z_{ij}}\right)\Theta\left(\frac{z_j - z}{z_{ij}}\right)\right\rangle$$
$$\beta p_T(z) = \rho(z) - \frac{1}{2A}\left\langle \sum_{i\neq j} \frac{x_{ij}^2 + y_{ij}^2}{2r_{ij}} \frac{\beta\phi'(r_{ij})}{|z_{ij}|} \Theta\left(\frac{z - z_i}{z_{ij}}\right)\Theta\left(\frac{z_j - z}{z_{ij}}\right)\right\rangle \quad (14.33)$$

where Θ is the Heaviside step function, $\phi' = d\phi/dr$ and $A = L_x L_y$ is the cross-sectional area of the rectangular simulation box. The product of the two Θ functions ensures that the configurational contribution is nonzero only between z_i and z_j.

The mid-eighties saw papers providing the corresponding tensorial expressions for the hard sphere [26] and square-well [32] potentials. Those papers (together with [33] which focuses on the wetting transition) present a large variety of sum rules that act, among other things, as convenient checks on the simulation results. In addition, they provide examples of the so-called *thermodynamic* route to the surface equations of state, specifically by integrating the Gibbs adsorption equation,

$$\frac{\partial\gamma}{\partial\mu} = -\Gamma \quad (14.34)$$

where the adsorption is given by

$$\Gamma = \int_0^\infty dz\, [\rho(z) - \rho]. \quad (14.35)$$

References

[1] Finney J L 2013 Bernal's road to random packing and the structure of liquids *Philosophical Magazine* **93** 3940–69

[2] Alder B J and Wainwright T E 1957 Phase transition for a hard sphere system *J. Chem. Phys.* **27** 1208

[3] Wood W W and Jacobson J D 1957 Preliminary results from a recalculation of the monte carlo equation of state of hard spheres *J. Chem. Phys.* **27** 1207

[4] Wood W W and Erpenbeck J J 1976 Molecular dynamics and monte carlo calculations in statistical mechanics *Ann. Rev. Phys. Chem.* **27** 319–48

[5] Allen M P and Tildesley D J 1987 *Computer Simulation of Liquids* (Oxford: Oxford University Press)

[6] Haile J M 1992 *Molecular Dynamics Simulation* (New York: Wiley)

[7] Smit B and Frenkel D 2002 Understanding *Molecular Simulation: From algorithms to Applications* (San Diego, CA: Academic)

[8] Ewald P P 1921 Die berechnung optischer und elektrostatischer gitter-potentiale *Ann. Phys.* **369** 253–87

[9] Lekner J 1989 Summation of dipolar fields in simulated liquid-vapour interfaces *Phys.* A **157** 826–38

[10] Lekner J 1982 Parametric solution of the van der waals liquid-vapor coexistence curve *Am. J. Phys.* **50** 161–3

[11] Tildesley D J 1993 *Computer Simulation in Chemical Physics* (Dortrecht: Kluwer)

[12] Woodcock L V 1981 Glass transition in the hard-sphere model and kauzmann's paradox *Ann. NY Acad. Sci.* **371** 274–98
[13] Rahman A 1964 Correlation in the motion of atoms in liquid argon *J. Phys.* **136** A405–11
[14] Verlet L 1967 Computer 'experiments' on classical fluids. i. thermodynamic properties of lennard-jones molecules *Phys. Rev.* **159** 98–103
[15] Ferrario M 1993 *Computer Simulation in Chemical Physics* (Dortrecht: Kluwer)
[16] Metropolis N, Rosenbluth A W, Rosenbluth M N, Teller A H and Teller E 1953 equation of state calculations by very fast computing machines *J. Chem. Phys.* **21** 1087–92
[17] Wood W W 1970 Nptâensemble monte carlo calculations for the hardâdisk fluid *J. Chem. Phys.* **52** 729–41
[18] Barker J A and Watts R O 1969 Structure of water: A Monte Carlo calculation *Chem. Phys. Lett.* **3** 144–5
[19] Moreira A G and Netz R R 2004 *Novel Methods in Soft Matter Simulations* (Berlin: Springer)
[20] Heyes D M 1981 The surface potential of point charge and point dipole lattices *Surf. Sci.* **110** L619–24
[21] Hautman J and Klein M L 1992 An ewald summation method for planar surfaces and interfaces *Mol. Phys.* **75** 379–95
[22] Widom B 1963 Some topics in the theory of fluids *J. Chem. Phys.* **39** 2808–12
[23] Widom B 1978 Structure of interfaces from the uniformity of the chemical potential *J. Stat. Phys.* **19** 563–74
[24] Leng C A, Rowlinson J S and Thompson S M 1976 *The gas-liquid surface of the penetrable sphere model Proc. Roy. Soc. Lond. A* 352 1–23
[25] Rowlinson J S and Widom B 1982 *Molecular Theory of Capillarity (Clarendon Press)* (Oxford: Univsersity Press)
[26] Henderson J R and van Swol F On the interface between a fluid and a planar wall theory and simulations of a hard sphere fluid at a hard wall *Mol. Phys.* **51** 991–1356
[27] Bellemans A 1962 Statistical mechanics of surface phenomena: I. a cluster expansion for the surface tension *Physica* **28** 493
[28] Schofield P and Henderson J R 1982 Statistical mechanics of inhomogeneous fluids *Proc. Roy. Soc. Lond.* A **379** 231–46
[29] Irving J H and Kirkwood J G 1950 The statistical mechanical theory of transport processes. iv. the equations of hydrodynamics *J. Chem. Phys.* **18** 817–29
[30] Harasima A 1958 title *Adv. Chem. Phys.* **1** 203
[31] Walton J P R B, Tildesley D J, Rowlinson J S and Henderson J R 1983 The pressure tensor at the planar surface of a liquid *Mol. Phys.* **48** 1357–68
[32] Henderson J R and van Swol F 1985 On the approach to complete wetting by gas at a liquid-wall interface exact sum rules, fluctuation theory and the verification by computer simulation of the presence of long-range pair correlations at the wall *Mol. Phys.* **56** 1313–68
[33] van Swol F and Henderson J R 1986 Wetting at a fluid-wall interface. computer simulation and exact statistical sum rules *J. Chem. Soc., Faraday Trans.* **2** 1685–99

IOP Publishing

Molecular Theory of Electric Double Layers

Dimiter N Petsev, Frank van Swol and Laura J D Frink

Chapter 15

Molecular simulation: applications

15.1 Background

In this chapter we will review the results of molecular simulation studies that impact electrolyte solutions and double layers. The primary goal is to present the state of the art in molecular simulation of electrolyte solutions and double layers.

Electrolyte systems span a wide range of fluids, from (molten) salts to (usually aqueous) electrolytes solutions. Ionic fluids, ionic systems that have low melting temperatures, such that many that are fluid at room temperature are included in this spectrum. They are closest to molten salts, in that there is no solvent. They are different in the sense that one of the ions (either the cation or the anion) is large, and not simply an element from the periodic table, Instead the large ion is a molecular compound. An early example is ethylammonium nitrate $(C_2H_5)NH_3^+NO_3^-$, which has a melting temperature of 12 C [1]. The large bulky ion can be thought of as a dielectric solvent molecule with an embedded ion.

15.2 One-component plasma

The first molecular simulation study that employed the Ewald summation was a Monte Carlo study reported by Brush, Sahlin and Teller [2] in the 1960s. These authors simulated a bulk plasma of heavy ions immersed in a uniform neutralizing background, known as a one-component plasma (OCP). They used systems containing from 32 to 500 particles, and employed periodic boundary conditions. The results of the study were presented in terms of a dimensionless parameter $\Gamma = (4\pi\rho/3)^{1/3}[(Ze)^2/k_BT]$, where ρ is the ion number density (particles per volume), T is the temperature, k_B is the Boltzmann constant, e is the electronic charge, and Z is the atomic number. Two different methods were used to determine the potential energy of a configuration. The first is the 'minimum image convention': each particle is allowed to interact only with each other particle in the basic cell, or with the nearest periodic image of each other particle if that image is closer. In the second method, the interaction of a particle with all the images of the other particles, and with the uniform

doi:10.1088/978-0-7503-2276-8ch15

background is taken into account by a technique similar to the Ewald procedure. It was found that both methods of determining the potential energy yield essentially the same results for the radial distribution function $g(r)$ for Γ values of 10 or less. For larger values of Γ the results given by the two methods differ significantly, indicating that the minimum image convention is inadequate for plasma systems at high densities and low temperatures. Energies and values of the pair distribution function are compared with predictions of various approximate theories for small Γ values. It is found that the nonlinear Debye–Hückel (DH) theory is in agreement with the MC results for values of Γ up to 0.1. At $\Gamma = 1.0$, significant deviations from the DH theory are observed. For $\Gamma = 1$, $g(r)$ is found to be monotonic and in close agreement with the integral equation using the Percus–Yevick equation closure (see section 4.2 of chapter 4). For values of $\Gamma > 2$, $g(r)$ begins to show oscillations. The system undergoes a fluid–solid phase transition in the vicinity of $\Gamma = 125$.

Moreira and Netz [3] have performed simulations that closely resemble the bulk OCP simulations. They performed simulations of counter ions next to a charged surface. The overall system is neutral. Whereas the OCP has a background charge uniformly throughout the 3D computational cell, the counter ion study has the uniform charge spread over the surface.

15.3 Molten salts

The first molecular simulation work on ionic systems was reported by Les Woodcock and Konrad Singer in 1971, who performed Monte Carlo and molecular dynamics calculations for molten potassium chloride, 108 K^+ ions mixed with an equal number of Cl^- ions. They employed the Ewald summation method discussed in an earlier chapter. The Coulombic potential by itself is insufficient, as opposite charges would end up overlapping given the divergence of the Coulombic interaction at zero separation. In some studies charged hard spheres were considered, but one is free to combine the Coulombic interaction with other short-range potentials. We describe selecting a Lennard–Jones potential to perform a DFT study of electrolyte solutions.

Other choices were made in the study of molten salts to better capture the thermodynamic properties. Thus, Woodcock and Singer [4], like others before them, employed the flexible so-called Born–Huggins–Mayer potential [5],

$$V_{ij} = \frac{q_i q_j e^2}{4\pi\epsilon\epsilon_0 r_{ij}} + A_{ij} e^{-\alpha_{ij} r_{ij}} - \frac{C_{ij}}{r_{ij}^6} + \frac{D_{ij}}{r_{ij}^8}. \tag{15.1}$$

The exponential term provides the needed short-ranged potential while the other two terms combine to provide the dispersion interactions (dipole–dipole and dipole-quadrupole). Another common potential is the so-called Pauling potential [6] where the short-ranged repulsion is represented by a power law term,

$$V_{ij} = \frac{q_i q_j e^2}{4\pi\epsilon\epsilon_0 r_{ij}} + \frac{A_{ij}}{r^n} - \frac{C_{ij}}{r_{ij}^6} + \frac{D_{ij}}{r_{ij}^8} \tag{15.2}$$

where n is typically taken to be 9. Both these potentials were originally developed to provide a description for the solid state.

The parameter values were provided by Tosi and Fumi [7]. The selection of KCl is convenient, as K^+ and Cl^- are isoelectronic with 18 electrons each, and both are isoelectronic with an often studied noble gas, Ar. Woodcock constructed a ball-and-spoke model of an instantaneous MD generated configuration, this is shown in figure 15.1. The state point makes this a molten salt, close to the triple point. A simple as it is, it immediately rules out the crystalline theories. They are not in agreement with the shown microstructure. The structure shows that ions are surrounded by the opposite ions, but occasionally like ions can meet. There is a distinct tendency for ion-pairing and clustering, and leaving considerable voids.

A quantitative analysis of the molten alkali-halide structure is obtained from the pair distribution functions, as shown in figure 15.2. For any binary ionic mixture there are three pair distribution functions, $g_{++}(r)$, $g_{+-}(r) = g_{-+}(r)$, and $g_{--}(r)$ but for the example at hand one finds that $g_{++}(r) \approx g_{--}(r)$. In figure 15.2 are shown the unlike, $g_u(r) = g_{+-}(r)$, and like, $g_l = (g_{++}(r) + g_{--}(r))/2$, radial distribution functions. There is a very clear charge ordering in the fluid that extends to large distances. The radii of charge neutrality (where the two distributions cross) occurs at regular intervals (i.e., at 0.4, 0.6, 0.8, and 1.0 nm). Finally, note that the like ions penetrate the first peak.

Figure 15.1. A ball-and-spoke model of an instantaneous configuration molten KCl due to Woodcock. The model was built using the ion coordinates of a molecular dynamics simulation of 216 ions at 1043 K and a density of 48.80 cm^3 mol^{-1}. The box size is 2.06 × 2.06 × 0.70 nm. The lighter sphere are the smaller K^+ ions, while Cl^- is represented by the white sphere. (Reproduced from [4] with permission of The Royal Society of Chemisty.)

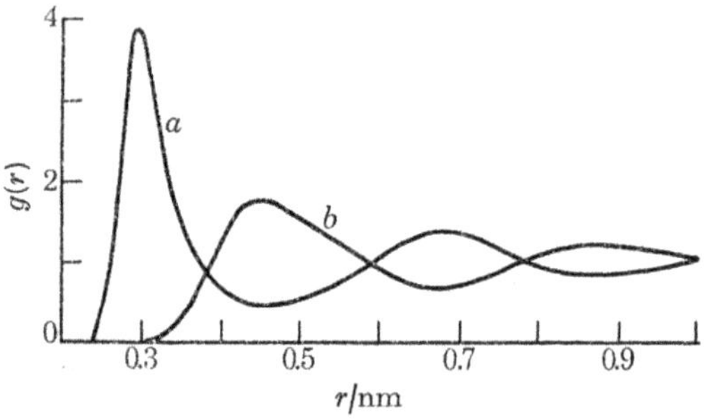

Figure 15.2. Unlike (labeled 'a') and like (labeled 'b') radial distribution functions for molten KCl. Two pair distributions curves are shown, the unlike $g_u(r) = g_{+-}(r)$ and the like $g_l = (g_{++}(r) + g_{--}(r))/2$. (Reproduced from [4] with permission of The Royal Society of Chemisty.)

15.4 Bulk electrolytes

15.4.1 Restricted primitive model

The thermodynamic behavior of bulk electrolytes has a long history, going back almost a century to 1923 when Peter Debye and Erich Hückel (DH) [8] published their now famous work on dilute electrolyte solutions in the Physikalische Zeitschrift. DH worked out their theory for charged hard spheres, or the so-called restricted primitive model or RPM for short. They used the Poisson equation and the Boltzmann approximation and linearization to arrive at the ionic distribution of ions in solution. As is usual for liquid state theories, DH theory comes about as a simplified mathematical model (i.e., RPM) that is then solved by making a number of approximations. We often have to accept that discrepancies between theory and experimental data cannot easily be designed to the model's simplifications or the subsequent approximations needed to solve equations. Molecular simulations, using the same model, are uniquely suited to ascertain the consequences of approximate methods.

Starting in the 1960s Monte Carlo calculations were performed for the RPM over different densities roughly spanning the range of molarities from 0.01 M to 2 M. The first MC simulations were reported by Vorontsov-Vel'yaminov *et al* [10] in 1966. Followed in 1970 by Valleau and co-workers. Both these groups avoided using the Ewald summation. In fact, Card and Valleau [9] reported using the minimum image convention (MIC) for calculating all the interactions.[1] Avoiding the Ewald method was successful because the concentrations were low and the central simulation box correspondingly large.

Both Vorontsov-Vel'yaminov *et al* [10] and Card and Valleau [9] studied a 1:1 electrolyte (with equal ionic diameters) and calculated thermodynamic properties (e.g., internal energy, equation of state, specific heat, osmotic coefficient and mean activity coefficient) as well as the like and unlike pair distribution functions. MC results for the distribution functions are shown in figure 15.3.

[1] To demonstrate accuracy they also extended the interaction calculations to 27 neighboring cells, and found little difference. Card and Valleau [9] also explored using a *spherical* cutoff as well. They showed that such an approach leads to bad results, as electroneutrality amongst the interacting particles is not guaranteed.

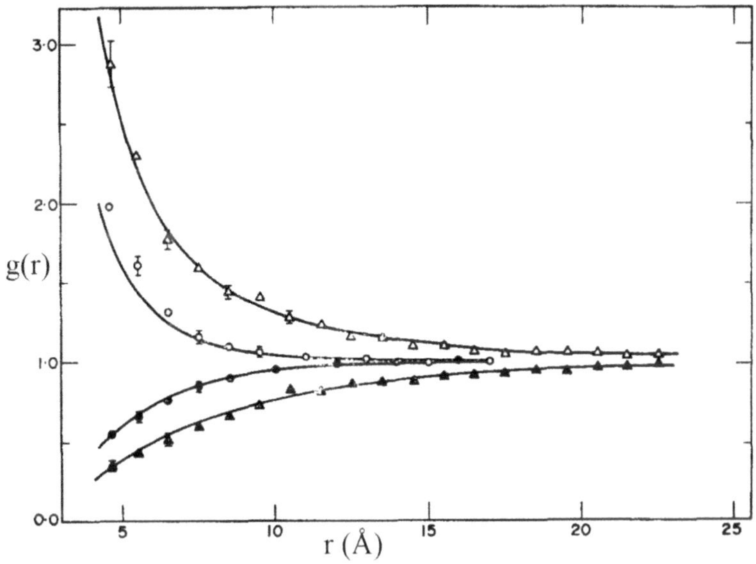

Figure 15.3. MC results for the unlike (open symbols) and like (filled symbols) radial distribution functions for RPM electrolyte ($d = 4.25$ Å) at 0.103 8M (triangles), and 1M (circles). The lines indicate the mDH pair distribution functions discussed in the text. (Reproduced from [9], with the permission of AIP Publishing.)

The solid lines in that figure correspond to a so-called modified DH (mDH) solution for the RPM model,

$$\begin{aligned} g_u(r) = g_l(r) &= 0 \qquad r < d \\ g_l(r) &= e^{-f(r)} \qquad r \geqslant d \\ g_u(r) &= e^{f(r)} \qquad r \geqslant d \end{aligned} \tag{15.3}$$

where d is the hard sphere diameter and the function $f(r)$ is given by:

$$f(r) = \frac{e^2}{\epsilon\epsilon_0 k_B T} \frac{e^{\kappa(d-r)}}{(1 + \kappa d)r} \tag{15.4}$$

were κ is the Debye screening length. With the distribution functions in hand one can now calculate the potential energy (U), osmotic coefficient(ϕ):

$$\frac{U}{Nk_BT} = \frac{\kappa^2}{4} \int_d^\infty drr[g_u(r) - g_l(r)] \tag{15.5}$$

$$\phi = 1 + \frac{U}{3Nk_BT} + \frac{\pi N d^3}{3V}[g_u(d) + g_l(d)]. \tag{15.6}$$

From these expressions we obtain the excess osmotic pressure (Π_{ex}) as,

$$\frac{\Pi_{ex} V}{NkT} = \phi - 1. \tag{15.7}$$

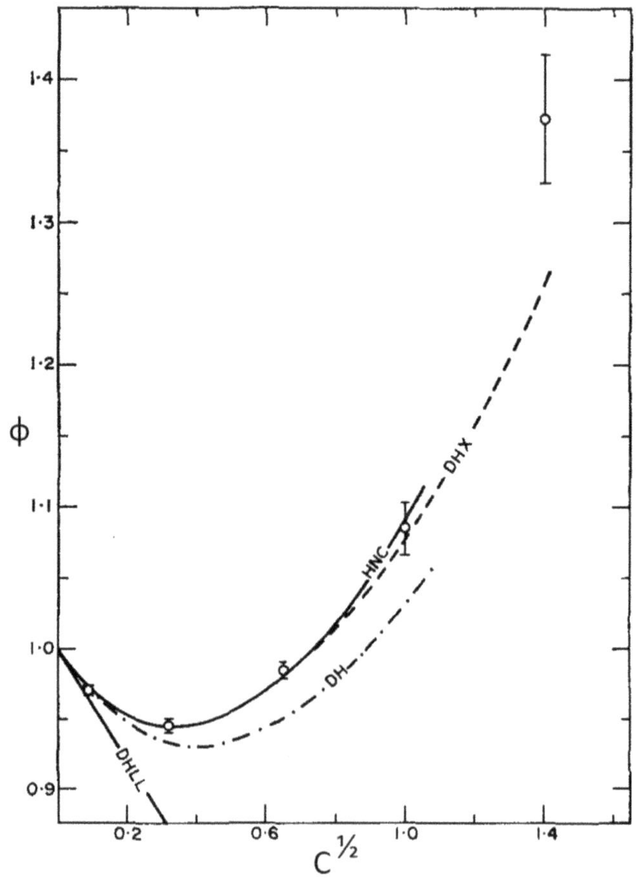

Figure 15.4. Results for the osmotic coefficient, ϕ as a function of square root of the concentration (i.e., $c^{1/2}$). MC simulation results are denoted by open circles. The results are compared to HNC integral equation theory results (solid line) of Rasaiah and Friedman [11]. The modified DH results (mDH) are labeled DHX. The traditional DH results are labeled DH. Traditional DH theory linearizes the pair distribution functions of equation (15.3) and uses these in equation (15.8). The limiting DH law is a straight line labeled DHLL. (Reproduced from [9], with the permission of AIP Publishing.)

Finally, the ionic activity coefficient ($\gamma_{\pm}$) is related to the osmotic coefficient through a Gibbs–Duhem equation, viz.,

$$cd(\ln \gamma_{\pm}) + d(c[1 - \phi]) = 0 \qquad (15.8)$$

where the concentration is denoted by c.

An example of the behavior of the osmotic coefficient as a function of concentration (i.e., $c^{1/2}$) is provided by figure 15.4, due to Card and Valleau [9].

15.4.2 Confined civilized electrolytes and water potentials

Qiao and Aluru [12] and, independently, Thompson [13] performed non-equilibrium molecular dynamics (NEMD) simulations of electrolytes in the presence of explicit solvent molecules. Their goal was not to study EDLs per se, but rather both papers addressed the limits of continuum models to describe electro-osmotic flow. Simply put, the continuum models cannot account for packing near a wall. As Thompson points out this a very relevant issue since the driving force for electro-osmotic is strongest close to the wall. This is in contrast to pressure-driven flow, where most of the applied force is located away from the wall. The details of explicit water representation is different for the two papers. Qiao and Aluru select the classical SPC/E (i.e., extended simple point charge model) water model due to Berendsen *et al* [14].

Water potentials

SPC/E is a *rigid* site-model of bonded hydrogen and oxygen atoms. There are (partial) charges at the center of the atoms, of magnitude 0.8476 e on the oxygen atom and half that amount on each of the hydrogens. The dipole moment of SPC/E water is 2.351D, larger than the experimental value of 1.85D.

Two water SPC/E molecules interact via the Coulombic partial charges as well as a single Lennard–Jones interaction. The latter is centered on the oxygen atom, and has a σ value of 0.3166 nm. This is much larger that the fixed oxygen-hydrogen distance of 0.1 nm. Thus, as far as the non-coulombic interaction is concerned the two hydrogen atoms are 'hidden' inside the oxygen atom.

Although SPC/E is an often used classical intermolecular potential, as might be expected, there are several commonly used *rigid* water interaction potentials proposed in the 1980s, namely SPC/E and TIP3P, TIP4P and later extended to include TIP5P and TIP6P. Where TIP stands for transferable intermolecular potential. All are so-called site interaction models. In general, the numbers in the names refer to the number of sites. The simplest are the (planar) three-site models (e.g., SPC/E and TIP3P). TIP4P adds a fourth non-atom (or dummy) site, *M*, to the atom sites, and places the negative charge on it instead of the oxygen atom, as shown in figure 15.5.

TIP5P replaces the M site by two L sites that are meant to represent the lone electron pairs on the oxygen atom (table 15.1). This makes the TIP5P model non-planar. The HOH and LOL planes are perpendicular to each other, and the LOL angle is the tetrahedral angle of 109.47 degrees. Finally, in the TIP6P model one employs both two L sites and the M site. Naturally, the computational cost of the water–water interaction goes up significantly as one moves from SPC/E (9 distances) to TIP6P (26 distances). The interactions between two water molecules i and j contains charge interactions between the partial charges k on i and partial charges l on molecule j. In addition there exists a LJ potential between the O atoms of i and j

$$U_{ij} = \sum_{k\in i}\sum_{l\in j}\frac{kq_k q_l}{r_{kl}} + \frac{A}{r_{OO}^{12}} - \frac{B}{r_{OO}^{6}}. \tag{15.9}$$

It should be pointed out that the potential listed here is just a part of a considerable field with many more potentials and/or updated parameter sets. Of course, the goal behind these efforts was (and is), to capture as much of the water properties and phase diagram as possible. Sometimes Ewald summations are used for these water models.

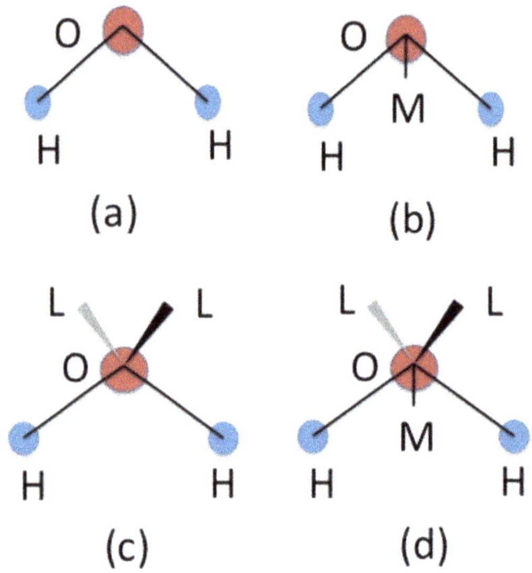

Figure 15.5. Schematics of four commonly used rigid water models, (a) SPC/E and TIP3P, (b) TIP4P; (c) TIP5P; and (d) TIP6P. In these models all the bond lengths and angles are held constant. Special site M is a dummy atom, it is massless and locates the position of the negative charge. M lies in the HOH-plane. In the case of the non-planar TIP5P and TIP6P models, the negative charges are located on the lone-pairs of the oxygen atom. They fall in a plane that is perpendicular to the HOH-plane.

Table 15.1. Parameters for a number of commonly used *rigid* water potentials. Each potential has coulombic interactions due to negative partial charges placed on (or near) the O-atom, and (equal) positive charges on the H-atoms. In addition, between two O atoms there is a LJ potential. The distances, r, are measured in nm, the partial charges in electron charge (e). A is in units of 10^3 kcal Å^{12} mol^{-1}, and B is in kcal Å^6 mol^{-1}.

	SPC/E	TIP3P	TIP4P	TIP5P
r_{OH}	0.1	0.09572	0.09572	0.09572
r_{OM}			0.015	0.015
r_{OL}			0.07	0.07
θ_{HOH}	109.47	104.52	104.52	104.52
θ_{LOL}				109.47
A	692.4	582.0	600.0	544.5
B	625.5	595.0	610.0	590.3
q_O	−0.84761	−0.834	−1.04	
q_H	0.423 8	0.417	0.52	0.241
q_L				−0.241

References

[1] Walden P 1914 Molecular weights and electrical conductivity of several fused salts *Bull. Acad. Sci. St. Petersburg* **1914** 405–22

[2] Brush S G, Sahlin H L and Teller E 1966 Monte carlo study of a one-component plasma. i *J. Chem. Phys.* **45** 2102–18

[3] Moreira A G and Netz R R 2004 *Novel Methods in Soft Matter Simulations* (Berlin: Springer)
[4] Woodcock L V and Singer K 1971 *Trans. Faraday Soc.* **67** 12
[5] Huggins M L and Mayer J E 1933 Interatomic distances in crystals of the alkali halides *J. Chem. Phys.* **1** 643–6
[6] Pauling L 1928 The influence of relative ionic sizes on the properties of ionic compounds *J. Am. Chem. Soc.* **50** 1036–45
[7] Fumi F G and Tosi M P 1964 Ionic sizes and born repulsive parameters in the NaCl-type alkali halides—i: The Huggins–Mayer and Pauling forms *J. Chem. Phys. Solids* **25** 31–43
[8] Debye P and Huckel E 1923 Zur theorie der elektrolyte. i. gefrierpunktserniedrigung und verwandte erscheinungen *Physikalische Zeitschrift.* **24** 185–206
[9] Card D N and Valleau J P 1970 Monte carlo study of the thermodynamics of electrolyte solutions *J. Chem. Phys.* **52** 6232–40
[10] Vorontsov-Vel'yaminov P N, El'yashevich A M and Kron A K 1966 *Elektrokhimiya* **2** 708
[11] Rasaiah J C and Friedman H L 1969 Integral equation computations for aqueous 1–1 electrolytes. accuracy of the method *J. Chem. Phys.* **50** 3965–76
[12] Qiao R and Aluru N R 2003 Ion concentrations and velocity profiles in nanochannel electroosmotic flows *J. Chem. Phys.* **118** 4692–701
[13] Thompson S M 1986 Private communication
[14] Berendsen H, Grigera J and Straatsma T 1987 The missing term in effective pair potentials *J. Phys. Chem.* **91** 6269–71

IOP Publishing

Molecular Theory of Electric Double Layers

Dimiter N Petsev, Frank van Swol and Laura J D Frink

Chapter 16

Numerical methods for classical-DFT

This chapter discusses several approaches for solving the systems of equations associated with classical density functional theories (c-DFTs) in general and electrolytes in particular. These approaches can also be applied in the context of c-DFT for polymer or chain fluids [1, 2], but those systems will not be discussed here. Numerical solutions in early c-DFT studies were most often accomplished with a straightforward procedure of successive substitution (using Picard iterations). Integrals present in the Euler–Lagrange equations were often represented using real space sums over a discretized space. During that time, virtually all work focused on systems with two degrees of symmetry including planar, cylindrical, and spherical surfaces.

Algorithms for numerically solving c-DFT systems of equations in two and three dimensions using a real space Newton's method approach were first developed in 2000 [3]. That work utilized parallel computing and computational matrix methods that had been previously developed in the context of solving partial differential equations (PDEs). These methods allowed access to new problem domains that required 2 or 3 dimensional calculations including adsorption at chemically patterned surfaces [4], small model ion channels [5], solvated molecules [6], charged nanoparticles near surfaces [7], molten salts [8], and membrane bound assemblies [9]. Furthermore, the approach allowed for application of bifurcation tools for the first time to investigations of a variety of problems in liquid state theory [10, 11]. Section 16.2 summarizes those bifurcation methods. Ultimately, those algorithms were released in the open source code Tramonto [12]. The bifurcation tools utilized in the Tramonto code are distributed in the Nox-Loca package [13] of the Trilinos library [14]. Trilinos also includes the parallel iterative solvers used to solve the c-DFTs in Tramonto. All of the results presented in chapter 7 were generated using these software packages. While powerful, the Achilles heel of the real space matrix based approach is the memory required for the matrix. As a result, large three-dimensional applications remain rare using this approach.

doi:10.1088/978-0-7503-2276-8ch16

A few years later, an alternate approach for solving the large systems was developed that applied (but did not store) a Jacobian matrix [15]. Coupled with fast Fourier transforms (FFTs) for convolutions, this approach allowed for significantly faster 3D calculations with reduced memory requirements. Since that time, FFTs have been generally accepted as the best way to compute the convolutions for neutral systems [16], ionic systems [17], and chain systems [18]. They have been recently applied in detailed 3D calculations of bubble gating in ion channel proteins [19] and arbitrary pore geometries [20], and they hold great promise for enabling studies in a wide range of large three-dimensional systems. Other recent significant developments in numerical methods for c-DFTs include a massively parallel GPU based acceleration of the convolution calculations [21] that allows for solution of larger problems on smaller platforms and a full utilization of available computing resources. Finally, we note that machine learning methods have just begun to be applied to c-DFTs [22, 23].

16.1 Solution methods

This chapter summarizes numerical methods for the systems of equations presented by classical density functional theories. In keeping with the spirit of this text, the focus is on electrolyte systems, but in some cases the discussion is restricted to neutral hard core fluids.

16.1.1 Review of system of equations

The equations to solve for electrolyte systems are Poisson's equation,

$$\nabla^2 \tilde{\psi}(\mathbf{r}) = -4\pi\lambda_B\left(\sum_i z_i\rho_i(\mathbf{r}) + Q_S(\mathbf{r})\right), \tag{16.1}$$

and an Euler–Lagrange equation written generally as

$$\rho_i(\mathbf{r}) = \rho_b \exp\left(-z_i\tilde{\psi} + \mu_i + \frac{\delta \mathcal{F}_{ex}}{\delta\rho_i(\mathbf{r})}\right). \tag{16.2}$$

For systems with steady state diffusion (see section 7.8), a transport equation,

$$\nabla\big(\rho_i(\mathbf{r})\nabla\big(\mu_i(\mathbf{r}) + q_i\psi(\mathbf{r})/kT\big)\big) = 0, \tag{16.3}$$

must also be solved.

16.1.2 Nonlinear solvers I: Picard iterations

The Picard iterative scheme has been used since the earliest days of c-DFT numerical investigations to solve the Euler–Lagrange equation, equation (16.2). This straightforward approach proceeds by writing the equations so that the various unknowns ($\mathbf{x} = [\{\rho_i\}, \psi, \{\mu_i\}]$) are expressed as

$$x_i = g_i(\{\mathbf{x}\}). \tag{16.4}$$

The Picard scheme iterates the previous equation so the $k + 1$ estimate of the solution vector is

$$x_i^{(k+1)} = g_i(\{\mathbf{x}^k\}). \tag{16.5}$$

For numerical stability, it is often necessary to take partial updates when using the Picard scheme so that

$$x_i^{(k+1)} = (1 - \alpha)x_i^{(k)} + \alpha g_i(\{\mathbf{x}^k\}) \tag{16.6}$$

where α is a parameter that may be tuned to achieve convergence.

16.1.3 Nonlinear solver II: Newton's method

Alternatively, the system of equations can be discretized and written in a matrix format as

$$\mathbf{Ax} = \mathbf{b} \tag{16.7}$$

where $\mathbf{x} = [x_i]$ are the unknowns of interest. If A can be easily inverted, then the solution x can be found from

$$\mathbf{x} = [\mathbf{A}]^{-1}\mathbf{b}. \tag{16.8}$$

However, a simple inversion is usually not possible, and so an iterative scheme is required. In a Newton's method scheme, the system of equations is posed as a set of residual equations, $\mathbf{R} = [R_i]$ with

$$\mathbf{R}(\mathbf{x}) = 0, \tag{16.9}$$

and the matrix problem is written

$$\mathbf{J}\Delta\mathbf{x} = -\mathbf{R}, \tag{16.10}$$

where the matrix is a Jacobian, $\mathbf{J} = [J_{ij}] = [\partial R_i/\partial x_j]$, and $\Delta\mathbf{x} = \mathbf{x}^{(k+1)} - \mathbf{x}^{(k)}$ where k is the iteration count. The goal is to solve equation (16.10) for $\Delta\mathbf{x}$ in order to then perform iterative updates. The Newton's method approaches in the Tramonto code were developed to address the sometimes slow convergence of the Picard scheme, to extend parallel iterative solvers developed for PDEs to c-DFT problems, to utilize bifurcation algorithms in studies of phase diagrams (see section 16.2.1), and to build tools that would be broadly applicable to neutral systems, electrolytes, and polymer fluids.

While the Picard and Newton's methods are distinct, it is possible to combine them. Edelmann and Roth recently developed a hybrid Picard–Broyden scheme of this type where the Broyden method is a quasi-Newton algorithm [24]. The combined scheme can improve stability of the solve.

It is important to note that solving PDEs (such as the elliptical Poisson's equation) has a long history in the numerical methods and engineering communities. Finite difference and finite element schemes are often employed to construct the matrix problem. Appendix E provides a brief procedural summary of those methods. In the Tramonto code, finite elements are used to construct the matrix for the Poisson and transport blocks of the matrix. The Euler–Lagrange equations are quite different from typical PDEs, and present unique challenges for a Newton's method solve.

16.1.4 Consequences of nonlocality in c-DFTs

For molecular models beyond the point charge approximation, a nonlocal Euler–Lagrange equation, equation (16.2), must be solved. The nonlocalities arise from finite size effects, cohesive forces, 2nd order correlation effects, and any other physical model that introduces a nonlocal density. Finite size models represented by the FMT functionals (section 7.3) introduce nonlocal summed densities with the general form

$$n_\gamma(\mathbf{r}) = \sum_i \int \rho_i(\mathbf{r}')w_{\gamma,i}(\mathbf{r}, \mathbf{r}', \lambda_i)dr', \tag{16.11}$$

where the various weight functions are variations on $\delta(\mathbf{r}, \mathbf{r}', R_i)$ and $\theta(\mathbf{r}, \mathbf{r}', R_i)$ functions, vector and tensor functions are sometimes used, and the range, $\lambda_i = R_i$, is set by the particle radii. Other models, including the inhomogeneous 2nd order correlation models discussed in section 7.4, introduce similar nonlocal densities written as

$$\bar{\rho}_i(\mathbf{r}) = \int \rho(\mathbf{r}')w_i(|\mathbf{r} - \mathbf{r}'|, \lambda_i)dr' \tag{16.12}$$

where most often $w_i = \theta(|\mathbf{r} - \mathbf{r}'|, \lambda_i)$. The range of the weight function may be $\lambda_i = R_i$ or $\lambda_i = d_i$. In some cases, it may depend on the properties of the system as in $\lambda_i(\bar{\rho}_i)$. Even the cohesive terms in the Euler–Lagrange equation (see section 7.5) can be written

$$\frac{\delta \mathcal{F}_{att}}{\delta \rho_i(\mathbf{r})} = \sum_j \int \rho_j(\mathbf{r}')u_{ij}(|r - r'|)w_{ij}(|r - r'|, \lambda_{ij})dr', \tag{16.13}$$

where $w_{ij} = \theta(|\mathbf{r} - \mathbf{r}'|, \lambda_{ij})$, and λ_{ij} is set by the cutoff distances for the pair potentials, $\mathbf{r}_{ij}^c$.

In all cases above, the convolution integrals extend well beyond the size of one element in real space, and change the nature of the matrix problem (from a PDE perspective). More specifically, the mesh spacing, h is usually set to be $d/100 < h < d/4$ for c-DFT calculations with the two extremes representing very refined and very coarse solutions. The sections below outline two different approaches for a Newton's iterative approach for the coupled system of equations presented by c-DFT functionals for electrolytes. Real space matrix based approaches are presented first followed by a matrix-free method that makes use of fast Fourier transforms (FFTs) for efficient calculation of convolutions.

16.1.5 A real space Newton solver for c-DFTs

This section summarizes the original real space implementation of equations (16.1)–(16.3) in the open source Tramonto code [3, 12] as well as later development of Schur complement solvers optimized for c-DFT applications [25].

Discretization and collocation

Cartesian rectangular (2D) or rectangular hexahedral (3D) elements with linear shape functions are used in the Tramonto code [12]. Finite element methods are often applied to much more complicated (and possibly dynamic) elements. However, there are advantages to simpler grid strategies as well. Nadal and coworkers discuss the efficiencies gained by using Cartesian mesh strategies in FE analysis while noting that up to 80% percent of some FE calculations is spent on geometry and meshing [26]. Adaptive Cartesian meshing has also been used in Poisson–Boltzmann solvers for computing surface electrostatic properties of large proteins [27]. In Tramonto, the base mesh is uniform, but systematic coarse graining of the mesh and/or matrix some distance away from surfaces is also possible [3]. Using a uniform grid allows for preliminary computation of both static FE stencils for the PDEs and static collocation stencils for the integrals in the Euler–Lagrange equations.

The FE stencils in Tramonto are constructed using standard techniques for linear shape functions as described briefly in appendix E. A collocation approach is used to compute the integrals in equations (16.11)–(16.13) on the discrete mesh with

$$\int f(\mathbf{r}')w^{(\gamma)}(\mathbf{r}-\mathbf{r}', \lambda)dr' \approx \sum_{i=1}^{N_{\text{sten}}^{(\gamma)}} C_i^{(\gamma)} f_i \tag{16.14}$$

where $N_{\text{sten}}^{(\gamma)}$ is the total number of points in the γth weight function stencil. The $C_i^{(\gamma)}$ are discretized weight functions differentiated by their type, $w = \delta$ or $w = \theta$, their range, and their prefactors. The discretized $C_i^{(\gamma)}$ are normalized so the sum is exact when $f(\mathbf{r}) = 1$. The uniform grid allows these collocation stencils to be precomputed for an arbitrary origin and translated as needed to any node in the domain.

The integration stencil coefficients are computed numerically by finding the contribution to a node located at the arbitrary origin from each element within the range of the integral, λ. In elements that fall entirely within the range of the stencil, a Gaussian quadrature is used to determine stencil weights for each element vertex. For elements only partially within the stencil range, a midpoint rule is used with numerous equally spaced and weighted quadrature points to numerically determine appropriate weights. Note that only δ function stencils in three-dimensional problems are solely composed of these edge elements. In systems that are uniform in one or two dimensions, equation (16.14) can be integrated analytically in the uniform dimensions [3]. Those integrals effectively collapse the δ function stencils to θ function stencils.

For a mesh with $h = d/10$, the FMT collocation stencils require O(10) points in 1D, O(100) points in 2D, and O(1000) points in 3D. Cohesive interactions require even longer range stencils (typically $\lambda = r_c \approx 2.5d - 5d$) with more collocation points. This leads to a matrix problem that is more dense than a typical sparse PDE/FE matrix as will be discussed below.

Matrix structure

For an N-component electrolyte, the solution vector $\mathbf{x}$ includes minimally the independent variables, $\mathbf{x}^T = [\rho_1, \ldots, \rho_N, \psi, \mu_1, \ldots, \mu_N]$. As mentioned above, a matrix problem is constructed as

$$\mathbf{A}\Delta\mathbf{x} = -\mathbf{R} \tag{16.15}$$

where the system is constructed as a set of residual equations,

$$\mathbf{R}(\mathbf{x}, \{\xi_i\}) = 0. \tag{16.16}$$

The ξ_i are a fixed set of state parameters that would be constant for a single c-DFT calculation. The matrix $\mathbf{A}$ is a Jacobian (or more properly Hessian since c-DFT is a minimization problem) matrix with entries $A_{ij} = \delta R_i/\delta x_j$, and $\Delta\mathbf{x} = \mathbf{x}^{i+1} - \mathbf{x}^i$ is the update vector (where i is an iteration count). Assigning $\mathbf{R}_\psi$, $\mathbf{R}_{\rho_i}$, and $\mathbf{R}_{\mu_i}$ to equations (16.1), (16.2), and (16.3) respectively, and expanding the matrix for the variables of an electrokinetic c-DFT problem results in

$$\begin{bmatrix} \frac{\delta \mathbf{R}_{\rho_i}}{\delta \rho_j} & \frac{\partial \mathbf{R}_{\rho_i}}{\partial \psi} & \frac{\partial \mathbf{R}_{\rho_i}}{\partial \mu_j} \\ \frac{\partial \mathbf{R}_{\psi}}{\partial \rho_j} & \frac{\partial \mathbf{R}_{\psi}}{\partial \psi} & 0 \\ \frac{\partial \mathbf{R}_{\mu_i}}{\partial \rho_j} & \frac{\partial \mathbf{R}_{\mu_i}}{\partial \psi} & \frac{\partial \mathbf{R}_{\mu_i}}{\partial \mu_j} \end{bmatrix} \begin{bmatrix} \Delta\mathbf{x}_{\rho_i} \\ \Delta\mathbf{x}_{\psi} \\ \Delta\mathbf{x}_{\mu_i} \end{bmatrix} = - \begin{bmatrix} \mathbf{R}_{\rho_i} \\ R_{\psi} \\ \mathbf{R}_{\mu_i} \end{bmatrix} \tag{16.17}$$

where $\partial R_\psi/\partial\mu_i = 0$ because there is no explicit dependence of Poisson's equation on the chemical potential. The total number of matrix rows (per node in the mesh) is set by the number of components, N. In a steady state diffusion problem with 3 component solvent primitive model, there are 7 unknowns (or matrix rows) per node. For the same model in an equilibrium situation, the block matrix above is reduced to a 2x2 since the μ_i are constant, and there will be 4 unknowns per node.

While most of the matrix entries, $\delta R_i/\delta x_j$ are straightforward, the nonlocal functionals make $\delta R_{\rho_i}/\delta\rho_i$ a bit thorny. Specifically, the hard sphere contribution to the first term

$$\frac{\delta R_{\rho_i}^{HS}}{\delta \rho_j(\mathbf{r})} = \frac{\delta^2 \mathcal{F}_{HS}}{\delta \rho_i(\mathbf{r})\delta\rho_j(\mathbf{r}')} = \int \sum_\gamma \sum_\epsilon \frac{\partial^2 \Phi}{\partial n_\gamma \partial n_\epsilon} w_{\gamma i}(\mathbf{r}, \mathbf{r}'', R_i) w_{\epsilon j}(\mathbf{r}', \mathbf{r}'', R_j) dr'', \tag{16.18}$$

where $R_i = d_i/2$ is a particle radius. Equation (16.18) is nonzero when two weight functions centered on $\mathbf{r}$ and $\mathbf{r}'$ overlap (see further discussion in [3]). This expensive operation can be avoided by including the nonlocal densities explicitly in the system of equations and treating them as independent variables so that

$x^T = [\{\rho_i\}, \{n_\gamma\}, \psi, \{\mu_i\}]$ [3]. A new residual equation, $\mathbf{R}_{n_\gamma}$ is introduced for each independent weight function based on equation (16.11), and the matrix problem becomes

$$\begin{bmatrix} \frac{\delta \mathbf{R}_{\rho_i}}{\delta \rho_j} & \frac{\delta \mathbf{R}_{\rho_i}}{\delta n_\epsilon} & \frac{\partial \mathbf{R}_{\rho_i}}{\partial \psi} & \frac{\partial \mathbf{R}_{\rho_i}}{\partial \mu_j} \\ \frac{\delta \mathbf{R}_{n_\gamma}}{\delta \rho_j} & \frac{\delta \mathbf{R}_{n_\gamma}}{\delta n_\epsilon} & 0 & 0 \\ \frac{\partial \mathbf{R}_{\psi}}{\partial \rho_j} & 0 & \frac{\partial \mathbf{R}_{\psi}}{\partial \psi} & 0 \\ \frac{\partial \mathbf{R}_{\mu_i}}{\partial \rho_j} & 0 & \frac{\partial \mathbf{R}_{\mu_i}}{\partial \psi} & \frac{\partial \mathbf{R}_{\mu_i}}{\partial \mu_j} \end{bmatrix} \begin{bmatrix} \Delta \mathbf{x}_{\rho_i} \\ \Delta \mathbf{x}_{n_\gamma} \\ \Delta \mathbf{x}_{\psi} \\ \Delta x_{\mu_i} \end{bmatrix} = - \begin{bmatrix} \mathbf{R}_{\rho_i} \\ \mathbf{R}_{n_\gamma} \\ \mathbf{R}_{\psi} \\ \mathbf{R}_{\mu_i} \end{bmatrix}. \tag{16.19}$$

In this case, the following straightforward matrix blocks take the place of equation (16.18):

$$\frac{\delta \mathbf{R}_{n_\gamma}}{\delta n_\epsilon} = \mathbf{I} \tag{16.20}$$

$$\frac{\delta \mathbf{R}_{n_\gamma}}{\delta \rho_j} = w^{(\gamma)}(\mathbf{r}, \mathbf{r}', \lambda_j) \tag{16.21}$$

$$\frac{\delta \mathbf{R}_{\rho_i}}{\delta n_\epsilon} = \sum_\gamma \frac{\partial^2 \Phi}{\partial n_\gamma n_\epsilon} w^{(\gamma)}(\mathbf{r}, \mathbf{r}', \lambda_i). \tag{16.22}$$

Of course, there are still ideal and possibly cohesive, correlations, or other contributions to $\frac{\delta R_{\rho_i}}{\delta \rho_j}$ depending on the fluid (electrolyte) model of interest. This approach simplifies construction of the matrix and works well in c-DFT problems [3]. In spirit, this method is akin to reducing the order of PDEs by introduction of variables.

Solving the linear system

Solving equation (16.15) generally proceeds with a parallel iterative solver. However, there are some unique features of the c-DFT matrix problems defined above that present challenges to solving the system of equations. To summarize, the main issues are:

- Stencil/Physics coupling: The integration stencils used for the Euler–Lagrange (or density) residuals depend on a physical parameter in the system. As the mesh size decreases, the number of stencil points in the collocation method increases. In contrast, a FE stencil for a given PDE will be the same size (and produce the same number of nonzeros in the matrix) independent of mesh refinement.

- Large number of DOFs: Compared with PDE problems, there are a large number of degrees of freedom per node, and this number grows with the number of fluid components. The DOFs also increase with the number of nonlocal density functions if the expanded unknowns approach is used (as in equation (16.19)).
- Matrix structure: The c-DFT matrices in equation (16.19) are banded diagonals where the width of the bands vary among the types of residual equations. The Poisson and transport equations yield typical sparse matrices with only a few nonzeros per row. The Euler–Lagrange (EL) and nonlocal density equations both result in banded diagonal blocks where the width of the band depends on physical parameters (particle sizes or interaction cutoffs). If the domain size is much larger than the defining physical parameter, these blocks may be significantly sparse (from a global perspective). In contrast for a very small confined system, these blocks can be completely filled.

The original implementation of Tramonto was designed to solve the matrix problem on platforms ranging from desktop workstations to large parallel computers [3]. The approach is based on a spatial decomposition of the problem with a heuristically weighted load-balancing scheme to partition nodes and the work associated with those nodes across the available processors. Each processor then has its unique set of local nodes. These local nodes are augmented by a larger set of nodes that contains the nodes necessary to allow the collocation integrals to be performed on the local nodes. By making the larger set of nodes rectangular, indexing is simplified. Tramonto uses parallel iterative solvers in the Trilinos library [14], and a a generalized minimum residual (GMRES) solver [28] works well for linear iterations. Convergence is often achieved in 10-20 nonlinear iterations. Between nonlinear iterations, processors communicate updates of the solution vector on their local nodes with other processors as needed.

Beyond utilizing standard matrix solvers, the specific structure of the c-DFT equations has led to the development a segregated Schur complement solver (SSS) specifically designed based on the structure of c-DFT problems [25]. In this approach, the matrix is segregated by variables that result in a diagonal block matrix that can be easily inverted (A_{11}) and a diagonal block matrix that is not easily inverted (A_{22}). The matrix from equation (16.19) is then rearranged to have the structure

$$\begin{bmatrix} \mathbf{A}_{11} & \mathbf{A}_{12} \\ \mathbf{A}_{21} & \mathbf{A}_{22} \end{bmatrix} \begin{bmatrix} \mathbf{x}_1 \\ \mathbf{x}_2 \end{bmatrix} = - \begin{bmatrix} \mathbf{b}_1 \\ \mathbf{b}_2 \end{bmatrix}. \tag{16.23}$$

The Schur complement of A with respect to A_{22} is [29]

$$\mathbf{S} = \mathbf{A}_{22} - \mathbf{A}_{21}\mathbf{A}_{11}^{-1}\mathbf{A}_{12}. \tag{16.24}$$

The x_2 unknowns are found by solving a smaller system,

$$\mathbf{S}\mathbf{x}_2 = (\mathbf{b}_2 - A_{21}A_{11}^{-1}\mathbf{b}_1), \tag{16.25}$$

with GMRES parallel iterative solvers [25]. Then, a straightforward calculation finds

$$\mathbf{x}_1 = A_{11}^{-1}(\mathbf{b}_1 - A_{12}\mathbf{x}_2). \tag{16.26}$$

The SSS method was shown to improve performance from general solvers by factors of 2–30 depending on the physical system (both hard sphere fluids and polymers fluid were considered), the number of processors, and the mesh density in the problem [25].

While these algorithms have been used to study a wide range of systems, there is no escaping the memory cost associated with the real space matrix which can become prohibitive for three-dimensional systems. The next section presents an alternative approach that exhibits improved performance for three-dimensional systems.

16.1.6 A matrix-free Newton solver for c-DFTs

Recognizing the cost of the real space matrix, an 'operational Jacobian' approach has been developed that applies the matrix without storing it [15]. A regular real space Cartesian mesh with periodic boundaries is used so that fast Fourier transform algorithms can be used to compute convolutions. The operational matrix method speeds up the linear iterations by taking advantage of the fact that the matrix vector multiply needed for the linear iterative solver (GMRES) is a convolution of similar structure as the convolutions that define the nonlocal densities, equation (16.11).

From an operational point of view, computing a convolution of a function with a spatial offset such as

$$\tilde{f}(\mathbf{r}) = \int f(\mathbf{r}')w(\mathbf{r} - \mathbf{r}')d\mathbf{r}' \tag{16.27}$$

proceeds as follows:

1. Use FFTs to transform $f(\mathbf{r})$ to $\hat{f}(k)$ in reciprocal space.
2. Compute the convolution in k space with the local operation $\hat{\tilde{f}} = \hat{w}(k)\hat{f}(k)$.
3. Use a reverse FFT to transform $\hat{\tilde{f}}(k)$ to $\tilde{f}(\mathbf{r})$.

Since the FMT weight functions, w_γ are based on δ and θ functions, the reciprocal space versions are known, and are detailed in table 16.1. Once $n_\gamma(\mathbf{r})$ are known on each real space node, the functions $\partial\Phi/\partial n_\gamma$ and $\partial^2\Phi/\partial n_\gamma \partial n_\epsilon$ are also straightforward local operations on the real space grid (see [15] for a complete enumeration of these functions for the Rosenfeld FMT).

Table 16.1. Real space and Fourier space weight functions for the Rosenfeld FMT functionals. Parameters are $R = R_i$, $k = |\mathbf{k}|$, and $w_{v1}(\mathbf{r}) = -\nabla w_3(\mathbf{r})/4\pi$.

γ	$w_\gamma(\|\mathbf{r} - \mathbf{r}'\|)$	$\tilde{w}_\gamma(k)$
0	$\delta(\|\mathbf{r} - \mathbf{r}'\| - R)/(4\pi R_i^2)$	$\sin(kR)/kR$
1	Rw_0	$R\tilde{w}_0(k)$
2	$4\pi R^2 w_0$	$4\pi R^2 \tilde{w}_0(k)$
3	$\theta(\|\mathbf{r} - \mathbf{r}'\| - R)$	$4\pi R^3(\sin(kR) - kR_i\cos(kR))/(kR)^3$
γ	$w_\gamma(\mathbf{r} - \mathbf{r}')$	$\tilde{w}_\gamma(\mathbf{k})$
v1	$\hat{\mathbf{r}}\delta(\|\mathbf{r} - \mathbf{r}'\| - R)/(4\pi R)$	$-i\mathbf{k}\tilde{w}_3(k)/(4\pi R)$
v2	$4\pi R_i w_{v1}$	$4\pi R\tilde{w}_{v1}(\mathbf{k})$

An operational Jacobian is then defined as

$$y_i(\mathbf{r}) = \int \sum_j A_{ij}(\mathbf{r}, \mathbf{r}')x_j(\mathbf{r}')dr'. \tag{16.28}$$

Note that it is not necessary to store $\mathbf{A}$ to perform this operation. In fact both the residual calculation and $y_i(\mathbf{r})$, can be found by applying fast Fourier transforms (FFTs) to the required convolutions.

The discussion here is restricted to a neutral mixture based on Rosenfeld FMT functionals for hard sphere interactions (see section 7.3). The densities are the only fields in the system, $x_i(\mathbf{r}) = \rho_i(\mathbf{r})$, the matrix problem for a multicomponent system is

$$A_{ij}(\mathbf{r}, \mathbf{r}')\rho_j(\mathbf{r}') = -R_i(\mathbf{r}), \tag{16.29}$$

where $R_i(\mathbf{r}) = R^{id}(\mathbf{r}) + R^{HS}(\mathbf{r}) + R^{att}(\mathbf{r}) - [\mu_i - V_i^{\text{ext}}(\mathbf{r})]$, and the matrix is

$$\begin{aligned} A_{ij} &= \frac{\delta R_i(\mathbf{r})}{\delta \rho_j(\mathbf{r}')} \\ &= \frac{\delta\left(R_i^{id}(\mathbf{r}) + R_i^{HS}(\mathbf{r}) + R_i^{att}(\mathbf{r})\right)}{\delta \rho_j(\mathbf{r}')} \\ &= \frac{\delta^2 \mathcal{F}_{id}}{\delta\rho_i(\mathbf{r})\delta\rho_j(\mathbf{r}')} + \frac{\delta^2 \mathcal{F}_{HS}}{\delta\rho_i(\mathbf{r})\delta\rho_j(\mathbf{r}')} + \frac{\delta^2 \mathcal{F}_{att}}{\delta\rho_i(\mathbf{r})\delta\rho_j(\mathbf{r}')}. \end{aligned} \tag{16.30}$$

The calculation of the various residual terms, $R^{id,\,HS,\,att}$, and the corresponding matrix operation, $y^{id,\,HS,\,att}$, are considered in turn. The simplest fluid contribution to $R_i(\mathbf{r})$ comes from the ideal free energy,

$$\mathcal{F}_{id} = kT \int \sum_i \rho_i(\mathbf{r})\left(\ln\left(\rho_i(\mathbf{r})\Lambda_i^3\right) - 1\right)dr. \tag{16.31}$$

Combining the ideal fluid term with the external field and the constant chemical potential constraint yields a local contribution to the residual,

$$R_i^{loc}(\mathbf{r}) = \ln(\rho_i(\mathbf{r})/\rho_b) + V_i^{ext}(\mathbf{r})/kT - \mu_i/kT, \tag{16.32}$$

and a diagonal contribution to the matrix

$$\delta R_i^{loc}(\mathbf{r})/\delta\rho_j(\mathbf{r}') = \delta_{ij}(\mathbf{r}, \mathbf{r}')/\rho_i(\mathbf{r}), \tag{16.33}$$

so that

$$y_i^{id}(\mathbf{r}) = \int \sum_j \left(\delta_{ij}(\mathbf{r}, \mathbf{r}')/\rho_i(\mathbf{r})\right) x_j(\mathbf{r}')dr'. \tag{16.34}$$

There is more complexity in the hard sphere term where the free energy functional is

$$\mathcal{F}_{HS} = \int \Phi(\{n_\gamma\})dr, \tag{16.35}$$

and the residual contribution is

$$R_i^{HS}(\mathbf{r}) = \sum_\gamma \int \frac{\partial \Phi}{\partial n_\gamma} w_{\gamma,i}(\mathbf{r} - \mathbf{r}')dr'. \tag{16.36}$$

The same process used to compute the convolutions that define n_γ can be used to compute R_i^{HS} where the function of interest, $f(\mathbf{r}) = \partial\Phi/\partial n_\gamma$, is computed on the real space grid at the start of a given nonlinear iteration. In reciprocal space,

$$\hat{R}_i^{HS}(k) = \sum_\gamma w_{\gamma,i}\hat{f}_k, \tag{16.37}$$

and $R_i^{HS}(\mathbf{r})$ is found with an inverse FFT operation applied to $\hat{R}_i^{HS}(k)$. The matrix operation for the hard sphere term has the form

$$\begin{aligned} y_i^{HS}(\mathbf{r}) &= \int \sum_j \left(\frac{\delta^2 \mathcal{F}_{HS}}{\delta\rho_i(\mathbf{r})\delta\rho_j(\mathbf{r}')} \right) x_j(\mathbf{r}')dr' \\ &= \int \sum_j \left(\int \sum_\gamma \sum_\epsilon \frac{\partial^2 \Phi}{\partial n_\gamma \partial n_\epsilon} w_{\gamma i}(\mathbf{r}, \mathbf{r}'', R_i) w_{\epsilon j}(\mathbf{r}', \mathbf{r}'', R_j)dr'' \right) x_j(\mathbf{r}')dr'. \end{aligned} \tag{16.38}$$

This matrix operation can be performed in the same fashion as other convolutions by defining [15]

$$y_i^{HS}(\mathbf{r}) = \int \sum_\gamma w_{\gamma i}(\mathbf{r}, \mathbf{r}'', R_i) z_\gamma(\mathbf{r}'')dr'', \tag{16.39}$$

where the function z_γ is defined to be

$$z_\gamma(\mathbf{r}'') = \sum_{\epsilon} \left.\frac{\partial^2 \Phi}{\partial n_\gamma \partial n_\epsilon}\right|_{\mathbf{r}''} t_\epsilon(\mathbf{r}'') \tag{16.40}$$

and

$$t_\epsilon(\mathbf{r}'') = \int \sum_j w_{\epsilon j}(\mathbf{r}', \mathbf{r}'', R_j) x_j(\mathbf{r}') dr'. \tag{16.41}$$

The following summarizes the procedure used to obtain R_i^{HS} and $y_i^{HS}(\mathbf{r})$ given a current iterate of the field $x_i(\mathbf{r})$:

1. Find $n_\gamma(\mathbf{r})$ using FFTs for the convolution in equation (16.11) where $f_i(\mathbf{r}') = x_i(\mathbf{r}')$.
2. Compute $\partial\Phi/\partial n_\gamma$ and $\partial^2\Phi/\partial n_\gamma \partial n_\epsilon$ on the real space grid.
3. Find $R_i^{HS}(\mathbf{r})$ using FFTs for the convolution in equation (16.36) where $f_\gamma(\mathbf{r}') = \partial\Phi/\partial n_\gamma|_{\mathbf{r}'}$.
4. Find $t_\epsilon(\mathbf{r}'')$ using FFTs for the convolution in equation (16.41) where $f_j(\mathbf{r}') = x_j(\mathbf{r}')$.
5. Calculate $z_\gamma(\mathbf{r})$ using equation (16.40). Use t_ϵ and $\partial^2\Phi/\partial n_\gamma \partial n_\epsilon$ previously computed on the real space mesh.
6. Find $y_i^{HS}(\mathbf{r})$ using FFTs for the convolution in equation (16.39) where $f_\gamma(\mathbf{r}') = z_\gamma(\mathbf{r}')$.

This method allows GMRES linear iterative updates without storing the full matrix. Reference [15] also discusses how cohesive interactions are treated with this method as well as options for preconditioning. Density profiles for planar surfaces, cylindrical surfaces, and spherical surfaces are in excellent agreement with the real space approach described in the previous section, but the code is significantly faster since it scales as $O(N^{1.2})$ while the real space code was found to scale as $O(N^{1.6})$-$O(N^{2.2})$ depending on the specific algorithm.

16.2 Algorithms for constructing phase diagrams

16.2.1 Arc-length continuation

In section 7.5, two-phase coexistence was found at slope discontinuities in the lowest free energy path as a state variable was varied. The state variable was the chemical potential in figure 7.13 and a wall–fluid interaction parameter in figure 7.15. The curves in those two figures also trace out solutions to the system of equations that are mathematically connected, but physically unstable. Metastable branches are the extensions of stable branches beyond the intersection (or coexistence) point. These metastable states have higher free energy than the competing alternate morphology, and they terminate at spinodals or turning points. In general, the rich state space that exists in interfacial fluid systems is accompanied by multiplicity (competing solutions) and hysteresis. The number of available states becomes increasingly large as the surface geometries become more complex (for a demonstration in a disordered porous media see [11]).

Continuation through metastable and unstable states can be achieved using the pseudo arc-length continuation (ALC) method of Keller [30], developed earlier in the bifurcation analysis community. Keller's approach tracks out families of solutions even if the solution branch exhibits hysteresis loops and solution multiplicity as is the case for inhomogeneous fluids. The ALC algorithms discussed here are part of the Nox-Loca package [13] distributed as part of the open source Trilinos library [14].

For a fixed geometry of interest and a grid with N nodes, the nonlinear system of equations that need to be solved can be represented as Newton residual equations as in

$$\mathbf{R}(\mathbf{x},\ \{\xi_i\}) = 0, \tag{16.42}$$

where x is the vector of length N of nodal unknowns (densities, electric field, etc) and $\{\xi_i\}$ are the state variables available to the system. These include both bulk conditions, such as temperature and chemical potentials, and microscopic parameters, such as surface charge, q_s and surface–fluid molecular interactions parameters.

The pseudo arc-length continuation algorithm takes steps in curve length and solves for both the solution vector, $\mathbf{x}$ and the parameter value, ξ_i, that are a given distance along the curve from the current step. While a branch can reach a spinodal point and end when parameterized by a state variable, ξ_i, it will continue when is parameterized by the distance along the curve, s. This algorithm alleviates the singularity seen at spinodal points by augmenting the linear system with an arc-length parameter s and an arc-length equation g. The augmented system is written

$$\begin{aligned}\mathbf{R}(\mathbf{x}(s),\ \xi_i(s)) &= 0\\ g(\mathbf{x}(s),\ \xi_i(s),\ s) &= 0.\end{aligned} \tag{16.43}$$

Following a solution curve requires the computation of x and ξ_i given the previous state x_p and $\xi_{i,p}$ and an increment in the arc-length parameter, Δs. The pseudo arc-length equation [30] is:

$$g = \left(\frac{\partial \mathbf{x}}{\partial s}\right)^{\mathrm{T}}(\mathbf{x} - \mathbf{x}_p) + \frac{\partial \xi_i}{\partial s}\left(\xi_i - \xi_{i,p}\right) - \Delta s = 0, \tag{16.44}$$

where the superscript T is the transpose operation. This formula is a linearization of the Euclidean distance formula in $N+1$ dimensional space: $\sum(\mathbf{x} - \mathbf{x}_p)^2 + (\rho - \rho_p)^2 = (\Delta s)^2$. The augmentation of the original system with this g equation frees the ξ_i parameter as an additional unknown and introduces the step size along the curve Δs as a parameter, allowing for continuation around spinodal points. There are additional practical issues with efficient solution of the augmented system in equation (16.44), appropriate scaling of the arc-length equation, and automated step size control that are addressed further in [10].

16.2.2 Binodal and spinodal tracking

A phase transition occurs when two phases with different order parameters (or morphology) have the same free energy at the same state point, $\{\xi_i\}$. The arc-length continuation algorithm described above is a powerful tool to find a single coexistence point. The phase transition can be located graphically by plotting the free energy, Ω, versus a state parameter of interest, ξ_i, to find the intersection of two branches. However, this approach can still be time consuming if the goal is to determine a complete phase diagram. In that case, phase transitions would need to be located for a range of state parameters. Locating each point on the phase diagram with ALC alone requires computing many (typically a few hundred) solutions connecting the two branches of interest. Therefore it is helpful to have different algorithms that can find the transition points on the phase diagram directly as the state parameters change. Phase transition tracking algorithms that can follow both binodal and spinodal points were also implemented in the Tramonto code [10, 12].

A phase transition (or binodal) tracking algorithm uses Newton's method to converge a 'bifurcation' parameter value, ξ_b, and two solution vectors, $\{\mathbf{x}_1, \mathbf{x}_2\}$, with a constraint of equal free energy, $\Omega(\mathbf{x}_1, \xi_b) = \Omega(\mathbf{x}_2, \xi_b)$ while simple steps are used to increment a second 'continuation' parameter, ξ_c. The algorithm then automatically tracks the phase transition in the ξ_b, ξ_c plane. It is not uncommon for a system to have many state parameters of interest, $\{\xi_i\}$. This algorithm allows for efficient, systematic study of one slice at a time in that multidimensional space.

The phase transition tracking algorithm requires solving the following set of $2N + 1$ equations:

$$\begin{aligned} \mathbf{R}_1(\mathbf{x}_1, \xi_b) &= 0 \\ R_2(\mathbf{x}_2, \xi_b) &= 0. \\ G = \Omega(x_1, \xi_b) - \Omega(x_2, \xi_b) &= 0 \end{aligned} \tag{16.45}$$

A full Newton method for this system has the form

$$\begin{bmatrix} \mathbf{J}_1 & 0 & \dfrac{\partial \mathbf{R}_1}{\partial \xi_b} \\ 0 & \mathbf{J}_2 & \dfrac{\partial \mathbf{R}_2}{\partial \xi_b} \\ \dfrac{\partial \Omega_1}{\partial \mathbf{x}_1} & -\dfrac{\partial \Omega_2}{\partial \mathbf{x}_2} & \dfrac{\partial G}{\partial \xi_b} \end{bmatrix} \begin{bmatrix} \Delta \mathbf{x}_1 \\ \Delta \mathbf{x}_2 \\ \Delta \xi_b \end{bmatrix} = - \begin{bmatrix} \mathbf{R}_1 \\ \mathbf{R}_2 \\ G \end{bmatrix}. \tag{16.46}$$

Here the subscript i on the variables $\mathbf{R}$, $\mathbf{J}$, and Ω represent evaluation with solution vector $\mathbf{x}_i$ ($i = 1$ or 2). Automated step controls allow a mostly automated tracking of the phase transition as state parameters vary [10]. Specifically, the step size of the continuation parameter, ξ_c, decreases near critical points where the solutions can become unstable and where the binodal transition ultimately disappears. The algorithm requires initial guesses for x_1, x_2 and ξ_i, which usually come from picking two solutions from near a phase transition that were found in an initial pseudo arc-length continuation calculation.

It should be noted that this algorithm does not give information on the phase transition with the minimum energy, and so can follow intersections of any combination of stable, metastable, and even unstable branches. A phase transition between two stable phases can pass through a triple point to a state point where both phases become metastable without any evidence from the phase tracking algorithm. For systems with more complicated phase behavior it may be necessary to calculate a few complete arc-length continuation calculations to correctly interpret the phase diagrams. Extending the method to three coexisting phases, the phase transition tracking approach can also be applied to triple points [10].

Finally, spinodal tracking has also been considered [10]. A spinodal point represents the limit of metastability, and is important in systems where metastable solutions are persistent such as in block copolymer self-assembly [1, 2]. The spinodal is the point where a solution to the c-DFT equation(s) switches from a local minimum to a saddle point. In the discrete system this is represented by $\det(\mathbf{J}) = 0$. The spinodal point is also known as a fold or turning point. A common method for locating a fold point was presented by Moore and Spence [31]. This formulation of $2N + 1$ equations,

$$\begin{aligned} \mathbf{R}(\mathbf{x}, \xi_i) &= 0 \\ \mathbf{J}\phi &= 0 \\ 1 - \mathbf{c}\cdot\phi &= 0 \end{aligned} \tag{16.47}$$

is used to solve for the solution vector $\mathbf{x}$, the null vector ϕ, and the state parameter ξ_i. The second vector equation forces $\det(\mathbf{J}) = 0$ for nontrivial ϕ. The third equation normalizes the null vector to unit length using a vector $\mathbf{c}$, which can be chosen almost arbitrarily. A full Newton method for this system can be solved using a bordering algorithm that requires four solves of $\mathbf{J}$ per iteration. As with the phase tracking algorithm, a simple continuation method in a second parameter computes a line of spinodal points in two-parameter (ξ_b, ξ_c) space [10].

References

[1] Frischknecht A L, Weinhold J D, Salinger A G, Curro J G, Frink L J D and McCoy J D 2002 Density functional theory for inhomogeneous polymer systems, i. numerical methods *J. Chem. Phys.* **117** 10385–97

[2] Frischknecht A L, Curro J G and Frink L J D 2002 Density functional theory for inhomogeneous polymer systems, ii. application to block copolymer thin films *J. Chem. Phys.* **117** 10398–411

[3] Frink L J D and Salinger A G 2000 Two- and three-dimensional nonlocal density functional theory for inhomogeneous fluids: I. Algorithms and parallelization *J. Comp. Phys.* **159** 407–24

[4] Frink L J D and Salinger A G 1999 Wetting of a chemically heterogeneous surface *J. Chem. Phys.* **110** 5969–77

[5] Frink L J D, Salinger A G, Sears M P, Weinhold J D and Frishknecht A L 2002 Numerical challenges in the application of density functional theory to biology and nanotechnology *J. Phys.: Cond. Matter* **14** 12167–87

[6] Frink L J D and Martin M G 2005 A combined molecular simulation-molecular theory method applied to a polyatomic molecule in a dense solvent *Cond. Matt. Phys.* **8** 271–80

[7] Sidhu V, Frischknecht A L and Atzberger P J 2018 Electrostatics of nanoparticle-wall interactions within nanochannels: Role of double-layer structure and ion-ion correlations *ACS Omega* **3** 11340–53

[8] Frischknecht A L, Halligan D O and Parks M L 2014 Electrical double layers and differential capacitance in molten salts from density functional theory *J. Chem. Phys.* **141** 054708

[9] Frink L J D and Frischknecht A L 2006 Computational investigations of pore forming peptide assemblies in lipid bilayers *Phys. Rev. Lett.* **97** 208701

[10] Salinger A G and Frink L J D 2003 Rapid analysis of phase behavior with density functional theory. i. novel numerical methods *J. Chem. Phys.* **118** 7457–65

[11] Frink L J D and Salinger A G 2003 Rapid analysis of phase behavior with density functional theory. ii. capillary condensation in disordered porous media *J. Chem. Phys.* **118** 7466–76

[12] The Tramonto Team The Tramonto Project Website: https://tramonto.github.io (accessed 20 March 2021)

[13] The Nox and Loca Team The Nox and Loca Project Website: https://trilinos.github.io/nox_and_loca.html (accessed 20 March 2021)

[14] The Trilinos Project Team The Trilinos Project Website: https://trilinos.github.io (accessed 20 March 2021)

[15] Sears M P and Frink L J D 2003 A new efficient method for density functional theory calculations of inhomogeneous fluids *J. Comp. Phys.* **190** 184–200

[16] Roth R 2010 Fundamental measure theory for hard-sphere mixtures: a review *J. Phys.: Cond. Matter* **22** 063102

[17] Knepley M G, Karpeev D A, Davidovits S, Eisenberg R S and Gillespie D 2010 An efficient algorithm for classical density functional theory in three dimensions: Ionic solutions *J. Chem. Phys.* **132** 124101

[18] Stierle R, Sauer E, Eller J, Theiss M, Rehner P, Ackermann P and Gross J 2020 Guide to efficient solution of pc-saft classical density functional theory in various coordinate systems using fast Fourier and similar transforms *Fluid Phase Equilibria* **504** 112306

[19] Gußmann F and Roth R Bubble gating in biological ion channels: A density functional theory study *Phys. Rev. E.*

[20] Bernet T, Piñeiro M M, Plantier F and C Miqueu 2020 A 3d non-local density functional theory for any pore geometry *Mol. Phys.* **118** e1767308

[21] Stopper D and Roth R 2017 Massively parallel GPU-accelerated minimization of classical density functional theory *J. Chem. Phys.* **147** 064508

[22] Lin S-C and Oettel M 2019 A classical density functional from machine learning and a convolutional neural network *SciPost Phys.* **6** 025

[23] Cats P, Kuipers S, de Wind S, van Damme R, Coli G M, Dijkstra M and van Roij R 2020 Machine-learning free-energy functionals using density profiles from simulations *APL Materials* **9** 031109

[24] Edelmann M and Roth R 2016 A numerical efficient way to minimize classical density functional theory *J. Chem. Phys.* **144** 074105

[25] Heroux M A, Salinger A G and Frink L J D 2007 Parallel segregated schur complement methods for fluid density functional theories *SIAM J. Sci. Comput.* **29** 2059–77

[26] Nadal E, Ródenas J J, Albelda J, Tur M, Tarancón J E and Fuenmayor F J 2013 Efficient finite element methodology based on cartesian grids: Application to structural shape optimization *Abstract and Applied Analysis* **2013** 953786

[27] Boschitsch A H and Fenley M O 2015 The adaptive cartesian grid-based Poisson-Boltzmann solver: Energy and surface electrostatic properties *Computational Electrostatics for Biological Applications* ed W Rocchia and M Spagnuolo (Cham: Springer) ch 3 pp 73–110

[28] Saad Y and Schultz M H 1986 Gmres: A generalized minimal residual algorithm for solving nonsymmetric linear systems *SIAM J. Sci. Stat. Comput.* **7** 856–69

[29] Saad Y 2003 *Iterative methods for sparse linear systems* 2nd edn (Philadelphia, PA: SIAM)

[30] Keller H B 1977 Numerical solution of bifurcation and nonlinear eigenvalue problems *Applications of Bifurcation Theory* ed P Rabinowitz (New York: Academic) 359–84

[31] Moore G and Spence A 1980 The calculation of turning points of nonlinear equations *SIAM J. Numerical Analysis* **17** 567–76

IOP Publishing

Molecular Theory of Electric Double Layers

Dimiter N Petsev, Frank van Swol and Laura J D Frink

Appendix A

Appendix: the Poisson–Nernst–Planck–Bikerman theory

While Poisson–Nernst–Planck (PNP) as discussed in section 7.8 is grounded in the simplest molecular model with point charges in a uniform solvent, recent extensions to the PNP approach have been pursued that incorporate finite size effects (including solvent), correlation effects, and solvent polarization [1–5] in a way that is different from the c-DFT development of chapter 7. A review of these extensions has been presented by Liu and Eisenberg [6]. Some of the main features of the model are presented here, but the original references are recommended for historical context, extra detail, and a discussion of mathematical limits.

This new approach presented by Liu and Eisenberg introduces a steric factor in the particles function (as compared with the Poisson–Boltzmann distribution) so that

$$\rho_i(\mathbf{r}) = \rho_i^{\text{bulk}} \exp\left(-z_i\tilde{\psi} + \frac{\nu_i}{\nu_0} S^{trc}(\mathbf{r})\right) \quad \text{(A.1)}$$

where S^{trc} is a steric potential, ν_i is the volume of a species i particle, and $\nu_0 = \sum_i(\nu_i)/(K+1)$ where K is the number of types of ions in the system, and the $K+1$ species is the solvent. The solvent has nonzero volume and so is explicitly included in this model. The steric potential is

$$S^{trc}(\mathbf{r}) = \ln\left(\Gamma(\mathbf{r})/\Gamma^b\right) \quad \text{(A.2)}$$

where $\Gamma(\mathbf{r})$ is a volume fraction of voids which can vary from its bulk values, Γ^b in an inhomogeneous fluid,

$$\Gamma(\mathbf{r}) = 1 - \sum_{i=1}^{K+1} \nu_i \rho_i(\mathbf{r}). \quad \text{(A.3)}$$

This approach emphasizes the voids in a collection of particles or molecules rather than the particles themselves (as in the c-DFT methods). Liu and Eisenberg have

doi:10.1088/978-0-7503-2276-8ch17

named equation (A.1) the Poisson–Bikerman distribution since Bikerman's first considered this steric factor in binary ionic liquids in 1942 [7, 8]. A recent review of steric effects in ionic systems has been presented by Gillespie [9]. The Poisson–Bikerman distribution has the properties (like a Fermi distribution) of saturating (reaching a maximal value of density) at high fields. This improves on the unbounded densities observed in the Poisson–Boltzmann distribution and discussed in section 7.2. In the limit of zero particle volume, the Poisson–Boltzmann distribution is recovered so the Poisson–Bikerman model is consistent with classical electrostatics in the proper limit [6].

To compute profiles in a system with diffusive gradients, the flux is now given by the Nernst–Planck–Bikerman equation,

$$J_i(\mathbf{r}, t) = 0 = -D_i\left[\nabla\rho_i(\mathbf{r}, t) + z_i\rho_i(\mathbf{r}, t)\nabla\tilde{\psi}(\mathbf{r}, t) - \frac{\nu_i}{\nu_0}\nabla S^{trc}(\mathbf{r}, t)\right] \tag{A.4}$$

where particle flow is in the direction of high to low concentrations or potentials, but low to high steric potential. The steric factor reflects the difficulty of a large particle moving into a region that is already packed and has little space despite the overall concentration or electrostatic potential driving forces.

To further improve on the classical electrostatics model, Liu and Eisenberg have also derived a modified Poisson's equation to include both a polarization field (to capture the dipolar nature of water) and correlations (both ion–ion and ion–dipole) [6]. They present a modified Poisson equation in an inhomogeneous electrolyte as

$$\epsilon_s[\lambda_c^2\nabla^2 - 1]\nabla^2\psi(\mathbf{r}) = \sum_i^K \rho_i(\mathbf{r}) \tag{A.5}$$

where $\epsilon_s = \epsilon_w\epsilon_0$, ϵ_w is the dielectric constant of water, and λ_c is a correlation length that reflects charge screening for ion interactions in an aqueous electrolyte. The expected screened interactions are given by the van der Waals potential kernel (Yukawa potential) as expected, with an effective pair interaction of

$$\phi_{ij}(\mathbf{r}, \mathbf{r}') = \frac{\exp(-|\mathbf{r} - \mathbf{r}'|/\lambda_c)}{|\mathbf{r} - \mathbf{r}'|/\lambda_c}. \tag{A.6}$$

In general nonideal situations, λ_c is an unknown functional of $\rho_i(\mathbf{r})$, and so has been fit to experiments or simulations [6]. Combining equations (A.5) and (A.1) results in the Poisson–Bikerman equation which is a 4th order partial differential equation (PDE). Finally, a 'dielectric operator', $\epsilon_s[\lambda_c^2\nabla^2 - 1]$, is defined that can be used to compute (as output) a variable dielectric constant that depends on both position and the state point of the fluids, $\hat{\epsilon}(\mathbf{r}, \{\rho_i^b\})$.

While the Poisson–Nernst–Planck–Bikerman approach is not a c-DFT method, focusing on the voids in the fluid rather than the particles results in an entirely different framework for treating the same electrolyte systems. It is a bit like an Escher symmetry drawing where the birds seem the obvious foreground subject until you notice the fish, and it becomes apparent that the drawing doesn't have a background, but rather two competing foreground image sets. Thus, the 'natural' approach for studying electrolytes is to some degree a matter of perspective. A researcher with a background in molecular modeling will likely find the c-DFT approach to electrolytes natural because it builds on molecular models while a researcher with an electronic device background find the PNPB approach more intuitive since PNP is used in modeling electronic devices. Most likely both approaches have insights to offer in the understanding of complex electrolyte interfaces.

References

[1] Liu J L and Eisenberg B 2014 Poisson–Nernst–Planck–Fermi theory for modeling biological ion channels *J. Chem. Phys.* **141** 22D532

[2] Liu J L and Eisenberg B 2015 Poisson–Fermi model of single ion activities in aqueous solutions *Chem. Phys. Lett.* **637** 1–6

[3] Liu J L and Eisenberg B 2015 Numerical methods for a Poisson–Nernst Planck–Fermi model of biological ion channels *Phys. Rev.* E **92** 012711

[4] Liu J L 2013 Numerical methods for the Poisson–Fermi equation in electrolytes *J. Comput. Phys.* **247** 88–99

[5] Chen J H, Chen R C and Liu J L 2018 A GPU Poisson–Fermi solver for ion channel simulations *J. Chem. Phys.* **229** 99–105

[6] Liu J L and Eisenberg B 2020 Molecular Mean-Field Theory of Ionic Solutions: A Poisson–Nernst–Planck–Bikerman Model *Entropy* **22** 550

[7] Bikerman J J 1942 Structure and capacity of electrical double layer *Philos. Mag.* **33** 384–97

[8] Bazant M Z, Kilic M S, Storey B D and Ajdari A 2009 Towards an understanding of induced-charge electrokinetics at large applied voltages in concentrated solutions *Adv. Coll. Interf. Sci.* **152** 48–88

[9] Gillespie D 2015 A review of steric interactions of ions: Why some theories succeed and others fail to account for ion size *Microfluid. Nanofluid* **18** 717–38

Appendix B

MSA: thermodynamic properties

Waisman and Lebowitz [1–3] worked out thermodynamic properties of charged hard spheres in the MSA approximation. Limiting ourselves to a binary mixture of hard spheres of identical diameter d and equal charge q they find for the energy density

$$\frac{\beta \mathscr{U}}{V} = \frac{\beta \mathscr{U}_{hs}}{V} - \frac{1}{4\pi d^3}\left(x + 1 - \sqrt{1 + 2x}\right)x \tag{B.1}$$

where, as before, x denotes the inverse Debye screening length, i.e., $x = \kappa d$. The Helmholtz free energy, $\mathscr{F}$, is found to be the sum of a hard sphere term $\mathscr{F}_{hs}$ and a term due to the soft charge interactions. The free energy density is,

$$\frac{\beta \mathscr{F}}{V} = \frac{\beta \mathscr{F}_{hs}}{V} - \frac{1}{12\pi d^3}\left(3x^2 - 2(1 + 2x)^{3/2} + 6x + 2\right). \tag{B.2}$$

Similarly, Waisman and Lebowitz obtained the Coulombic contribution to the activity as,

$$\ln \gamma = -\frac{1}{4\pi \rho d^3}\left(x^2 - x\sqrt{1 + 2x} + x\right). \tag{B.3}$$

Finally, the osmotic coefficient, ϕ (not to be confused with the pair potential), is the sum of a hard sphere term and an ionic term

$$\phi = \phi_{hs} - \frac{1}{12\pi \rho d^3}\left(3x + 3x\sqrt{(1 + 2x)} - 2(1 + 2x)^{3/2} + 2\right) \tag{B.4}$$

where $\phi_{hs} = \beta p_{hs}/\rho$, the hard sphere compressibility factor.

References

[1] Waisman E and Lebowitz J L 1972 Mean spherical model integral equation for charged hard spheres. II. Results *J. Chem. Phys.* **56** 3093

[2] Waisman E and Lebowitz J L 1970 Exact solution of an integral equation for the structure of a primitive model of electrolytes *J. Chem. Phys.* **52** 4307

[3] Waisman E and Lebowitz J L 1972 Mean spherical model integral equation for charged hard spheres. I. Method of solution *J. Chem. Phys.* **56** 3086

Dimiter N Petsev, Frank van Swol and Laura J D Frink

Appendix C

Some conventions: dimensionless quantities

As stated before hard particle systems have phase diagrams that consist of just a single isotherm, depending only on the (number) density. Thus for monodisperse hard spheres all properties depend on the density $\rho = N/V$, where N is the number of particles and V is the volume of the MD box. It is customary to select the sphere diameter, σ, as the unit of length and to define a dimensionless density (sometimes denoted as the reduced density ρ^*) which in 3D is $\rho\sigma^3 = N\sigma^3/V$. A derived quantity is the packing fraction, η, (or in some parts of the literature ϕ). For monodisperse hard spheres we have

$$\eta = \frac{N}{V}\frac{\pi\sigma^3}{6}. \tag{C.1}$$

The packing fraction is, as the name suggest, the fraction of the volume that is occupied by the spheres. Recall that $\pi\sigma^3/6$ is the volume of a sphere of diameter σ. This interpretation assumes, of course, that the spheres are nonoverlapping. Otherwise it is an upper bound.

To preserve the above interpretation of packing fraction, for an M-component mixtures of spheres the expressions must become

$$\begin{aligned} \eta &= \frac{1}{V}\sum_i^M N_i\frac{\pi\sigma_i^3}{6} \\ &= \frac{\sigma_1^3}{V}\sum_i^M N_i\frac{\pi R_i^3}{6} \end{aligned} \tag{C.2}$$

where N_i denotes the number of spheres of species i of diameter σ_i. The diameter ratios are σ_i/σ_1.

Analogously, for nonspherical particles we have, in terms of the volume, v_i, of species i,

$$\eta = \frac{1}{V}\sum_{i}^{M} N_i v_i. \tag{C.3}$$

It is instructive to re-arrange expression (C.2), and rewrite it with the help of the molefractions $x_i \equiv N_i/N$ as:

$$\eta = \frac{N}{V}\frac{\pi}{6}\sum_{i}^{M} x_i \sigma_i^3. \tag{C.4}$$

This allows a straightforward generalization to polydisperse mixtures. First, we recognize that the molefraction x_i is essentially the probability density, $f(\sigma_i)$ for finding a sphere of diameter σ_i. Then, upon re-expressing the packing fraction more formally in terms of the probability density distribution, we obtain an expression that is valid for any mixture, polydisperse or discrete

$$\begin{aligned} \eta &= \frac{N}{V}\frac{\pi}{6}\int_0^{\infty} d\sigma\, f(\sigma)\sigma^3 \\ &= \frac{N}{V}\frac{\pi}{6}\left\langle \sigma^3 \right\rangle \end{aligned} \tag{C.5}$$

where $\langle \sigma^3 \rangle$ is the third moment of the particle size distribution (to be distinguished from $\langle \sigma \rangle^3$, the average diameter cubed).

Appendix D

Details of Lekner's mathematical transformations

Lekner uses three transformations that make the summations of equations (14.17) rapidly convergent. To start, Lekner introduces the Euler transformation

$$[(x+l)^2+(y+m)^2+z^2]^{-3/2} = 2\pi^{-1/2}\int_0^\infty dt\ \sqrt{t}\, e^{-t[(x+l)^2+(y+m)^2+z^2]}. \tag{D.1}$$

The term on the left hand side appears under a double summation in equations (14.17). The resulting summations of the exponential of equation (D.1) can be transformed using the so-called Poisson–Jacobi identity, which reads

$$\sum_{-\infty}^{\infty} e^{-(x+l)^2 t} = \sqrt{\pi/t}\sum_{-\infty}^{\infty}\cos(2\pi l x)e^{-\pi^2 l^2/t}. \tag{D.2}$$

Once this identity is substituted, the resulting integral is a representation of K_ν, a Bessel function of the second kind, viz.,

$$\int_0^\infty dt\ e^{-\pi^2 l^2/t - m^2 t} = 2\left(\pi\left|\frac{l}{m}\right|\right)^\nu K_\nu(2\pi|lm|). \tag{D.3}$$

Together these transformations produce

$$\begin{aligned} F_x(x, y, z) = &\frac{2\pi \sinh 2\pi x}{\cosh 2\pi z - \cos 2\pi x} \\ &+ 8\pi\sum_{l=-\infty}^{\infty}\frac{x+l}{[(x+l)^2+z^2]^{1/2}}\sum_{m=1}^{\infty} m\cos 2\pi m y \\ &\times K_1\left(2\pi m\sqrt{(x+l)^2+z^2}\right) \end{aligned} \tag{D.4}$$

and the expression for $F_z(x, y, z)$ as given on the second line of equation (14.18).

Finally, the expressions containing the second modified Bessel function K_0 can lead to numerical problems for small arguments where K_0 diverges. This problem was resolved by Sperb [1], who found the following transformation[1]

$$
\begin{aligned}
4\sum_{l=1}^{\infty}\cos(2\pi lx)K_0(2\pi l\rho) = &-1.386\,294 + 2\ln(\rho) + \frac{1}{\left(\sqrt{x^2+\rho^2}\right)} - \\
&\Psi(1+x) - \Psi(1-x) + \\
&\sum_{l=1}^{\infty}\binom{-1/2}{l}\rho^{2l} \times \\
&[Z(2l+1,\, 1+x) + Z(2l+1,\, 1-x]
\end{aligned}
\tag{D.5}
$$

where $\rho = \sqrt{(y+m)^2 + z^2}$ and Ψ denotes the digamma function, which is defined as

$$
\Psi(x) = \frac{d}{dx}\ln(\Gamma(x)) = \frac{\Gamma'(x)}{\Gamma(x)}. \tag{D.6}
$$

Z is the so-called Hurwitz zeta function,

$$
Z(m,\, x) = \sum_{k}^{\infty}\frac{1}{(k+x)^m} \tag{D.7}
$$

which reduces to the more familiar Riemann zeta function when $x = 1$.

Finally, he fractional binomial coefficient, $\binom{-1/2}{k}$ is given by

$$
\begin{aligned}
\binom{-1/2}{k} &= \left(-\frac{1}{4}\right)^k\binom{2k}{k} \\
&= \left(-\frac{1}{2}\right)\left(-\frac{3}{2}\right)\left(-\frac{5}{2}\right)\ldots\left(-\frac{2k-1}{2}\right)
\end{aligned}
\tag{D.8}
$$

conveniently evaluated as $(-1/2)(-3/2)(-5/2)\ldots-(2k-1)/2$.

References

[1] Sperb R 1998 Research Report No 96–18

[1] Sperb [1] states that the value of the constant, $-1.386\,294$, in (D.5) was determined by evaluation of a number of special cases. We point out that, by accident, we found that the constant appears to be equal to $-2\ln 2$ to all significant figures. We have no further explanation for this observation.

Dimiter N Petsev, Frank van Swol and Laura J D Frink

Appendix E

Finite difference and finite element methods for PDEs

Solving elliptical PDE's such as Poisson's equation has a long history in the numerical methods community. Both finite difference and finite element methods have been developed to solve these kinds of problems on discretized domains, and discussions of the methods can be found in many standard texts (see [1] and [2] for two examples with introductory material). These methods are frequently used when the electrolyte is treated with the ideal point particle model. Two important domains where the Poisson–Boltzmann electrolytes are often applied involve solvation of quantum systems [3] and solvation of large biological molecules [4].

Briefly, a finite difference (FD) approach for PDEs requires the following steps:

1. Discretize the domain of interest with a grid (or mesh).
2. Replace derivatives in equation (16.1) and/or equation (16.3) with discrete representations. This results in a system of algebraic equations with one equation per grid point.
3. Construct a matrix for the linear algebraic equations corresponding to the discrete representation of the system of equations. The matrix will be a sparse banded diagonal (often tridiagonal).
4. Perform numerical iterations of the matrix problem from an initial guess to find successive approximations to the unknowns in the discretized system of equations.
5. Perform convergence checks to stop the iterative loop when the problem is converged.

A finite element (FE) approach for PDEs is a bit more involved, but has been widely applied due to its flexibility for complex geometries when nonuniform mesh are needed. Briefly, the methods first proposed by Galerkin in 1915 proceeds as follows [5]:

1. Select the geometry for the elements in the problem. Some typical choices are: 1D (linear element), 2D or boundary elements (triangle or quadrilateral),

doi:10.1088/978-0-7503-2276-8ch21

and 3D (tetrahedron or hexahedron). In general, the edges of the elements may be straight or curved.

2. Discretize the domain of interest with the chosen elements. Element vertices are the nodes in the mesh where the PDE will be solved.
3. Choose the type of interpolation that will define the parameter values in the interior of the element, and derive shape function coefficients, ϕ_s. Each element will have N_v shape functions, where N_v is the number of vertices in the element. Each shape function is $\phi_s = 1$ at one node and $\phi_s = 0$ at all other nodes in the element. In a simple linear interpolation scheme, any variable $f(x, y, z)$ within an element will be represented as the sum

$$f(x, y, z) = \sum_{j=1}^{N_v} \phi_{s,j} f_j. \tag{E.1}$$

Linear shape functions in a 1D element of size l (with nodes 1 and 2) are

$$\phi_{s,i} = \begin{cases} \dfrac{x_2 - x}{l} & \text{for } i = 1 \\ \dfrac{x - x_1}{h} & \text{for } i = 2 \end{cases}. \tag{E.2}$$

In 2D, a bilinear rectangular element with mesh spacing h_x in the x direction and h_y in the y direction that has the local origin in the center of the element and is numbered starting with $i = 1$ in the lower left corner and then incrementing in the counterclockwise direction will have

$$\phi_{s,i} = \begin{cases} \dfrac{(h_x - x)(h_y - y)}{4h_x h_y} & \text{for } i = 1 \\ \dfrac{(h_x + x)(h_y - y)}{4h_x h_y} & \text{for } i = 2 \\ \dfrac{(h_x + x)(h_y + y)}{4h_x h_y} & \text{for } i = 3 \\ \dfrac{(h_x - x)(h_y + y)}{4h_x h_y} & \text{for } i = 4 \end{cases}. \tag{E.3}$$

Not surprisingly, a 3D hexahedral element has eight shape functions.

4. Recast the PDE as a weak-form variational problem. Integrate the PDE multiplied by the trial functions, $\nu(\mathbf{r})$ where $\nu = 0$ at the boundary nodes. Then integrate by parts to reduce the Laplace operator to a gradient operator. Since $\nu = 0$ at the boundaries, the surface terms vanish. For Poisson's equation (within constant factors) in the 1D case the result is

$$
\begin{aligned}
&0 = \nabla^2\psi(x) + q(x) \\
&0 = \int_V \nabla^2\psi(x)\nu(x) + \int_V q(x)\nu(x) \\
&0 = \nabla\psi(x)\nu(x)\Big|_0^l - \int_V \nabla\psi(x)\nabla\nu(x)dx + \int_V q(x)\nu(x)dx = 0 \\
&0 = \int_0^l \nabla\psi(x)\nabla\nu(x)dx - \int_V q(x)\nu(x)dx.
\end{aligned}
\tag{E.4}
$$

5. Construct a discrete form of the equations by using the shape functions to define the test function, $\nu = \phi_s$
6. Assemble the matrix with the number of rows (and columns) equal to the number of nodes multiplied by the number of unknowns per node in the problem. Each element contributes to the rows corresponding to the nodes of that element. The contributions are given by the shape functions
7. Perform numerical iteration of the matrix problem from an initial guess to find successive approximations to the unknowns in the discretized system of equations.
8. Perform convergence checks to stop the iterative loop when the problem is converged.

References

[1] Li Z, Quao Z and Tangm T 2018 *Numerical Solution of Differential Equations: Introduction to Finite Difference and Finite Element Methods* (Cambridge: Cambridge University Press)

[2] Cardoso J R 2017 *Electromagnetics Through the Finite Element Method* (Boca Raton, FL: CRC Press)

[3] Fisicaro G, Genovese L, Andreussi O, Marzari N and Goedecker S 2016 A generalized Poisson and Poisson–Boltzmann solver for electrostatic environments *J. Chem. Phys.* **144** 014103

[4] Liu J L and Eisenberg B 2015 Numerical methods for a Poisson–Nernst Planck–Fermi model of biological ion channels *Phys. Rev.* E **92** 012711

[5] Galerkin B G 1915 Rods and plates: series in some questions of elastic equilibrium of rods and plates *Vestnik Inzhenerov i Tekhnikov* **19** 897–908

www.ingramcontent.com/pod-product-compliance
Ingram Content Group UK Ltd.
Pitfield, Milton Keynes, MK11 3LW, UK
UKHW051903100926
8556IPUK00005B/741